Risk Identification, Rapid Detection, Prevention and Control of Foodborne Pathogenic Microorganisms

Risk Identification, Rapid Detection, Prevention and Control of Foodborne Pathogenic Microorganisms

Editor

Xinjun Du

MDPI

Basel • Beijing • Wuhan • Barcelona • Belgrade • Novi Sad • Cluj • Manchester

Editor
Xinjun Du
Tianjin University of Science
and Technology
Tianjin
China

Editorial Office
MDPI
St. Alban-Anlage 66
4052 Basel, Switzerland

This is a reprint of articles from the Special Issue published online in the open access journal *Foods* (ISSN 2304-8158) (available at: https://www.mdpi.com/journal/foods/special_issues/93S62WSZ09).

For citation purposes, cite each article independently as indicated on the article page online and as indicated below:

Lastname, A.A.; Lastname, B.B. Article Title. *Journal Name* **Year**, *Volume Number*, Page Range.

ISBN 978-3-7258-0359-0 (Hbk)
ISBN 978-3-7258-0360-6 (PDF)
doi.org/10.3390/books978-3-7258-0360-6

Contents

About the Editor . . . vii

Mohamed-Yousif Ibrahim Mohamed and Ihab Habib
Pathogenic *E. coli* in the Food Chain across the Arab Countries: A Descriptive Review
Reprinted from: *Foods* **2023**, *12*, 3726, doi:10.3390/foods12203726 . . . 1

Yijia Deng, Rundong Wang, Yuhao Zhang, Jianrong Li and Ravi Gooneratne
Effect of Amino Acids on *Fusarium oxysporum* Growth and Pathogenicity Regulated by TORC1-*Tap42* Gene and Related Interaction Protein Analysis
Reprinted from: *Foods* **2023**, *12*, 1829, doi:10.3390/foods12091829 . . . 19

Anna Austrich-Comas, Cristina Serra-Castelló, Maria Viella, Pere Gou, Anna Jofré and Sara Bover-Cid
Growth and Non-Thermal Inactivation of *Staphylococcus aureus* in Sliced Dry-Cured Ham in Relation to Water Activity, Packaging Type and Storage Temperature
Reprinted from: *Foods* **2023**, *12*, 2199, doi:10.3390/foods12112199 . . . 37

Yufei Fan, Ping Li, Dongdong Zhu, Chumin Zhao, Jingbo Jiao, Xuemeng Ji and Xinjun Du
Effects of *ESA_00986* Gene on Adhesion/Invasion and Virulence of *Cronobacter sakazakii* and Its Molecular Mechanism
Reprinted from: *Foods* **2023**, *12*, 2572, doi:10.3390/foods12132572 . . . 54

Honggang Lai, Yuanyue Tang, Fangzhe Ren, Xin-an Jiao and Jinlin Huang
Evaluation of Hygiene Practice for Reducing *Campylobacter* Contamination on Cutting Boards and Risks Associated with Chicken Handling in Kitchen Environment
Reprinted from: *Foods* **2023**, *12*, 3245, doi:10.3390/foods12173245 . . . 71

Dongdong Zhu, Yufei Fan, Xiaoyi Wang, Ping Li, Yaping Huang, Jingbo Jiao, Chumin Zhao and et al.
Characterization of Molecular Chaperone GroEL as a Potential Virulence Factor in *Cronobacter sakazakii*
Reprinted from: *Foods* **2023**, *12*, 3404, doi:10.3390/foods12183404 . . . 84

Monika Garbowska, Anna Berthold-Pluta, Lidia Stasiak-Różańska, Antoni Pluta, Stephen Forsythe and Ilona Stefańska
The Genotyping Diversity and Hemolytic Activity of *Cronobacter* spp. Isolated from Plant-Based Food Products in Poland
Reprinted from: *Foods* **2023**, *12*, 3873, doi:10.3390/foods12203873 . . . 100

Xinjie Song, Wei Li, Li Wu, Tianfeng Lv, Yao Zhang, Juan Sun, Xuping Shentu and et al.
Detection of *Vibrio parahaemolyticus* Based on Magnetic and Upconversion Nanoparticles Combined with Aptamers
Reprinted from: *Foods* **2023**, *12*, 4433, doi:10.3390/foods12244433 . . . 110

Mao Mao, Zilei Zhang, Xuchong Zhao, Haoran Geng, Liang Xue and Danlei Liu
Spatial Distribution and Enrichment Dynamics of Foodborne Norovirus in Oyster Tissues
Reprinted from: *Foods* **2024**, *13*, 128, doi:10.3390/foods13010128 . . . 122

Jan Bernd Hinrichs, Antonia Kreitlow, Madeleine Plötz, Ulrich Schotte, Paul Becher, Nele Gremmel, Roger Stephan and et al.
Development of a Sensitive and Specific Quantitative RT-qPCR Method for the Detection of Hepatitis E Virus Genotype 3 in Porcine Liver and Foodstuff
Reprinted from: *Foods* **2024**, *13*, 467, doi:10.3390/foods13030467 . . . 132

About the Editor

Xinjun Du

Xinjun Du, Ph.D., Professor, is currently the secretary of the party committee of college of Food Science and Engineering of Tianjin University of Science and Technology. In 2017, he was selected as the leading talent of Tianjin discipline. In 2020, he was selected as a young leader in scientific and technological innovation in the Tianjin Innovative Talents Promotion Plan and was selected as a special professor in Tianjin. Dr. Du has been engaged in food quality and safety assurance and functional food development work. In the past five years, he has presided over 9 national projects such as the National Key R & D Program and the National Natural Science Foundation of China, 4 provincial and ministerial projects such as the Natural Science Foundation of Tianjin, and more than 10 projects entrusted by enterprises. He published more than 100 scientific papers with the first or corresponding author, developed more than 10 kinds of functional foods, developed more than 10 kinds of rapid detection methods for food safety, applied for more than 10 invention patents, revised 2 national standards and 2 industry standards, and undertook 1 task of research and development of food reference materials.

foods

MDPI

Review

Pathogenic *E. coli* in the Food Chain across the Arab Countries: A Descriptive Review

Mohamed-Yousif Ibrahim Mohamed [1,*] and Ihab Habib [1,2,3,*]

1 Department of Veterinary Medicine, College of Agriculture and Veterinary Medicine, United Arab of Emirates University, Al Ain P.O. Box 1555, United Arab Emirates
2 Department of Environmental Health, High Institute of Public Health, Alexandria University, Alexandria P.O. Box 21511, Egypt
3 ASPIRE Research Institute for Food Security in the Drylands (ARIFSID), United Arab Emirates University, Al Ain P.O. Box 15551, United Arab Emirates
* Correspondence: mohamed-yousif-i@uaeu.ac.ae (M.-Y.I.M.); i.habib@uaeu.ac.ae (I.H.)

Abstract: Foodborne bacterial infections caused by pathogens are a widespread problem in the Middle East, leading to significant economic losses and negative impacts on public health. This review aims to offer insights into the recent literature regarding the occurrence of harmful *E. coli* bacteria in the food supply of Arab countries. Additionally, it aims to summarize existing information on health issues and the state of resistance to antibiotics. The reviewed evidence highlights a lack of a comprehensive understanding of the extent to which harmful *E. coli* genes are present in the food supply of Arab countries. Efforts to identify the source of harmful *E. coli* in the Arab world through molecular characterization are limited. The Gulf Cooperation Council (GCC) countries have conducted few surveys specifically targeting harmful *E. coli* in the food supply. Despite having qualitative data that indicate the presence or absence of harmful *E. coli*, there is a noticeable absence of quantitative data regarding the actual numbers of harmful *E. coli* in chicken meat supplies across all Arab countries. While reports about harmful *E. coli* in animal-derived foods are common, especially in North African Arab countries, the literature emphasized in this review underscores the ongoing challenge that harmful *E. coli* pose to food safety and public health in Arab countries.

Keywords: pathogenic *E. coli*; food safety; foodborne infection; middle east; zoonoses; one health

check for updates

Citation: Mohamed, M.-Y.I.; Habib, I. Pathogenic *E. coli* in the Food Chain across the Arab Countries: A Descriptive Review. *Foods* **2023**, *12*, 3726. https://doi.org/10.3390/foods12203726

Academic Editor: Dario De Medici

Received: 9 September 2023
Revised: 29 September 2023
Accepted: 9 October 2023
Published: 11 October 2023

1. Introduction

Foodborne pathogens continue to pose a significant challenge to both food safety and global trade [1]. Recent times have seen a noticeable increase in the occurrence of harmful *E. coli* strains in the United States of America (USA) [2]. The Centre for Disease Control (CDC) has reported a consistent rise in the number of hospitalizations linked to foodborne illnesses in the USA [3]. In Europe, foodborne infections have emerged as a prominent public health concern [4]. Reports from the European Food Safety Authority (EFSA) and the European Centre for Disease Prevention and Control (ECDC) have designated *Salmonella*, *Campylobacter*, *Listeria*, and Shiga toxin-producing *E. coli* as high-priority pathogens at the European Union (EU) level [5]. The World Health Organization (WHO) [6] has reported that the Middle East and North Africa (MENA) region ranks as the third highest in terms of the burden of foodborne diseases per population, following closely behind the Southeast Asian and African regions. The WHO report also indicates that approximately 70% of foodborne diseases in the Middle East and North Africa region are attributed to *E. coli*, *Campylobacter*, non-typhoidal *Salmonella* (NTS), and Norovirus, underscoring the significant threat posed by these disease-causing agents [6]. Accurately assessing the true prevalence of foodborne infections in the Middle East is a complex task due to limited epidemiological surveillance efforts aimed at identifying individual cases and outbreaks, as well as providing isolates suitable for determining sources and estimating risks at both national and regional levels [7,8]. Moreover, comprehensive data on the extent of antimicrobial resistance in foodborne bacteria responsible for infections in the Middle East region are lacking. While some reports hint at various trends in the relationship between human health and food, a complete understanding of this aspect remains elusive [1,9].

This review offers an updated analysis of the epidemiology of common pathogenic *E. coli* groups that are responsible for causing foodborne illnesses, both in the Arab world and worldwide. The methodology used in this review is a descriptive approach, which involves systematically identifying, evaluating, and compiling a body of literature related to a specific research question. The goal is to identify noticeable patterns or trends related to the research inquiry. Essentially, the descriptive study serves as a crucial part of the analysis, using the collective published information as a comprehensive dataset to draw overarching conclusions regarding the research question [8]. Researchers conducting descriptive review studies extract relevant details from each study, including research methods, publication dates, research findings, and data collection techniques (such as positive, negative, or inconclusive results). These details are then subjected to a recurrence analysis to generate quantitative findings [9]. In conducting this review, various databases, including PubMed, Scopus, Science Direct, Google Scholar, and Web of Science, were employed to gather available research materials on foodborne infections across a wide range of food types in the Arab world over the past two decades. Additionally, efforts were made to identify relevant resources, such as national reports that provide insights into the presence of pathogenic *E. coli* in the food supply chain within the region.

The Arab world encompasses 22 countries situated in the Middle East and North Africa (MENA) region. These countries include Algeria, Saudi Arabia, Sudan, Iraq, Bahrain, Comoros, Djibouti, Egypt, Jordan, Kuwait, the United Arab Emirates, Lebanon, Libya, Mauritania, Morocco, Oman, Palestine, Qatar, Somalia, Syria, Tunisia, and Yemen. Geographically, this region stretches from the Zagros Mountains in Southwest Asia to the Atlantic Ocean, covering a total land area of 14,291,469 square kilometers, which accounts for approximately 10.2% of the world's total landmass. Within this area, 27.5% is located in Asia, while the remaining 72.5% is in Africa. The region is primarily characterized by dry sub-humid, arid, and semiarid zones, with approximately 90% of the Arab region having limited water resources and arable land, which contributes to its unique ecological profile [10].

In this descriptive review, we aim to elucidate (i) background knowledge and (ii) recent updates, based on published research in the past twenty years, on the prevalence of the concerned pathogenic *E. coli* groups in the food chain in the Arab countries (iii) and the status of antimicrobial resistance (AMR).

Escherichia coli is a rod-shaped bacterium with a Gram-negative cell wall structure with dimensions ranging from 1.1 to 1.5 μm in width and 2.0 to 6.0 μm in length; it is a facultative anaerobe that swiftly colonizes the gastrointestinal tracts of both humans and animals shortly after birth, benefiting both the host and bacterium. It belongs to the family Enterobacteriaceae and falls under the *Escherichia* genus. When grown under aerobic conditions at a temperature of 37 °C, it exhibits robust growth on both general and selective agar media. This growth results in the formation of distinct round colonies that produce indole [11]. *E. coli* is typically oxidase-negative, catalase-positive, capable of reducing nitrate to nitrite, shows motility, lacks acid-fast properties, and does not form spores. The identification of specific strains of *E. coli* has traditionally relied on serotyping, a method that involves characterizing the presence of O (somatic), H (flagellar), K (capsular), and F (fimbriae) antigens [12].

In the late 19th century, pediatrician Theodore Escherich discovered *E. coli* and initially referred to it as normal intestinal flora, naming it "*Bacterium coli commune*". Later, it was officially renamed *E. coli* [13]. This bacterium belongs to the Enterobacteriaceae family and exhibits facultative anaerobic characteristics. *E. coli* can be motile, often utilizing flagella, but can also be non-motile, and it can thrive in both aerobic and anaerobic environments [14]. *Escherichia coli* is one of the most frequently encountered bacteria in clinical samples [15].

In healthy hosts, most *E. coli* strains are non-pathogenic and contribute significantly to the commensal population residing in the host's intestinal tract, primarily within the mucosal layer of the colon [13]. However, *E. coli* can become pathogenic under certain circumstances, such as when the host's immune system is suppressed. Some strains are inherently pathogenic and can cause gastrointestinal and urinary tract infections [16]. *Escherichia coli* can endure for extended periods in environmental settings like soil and water [17]. The presence of *E. coli* in food or water can signal inadequate cleaning and careless handling, or it may suggest the potential presence of enteric pathogens [18]. Based on genetic and clinical criteria, *E. coli* can be broadly classified into three major groups: commensal *E. coli*, intestinal pathogenic (diarrheagenic) *E. coli*, and extraintestinal pathogenic *E. coli* (ExPEC). Additionally, various molecular typing techniques, including PCR (polymerase chain reaction) and PFGE (pulsed-field gel electrophoresis), can be employed to differentiate between *E. coli* strains.

The polymerase chain reaction (PCR) has become a cornerstone technique in molecular biology and microbiology for its ability to selectively amplify specific DNA sequences. It is extensively utilized in the detection of genes and their variants within various organisms, including *E. coli*. However, the absence of a PCR signal for a particular gene of interest in *E. coli* isolates can pose challenges and uncertainties. It is crucial to explore the reasons behind such failures and consider the implications they may have on research outcomes and interpretations.

Most *E. coli* found in the environment are non-pathogenic; however, some groups are pathogenic [19,20]. Generally, pathogenic *E. coli* is broadly classified into two major categories: diarrheagenic *E. coli* and extraintestinal pathogenic *E. coli*. The intestinal or diarrheagenic pathogenic strains of *E. coli* are rarely found among the intestinal flora of healthy mammals [18,21]. Based on the virulence factors, six different pathogenic classes of intestinal pathogenic *E. coli* have been identified, namely, enterotoxigenic *E. coli* (ETEC), enteropathogenic *E. coli* (EPEC), enterohemorrhagic *E. coli* (EHEC)/Shiga toxin-producing *E. coli* (STEC), enteroinvasive *E. coli* (EIEC), enteroaggregative *E. coli* (EAEC), and diffusely adherent *E. coli* (DAEC) [13,22]. Extraintestinal pathogenic *E. coli* is phylogenetically and epidemiologically different from diarrheagenic *E. coli*. They could inhabit a range of anatomical locations and cause various infections outside the gastrointestinal tract, among which urinary tract infections are the most common [22]. The *E. coli* strains causing extraintestinal infections have been collectively called extraintestinal pathogenic *E. coli* (ExPEC), which includes two major pathotypes: uropathogenic *E. coli* (UPEC) and neonatal meningitis *E. coli* (NMEC). Several strains of *E. coli* can cause disease in the GIT by toxin production, which includes enterohemorrhagic, enterotoxigenic, enteroinvasive, and enteroaggregative *E. coli* [23,24]. In developing countries, ETEC is the causative agent of travelers' diarrhea (watery diarrhea without fever). In humans, EIEC is the causative agent of the invasive, dysenteric form of diarrhea because of its ability to invade the colonic mucosa. EHEC produces Vero or Shiga toxins and is the causative agent of hemorrhagic colitis and bloody diarrhea [25,26].

2. Epidemiology of Pathogenic *E. coli* Groups

2.1. Epidemiology of Enteropathogenic E. coli *(EPEC)*

Typical enteropathogenic *E. coli* (EPEC) strains are a leading cause of infantile diarrhea in developing countries, whereas they are rare in industrialized countries, where atypical EPEC seems to be a more important cause of diarrhea [27,28]. Also, they are among the most important food-borne pathogens worldwide [29]. EPEC infection results in an excessive loss of water and electrolytes from the body, leading to dehydration and death. However, the underlying molecular mechanisms are not completely understood. EPEC has been reported to disrupt the ion transporters and channels as well as tight junctions in the intestinal epithelial cells leading to the rapid onset of diarrhea. EPEC directly injects virulence factors into the host cells that target multiple signaling pathways of which some have been linked to tight junction disruption [28].

Atypical EPEC strains were obtained from chicken suggesting that it could be a reservoir of these bacteria [29,30]. Their presence in chicken products is evidence that contamination can occur during the slaughtering and manufacturing processes, thereby representing a risk for humans. These results highlight the need for more molecular characterization studies to detect the EPEC genotypes and compare them with human diarrhea.

EPEC harbors the *eaeA* gene for attaching and effacing or causing A/E lesions on intestinal cells, do not possess the Shiga toxin gene, but may possess other genes such as *bundle-forming pili* (*bfpA*), the intimate adhesin intimin gene. Typical EPEC is *eaeA*-positive and *bfpA*-positive (humans are the reservoirs), while atypical EPEC is only *eaeA*-positive (both humans and animals can be reservoirs).

In Algeria, Ferhat et al. [31,32] successfully isolated the *eaeA* gene from *E. coli* isolates obtained from ovine carcasses in slaughterhouses located in the city of Algiers. Similarly Chahed et al. [33] and Mohamed et al. [34] also managed to isolate the *eaeA* gene, but this time from bovine carcasses in Algiers slaughterhouses. Furthermore, Dib et al. [35] made a significant discovery, detecting a notable prevalence of the *eaeA* gene in sardines (14.3%, *n* = 32 *E. coli* isolates) and shrimps (33.3%, *n* = 66 *E. coli* isolates). These findings raise awareness about the potential role of sardines and shrimp in the dissemination of the *eaeA* gene within the Algerian context. Interestingly, studies focused on chicken meat in Algeria, such as those conducted by Benameur et al. [36] and Laarem et al. [37], did not find any presence of the *eaeA* gene. These results collectively provide compelling evidence for the contamination of widely consumed bovine and ovine carcasses, as well as fish and seafood (specifically sardines and red shrimp), with EPEC in Algeria.

In Egypt, several studies have shed light on the role of fresh fish as a potential source of the *eaeA* gene. For example, Galal et al. [38] conducted research in Kafr El-Shikh and identified the *eaeA* gene in 57.1% of 45 *E. coli* isolates obtained from fish samples. Similarly, Saqr et al. [39] found the gene in 83.3% of 6 *E. coli* isolates obtained from Nile tilapia. Fresh beef meat has also been implicated as a source of the *eaeA* gene, with Merwad et al. [40] and Mohammed et al. [41] reporting its presence in 18% (*n* = 27) and 20.7% (*n* = 87) of their respective *E. coli* isolates samples. Additionally, Merwad [40] detected the *eaeA* gene at a rate of 19.1% out of 120 *E. coli* isolates obtained from raw milk samples. The use of poultry waste, sewage, and cow dung as fertilizers for fishponds has been identified as a hazardous source of contamination for water and fish. This contamination poses a direct threat to public health. Detecting EPEC in *E. coli* isolated from fish samples highlights potential risks, as these bacteria are known to cause food poisoning and hemorrhagic enterocolitis in humans who consume improperly processed fish. It is important to note that not all food sources have been found to carry the *eaeA* gene. Hamed et al. [42] were unable to identify the presence of the gene in *E. coli* strains obtained from luncheon and sausage. Similarly, Sahar et al. [43] did not discover the *eaeA* gene in *E. coli* strains collected from a wide range of sources, including minced meat, steaks, sausage, kofta, burgers, luncheon, liver, chicken livers, lambs, oysters, calamari, bivalves, raw milk, yogurt, and cheese.

In the city of Duhok, Iraq, a study conducted by Taha and Yassin [44] examined various food samples. Out of 120 beef carcass samples, eight isolates tested positive for the presence of the *eaeA* gene. Similarly, from 120 imported chicken carcass samples, two isolates were found to carry the *eaeA* gene. These findings suggest that beef carcasses and imported chicken carcasses could potentially serve as sources for the *eaeA* gene in this region. However, the study did not detect the *eaeA* gene on fish surfaces or in *E. coli* isolated from samples of imported and local raw burgers, local raw ground meat, and local raw milk. These results indicate the absence of the *eaeA* gene in these specific food items. The implications of this study point to beef carcasses and imported chicken carcasses as potential contributors to the dissemination of the *eaeA* gene in the studied area.

In Jordan, a study conducted by Swedan and Alrub [45] revealed the presence of the *eaeA* gene in 2.8% of 109 isolates obtained from different drinking water sources within Amman city. Additionally, Tarawneh et al. [46] detected the *eaeA* gene in 6% of 50 isolates collected from six slaughterhouses situated in the southern region of Jordan. The detection of the *eaeA* gene within *E. coli* strains from water sources suggests a potential mechanism for the dissemination of virulence genes among various animal species in the environment. This finding underscores the importance of understanding and monitoring the presence of such genes in water sources, as they can contribute to the transmission of virulence factors and impact both animal and public health. Continued research in this area is crucial to comprehending the dynamics of gene spread and potential health risks associated with contaminated water sources in Jordan.

In Lebanon, certain widely consumed dairy products have been identified as potential sources of public health risks due to their role in transmitting the *eaeA* gene. Saleh et al. [47] conducted a study in which they isolated *E. coli* bacteria from a total of 340 dairy products, including Shankleesh, Kishk, and Baladi. The *eaeA* gene was found to be present in 102 *E. coli* isolated obtained from these products. Specifically, the *eaeA* gene was detected in *E. coli* isolated from Kishk and Baladi at rates of 13.5% and 2.7%, respectively. These findings underscore the importance of monitoring and addressing potential contamination of dairy products with pathogenic genes such as *eaeA*. The presence of this gene in certain dairy items highlights the need for robust quality control measures and strict hygiene practices within the dairy production and processing industry. These measures are essential to ensure the safety of these products and protect public health.

In Morocco, the presence of the *eaeA* gene has been identified in various food products, including ground beef, sausage, and turkey. Badri et al. [48] reported that 12.5%, 2.8%, and 2.8% of *E. coli* isolates from 140 ground beef samples, 200 turkey samples, and 120 sausage samples, respectively, tested positive for the *eaeA* gene. The authors emphasized the importance of gaining a better understanding of the molecular characteristics of potentially *eaeA*-positive *E. coli* strains and their role in causing diseases. This understanding is crucial for assessing the pathogenic potential of these strains in human patients.

In Libya, a study conducted by Garbaj et al. [49] identified the presence of the *eaeA* gene in raw cow milk and raw camel milk, but notably, it was not detected in raw goat milk. These findings highlight the importance of improving and implementing rigorous hygienic practices within the dairy production sector. The authors of the study strongly recommended the adoption and application of Libyan standards for dairy products. This is essential to ensure effective monitoring throughout the entire dairy production process, from the farm to the point of delivery to consumers.

In Palestine, a study conducted by Adwan et al. [50] reported the absence of the *eaeA* gene in *E. coli* isolated from beef, chicken meat, and turkey samples obtained from the Jenin district. This absence of the gene in these meat products suggests a lower risk of contamination with EPEC carrying the *eaeA* gene in this specific area during the time of the study. It is important to note that the absence of the gene in these isolates contributes to our understanding of the safety of these meat products in the region. However, food safety standards and monitoring should still be maintained to ensure ongoing consumer protection.

In Khartoum, Sudan, a study conducted by Adam [51] did not detect the presence of the *eaeA* gene in *E. coli* isolated from drinking water samples. However, it is important to note that water quality can vary over time and across different sources, so continued monitoring of water sources is essential to ensure the safety of drinking water for the population. While the study did not find the *eaeA* gene in the *E. coli* isolates obtained from samples, maintaining high standards of water treatment and hygiene practices is crucial for public health.

In Qatar, a comprehensive survey conducted by Johar et al. [52] uncovered a high prevalence of the *eaeA* gene in *E. coli* O157:H7 isolated from beef, mutton, and chicken. These findings emphasize the substantial presence of the *eaeA* gene in these food sources, which raises concerns about its potential transmission to humans through contaminated food. This underscores the importance of implementing specialized monitoring programs to detect and control the presence of the *eaeA* gene in food production processes in Qatar.

In Saudi Arabia, a study by Al-Zogibi et al. [53] reported that the primary sources of the *eaeA* gene were milk and raw meat. They isolated *E. coli* from milk, finding it in 15.9% of

540 samples, and detected the *eaeA* gene in 44.2% of the *E. coli* isolates from milk. Similarly, *E. coli* was isolated from raw meat, present in 11.3% of 150 samples, and the *eaeA* gene was detected in 58.8% of the *E. coli* isolates from raw meat. However, it is worth noting that other studies conducted by Hessain et al. [54] and Abu-Duhier [55] did not detect the *eaeA* gene in *E. coli* isolated from beef meats, chicken meats, fresh vegetables, and fruits. These variations in findings could be due to differences in the samples tested, the sampling methods, or regional factors. While the *eaeA* gene was not found in *E. coli* isolated from these specific samples in the studies mentioned, ongoing monitoring and adherence to food safety practices are essential to ensure consumer protection and the safety of food products in Saudi Arabia.

In the United Arab Emirates, the *eaeA* gene was reported in *E. coli* O157 isolated from camel meat, goat meat, cattle meat, and sheep meat obtained from slaughterhouses. This finding indicates the presence of the *eaeA* gene in *E. coli* isolated from animals being processed for meat production. This raises concerns about the potential contamination of meat products intended for human consumption [56] (Table 1).

These studies underscore the importance of continued surveillance and the implementation of hygiene measures in the handling and processing of these meats to mitigate potential risks to public health.

2.2. *Epidemiology of Enteroinvasive E. coli (EIEC)*

Enteroinvasive *E. coli* (EIEC) infections in humans appear to be primarily sourced from infected individuals, as no animal reservoirs have been identified. The main mode of transmission is through the oral–fecal route. While EIEC infections can be found worldwide, they are particularly prevalent in low-income countries where poor general hygiene facilitates their spread. The incidence of Enteroinvasive *E. coli* varies by region [57], and there may be discrepancies in some reports, likely due to the challenge of distinguishing between Shigella and EIEC. In certain countries in Latin America and Asia, such as Chile, Thailand, India, and Brazil, EIEC is frequently identified as a causative agent of diarrhea, with frequent reports of asymptomatic individuals excreting the pathogen. In industrialized countries, EIEC infections are often linked to travel, reported mainly in returning travelers from high-incidence countries. Occasionally, food and water sources have been identified as vehicles of infection, but this is typically traced back to secondary contamination from a human source [58].

Enteroinvasive *E. coli* carrying the *ial* gene can cause sporadic infections but have also been implicated in outbreaks, sometimes affecting a significant number of individuals. In the 1970s, a major outbreak occurred in the United States, impacting 387 patients, and was linked to cheese contaminated with an O124 *E. coli* strain [59]. Europe has also observed an increase in the number of infection cases associated with an emerging EIEC clone. In 2012, Italy reported a large and severe outbreak of bloody diarrhea involving more than 100 individuals [60,61]. During this outbreak, an EIEC O96:H19 strain, a serotype never previously described for EIEC, was isolated, and the suspected source of infection was traced back to cooked vegetables [60]. In the course of the outbreak investigation, an EIEC O96:H19 strain was also found in two asymptomatic food handlers working in the canteen linked to the outbreak. This supported the hypothesis of secondary contamination of the vegetables during post-cooking handling procedures [60]. In 2014, the United Kingdom experienced two interconnected outbreaks of gastrointestinal diseases, affecting more than 100 cases. One of these episodes was associated with the consumption of contaminated salad vegetables, and once again, an O96:H19 EIEC strain was isolated from some of the patients and from vegetable samples [62].

In Arab countries, the presence of the *ial* gene has been a subject of investigation in numerous studies [41,48].

In Egypt, Mohammed et al. [41] detected the *ial* gene in only two *E. coli* O157:H7/H- isolates obtained from beef meat products in Mansoura city. Their study brought attention to the contamination of meat products, especially beef burgers, with various non-O157 STEC and EIEC serotypes. This contamination raised significant concerns regarding potential health risks for consumers of these products.

In Morocco, a comprehensive study conducted by Badri et al. [48] found no detection of the *ial* gene in *E. coli* isolates from ground beef (n = 140), sausage (n = 120), turkey (n = 200), and well water (n = 50) (Table 1).

Nevertheless, despite these findings, additional research is warranted to gain a deeper understanding of the prevalence and significance of the *ial* gene markers in *E. coli* strains found in food throughout Arab countries. Such research efforts would contribute to a more comprehensive understanding of the potential implications of the ial gene in public health within the region.

2.3. *Epidemiology of Enterotoxigenic E. coli (ETEC)*

Enterotoxigenic *E. coli* (ETEC) is a major contributor to the global health burden, responsible for an estimated 400 million cases of diarrhea and nearly 400,000 deaths annually among children under 5 years of age in low and middle-income countries [3]. It is also a prevalent cause of travelers' diarrhea. ETEC is characterized by its ability to produce specific toxins, including heat-labile toxin (LT) and heat-stable toxin (ST), which further comprise two subtypes: STh and STp. ETEC infections are linked to the presence of genes encoding these toxins, namely *elt (encoding LT)*, *esth (encoding STh)*, and *estp (encoding STp)* [63]. These toxins play a crucial role in causing the diarrheal symptoms and gastrointestinal distress observed in infected individuals. Importantly, ETEC strains can be distinguished by their production of either LT, ST, or both, as well as the specific combination of toxin genes they carry. Additionally, ETEC strains utilize various colonization factors to adhere to the intestinal lining, facilitating the establishment of infection. These factors enhance the bacteria's ability to colonize and thrive within the host. To date, researchers have identified at least 25 distinct colonization factors associated with human ETEC strains, as detailed in a study by Von Mentzer et al. [64]. Understanding the genetic and molecular factors contributing to ETEC infections is of paramount importance for the development of effective preventive measures. This includes the development of vaccines and the promotion of improved hygiene practices, with the aim of reducing the substantial disease burden caused by ETEC in vulnerable populations.

In Arabic countries, the *elt*, *esth*, and *estp* genes were detected in *E. coli* strains from food samples, specifically in the city of Duhok, Iraq. These *E. coli* strains were isolated from various sources, including beef carcasses (50/20, 41.6%), imported chicken carcasses (52/120, 43.3%), fish surfaces (47/120, 39.1%), imported and local raw burgers (15/120, 37.5%), and local raw ground meat (46/120, 38.3%). The presence of these genes was detected at rates of 34.6%, 91.3%, 100%, 71.4%, and 100%, respectively [44]. This study revealed a high level of food items contaminated with ETEC in the specified area. The authors attributed this high contamination rate to poor hygienic conditions during the slaughtering process, suboptimal food management and storage practices at retail shops, or potential cross-contamination occurring during these processes. Such contamination poses significant community health hazards to the local population and travelers in the region. Therefore, the authors strongly recommend the implementation of strict hygienic practices throughout all stages of food production, handling, and storage to reduce the risk of contamination and subsequent outbreaks of diarrhea. These data provide valuable baseline information for the ongoing monitoring and assessment of ETEC in food products across Arab countries, ultimately contributing to improved food safety and public health in the region (Table 1).

Indeed, various studies have detected the *elt* gene in beef products in different Arabic countries, including Algeria [34] and Egypt [41]. These findings suggest the potential presence of ETEC in beef products in these countries. However, it is worth noting that in some other Arabic countries, there is a lack of published data on the prevalence of ETEC strains carrying the *elt*, *esth*, and *estp* genes in beef products. This underscores the need for further research and surveillance in these regions to better understand the extent of

ETEC contamination in various food sources and to implement measures to ensure food safety and public health. Such studies can contribute to a more comprehensive assessment of the situation and help develop strategies for preventing ETEC-related health risks in these areas.

2.4. Epidemiology of Enterohemorrhagic *E. coli* (EHEC)

Enterohemorrhagic *E. coli* (EHEC) carrying the *stx1*, *stx2*, *eaeA* genes, or Hemolysin (*hlyA*) is a subset of pathogenic *E. coli* capable of causing diarrhea or hemorrhagic colitis in humans [65]. In some cases, hemorrhagic colitis can progress to hemolytic uremic syndrome (HUS) [66], a condition that can lead to acute renal failure in children and result in significant morbidity and mortality in adults [67]. While this has been recognized as a cause of these syndromes since the 1980s, clinical cases and outbreaks attributed to other EHEC serogroups are increasingly being identified [68]. In certain regions, non-O157 EHEC strains may account for a higher number of cases than EHEC O157:H7. What all *E. coli* strains associated with HUS appear to share is the capacity to produce verotoxins and the ability to adhere to and colonize the human intestines. Since verotoxin genes can be transferred between bacteria, it is possible that additional *E. coli* pathotypes linked to HUS could be discovered in the future [69]. According to Yamasaki et al. [70], in Japan, the quantitative detection of shiga toxins directly from stool specimens of patients played a crucial role in identifying an outbreak of EHEC.

The detection of *stx1*, *stx2*, and *eaeA* genes in food samples in various Arabic countries suggests the potential presence of EHEC capable of causing severe health problems. Numerous studies have underscored the presence of EHEC in food products across different regions of the Arab world. Mohammed et al. [41] identified EHEC in ground beef in the city of Mansoura, Egypt, with the presence of EHEC-like strains (*eaeA* + *stx1* or *stx2*). Mohammed et al. [71] detected EHEC in beef products (9.4% of *E. coli* isolates) in Egypt. Merwad et al. [40] found EHEC (5%) in cow milk samples in Egypt. In Iraq, Taha and Yassin [44] discovered EHEC in 30.7% of *E. coli* isolated from 120 beef carcasses. Klaif et al. [72] detected EHEC in camel meat samples from Iraq (45% of *E. coli* isolates). Saleh et al. [47] identified EHEC in dairy products (Kishk and Baladi) in Lebanon at 8.1% and 5.4%, respectively. In Libya, EHEC was detected in *E. coli* O157 from milk and dairy product samples (25%) [49]. In Saudi Arabia, Hessain et al. [54] detected EHEC in raw beef, raw mutton, and raw chicken samples (1%, 2.5%, and 2.5%, respectively). Collectively, these studies reveal a significant contamination of various meat and dairy products with EHEC strains in Arabic countries. The presence of EHEC in these food items underscores a potential health risk for consumers. Consumption of such products can lead to gastrointestinal issues and, in severe cases, conditions like HUS. To mitigate the risk of EHEC contamination and related health hazards, it is essential to ensure strict hygiene practices throughout the entire food production, handling, and distribution processes (Table 1).

2.5. Epidemiology of Shiga Toxin-Producing *E. coli* (STEC)

Shiga toxin-producing *E. coli* (STEC) carrying *Stx1* and *Stx2* are significant pathogens with global implications, known for their association with various human illnesses. These illnesses include diarrhea, bloody diarrhea, hemorrhagic colitis, and hemolytic uremic syndrome (HUS) [73]. Ruminants, particularly cattle, are recognized as the primary reservoir for STEC, and it can spread to humans through contaminated food and water sources [74]. The severity of STEC infections is influenced by a wide array of virulence factors. One of the key virulence factors is Shiga toxin, which plays a pivotal role in the development of severe symptoms like HUS. Shiga toxin can be categorized into two primary types: Shiga toxin 1 (*stx1*) and Shiga toxin 2 (*stx2*) [73]. These toxins are central to the pathogenicity of STEC and the associated illnesses in humans.

In Algeria, although limited data are available regarding STEC isolated from food sources, several studies have provided valuable insights into the presence and characteristics of STEC strains in different food products. Salih et al. [75] detected a single STEC isolate from frozen bovine meat in Algeria. Dib et al. [35] identified STEC strains in seafood, including three from sardines and three from shrimps. Ferhat et al. [31] conducted a study involving 116 sheep carcasses in an Algiers slaughterhouse. Among the *E. coli* strains isolated from these carcasses, five strains (17.2%) were classified as STEC (Table 1). While the available data are limited, these studies suggest the presence of STEC strains in various food products in Algeria. The detection of STEC in frozen bovine meat, seafood, and sheep carcasses highlights the importance of monitoring and understanding the prevalence of these pathogens in the food supply chain. Continuous research and surveillance efforts are crucial to assess the potential risks associated with STEC contamination in Algerian food sources and to implement appropriate measures to ensure food safety and protect public health.

2.6. Epidemiology of Enteroaggregative E. coli (EAEC)

Enteroaggregative *E. coli* (EAEC) infections have been increasingly recognized as important enteropathogens since their initial discovery by patterns of adherence to HEp-2 cells in *E. coli* isolates from Chilean children with diarrhea [76]. EAEC have since been associated with foodborne outbreaks of diarrhea [77], traveler's diarrhea [78], diarrhea in adults with HIV infection [79], and endemic diarrhea in cities in the USA [80]. A meta-analysis of 41 studies found EAEC to be significantly associated with acute diarrheal illness among both children and adults in developing regions [80]. However, because EAEC are also a highly common infection among children without overt diarrhea in low-resource settings, they have not been found to be a major cause of diarrhea in some endemic settings [81]. Regardless, EAEC, independent of diarrheal symptoms, have been associated with other poor health outcomes in children, such as growth failure [82] and mild-to-moderate intestinal inflammation [76].

Table 1. The occurrence of virulence *E. coli* genes in foods in Arab countries.

Country	Tested Food Samples (Total Number)	% of *E. coli*-Positive Samples or Isolates No.	Virulence Genes (%) (Out of Total Number) of *E. coli*-Positive Samples	References
Algeria	Sheep carcasses (*n* = 363)	ND	*eaeA* (9.92)	[31]
	Bovine carcasses (*n* = 230)	66	*eaeA* (21.2); *stx*1 (10.6); *stx*2 (12.1); *eaeA*, *stx* (4.5)	[33]
	Sardines (*n* = 100)	32	*eaeA* (14.3); *eae*, *stx*1 (14.3); *stx*2 (42.9); *stx*1, *stx*2 (14.3)	[35]
	Shrimps (*n* = 50)	66	*eaeA* (33.3); *stx*1, *stx*2 (16.7); *stx*2 (16.7)	
	Chicken samples (*n* = 32)	56.3	*stx*2 (5.6); *eaeA* (0); *rfbE* (0); *fliC* (0)	[36]
	Retail chicken meat (*n* = 33)	87.8	*stx*1 (6.9); *stx*2 (3.4); *ehxA* (3.4)	[37]
	Frozen beef liver (*n* = ND)	92 isolates	*iss* (85.9); *hylF* (82.6); *ompT* (80.4); *iroN* (87); *fimC* (70.7); *iutA* (90.2); *elt* (5.4); *stx* (2.2); *ipaH* (2.2); *eaeA* (0); *aggR* (0)	[34]
	Chicken samples (*n* = ND)	17 isolates	*iss* (82.4); *hlyF* (52.9) *ompT* (76.5); *iroN* (52.9); *iutA* (52.9); *fimC* (88.2)	[83]
	Frozen bovine meat (*n* = 756)	Five *E. coli* O157:H7 isolates	*stx*1 (20); *stx*2 (100); *eae* (80); *ehxA* (100)	[75]
	Ovine carcasses (*n* = 151)	13 *E. coli* O157:H7 isolates	*eae* (69.2); *stx*1 (7.7); *stx*2 (76.9)	[32]

Table 1. *Cont.*

Country	Tested Food Samples (Total Number)	% of *E. coli*-Positive Samples or Isolates No.	Virulence Genes (%) (Out of Total Number) of *E. coli*-Positive Samples	References
Egypt	Fresh fishes ($n = 45$)	15.6	*eaeA* (57.1); *stx1* (42.9); *stx2* (0); *hylaA* (57.1); *sta* (57.1); *Stb* (42.8)	[38]
	Drinking water ($n = 46$)	91	*stx1* (24.4); *stx2* (2.4); *eae* (0); *hly* (4.8); *fliCh7* (0)	[84]
	Freshwater canal (n = ND)	49 isolates	*eae* (2); *stx1* (2); *stx2* (0); *hlyA* (0); *hly* (0)	
	Broiler meats (n = ND)	11 isolates	*iroN* (90.9); *ompA* (81.8); *iss* (100); *tsh* (81.8); *papC* (81.8)	[85]
	Karish cheese ($n = 55$)	74.5	*stx* (2.3); *eaeA* (0); *astA* (4.5); *ehaA* (34.8); *lpfA* (33.7); (3.4); *iha* (2.3); *hlyA* (0); *cdt* *cnf* (0)	[86]
	Ras cheese ($n = 60$)	21.7	*stx* (0); *eaeA* (0); *astA* (9.1); *ehaA* (36.4); *lpfA* (45.5); (0); *iha* (0); *hlyA* (4.6); *cdt* *cnf* (4.6)	
	Raw milk ($n = 120$)	19.1	*stx1*(21.7); *stx2* (34.8); *eaeA* (17.3); *ehxA* (17.3)	[40]
	Fresh beef ($n = 27$)	100	*eae* (18); *ipaH* (18); *stx1* (18); *stx2* (10)	[41]
	Beef meat products ($n = 218$)	18.3	*eae* (30); *ipaH* (18); *stx1*(18); *stx2* (18); *eltB* (8); *estA* (8); *ial* (2)	
	Nile tilapia (Oreochromis niloticus) (n = ND)	Six isolates	*eaeA* (83.3); *stx2* (50); *aadA2* (50)	[39]
	Minced meat ($n = 50$)	Eight	*eaeA* (12.5); *stx1* (25); *stx2* (12.5)	[42]
	Luncheon ($n = 50$)	Four	*eaeA* (0); *stx1* (0); *stx2* (0)	
	Beef burgers ($n = 50$)	Two	*eaeA* (100); *stx1* (0); *stx2* (0)	
	Sausage ($n = 50$)	10	*eaeA* (0); *stx1* (20); *stx2* (0)	
	Karish cheese ($n = 60$)	3.3	*eaeA* (50); *stx1* (0); *stx2* (50)	
	Raw bovine milk ($n = 121$)	13.2	*stx1* (12.5); *stx2* (18.8); *Sta* (12.5); *lt* (0)	[87]
	Meat products ($n = 100$)	32	*lt* (15.6); *eae* (12.5); *stx1* (6.3); *stx2* (9.4); *bfpA* (3.1); *ipaH* (3.1)	[71]
	Drinking water ($n = 300$)	5.3	*lt* (25); *st* (12.5); *stx1* (18.8); *stx2* (6.3); *eaeA* (31.3)	[88]
	Raw beef ($n = 100$)	ND	*stx1* (6); *stx2* (6)	[89]
	Raw milk ($n = 100$)	ND	*stx1* (7); *stx2* (7)	
	Sausages ($n = 8$)	25	*eae* (0); *stx1* (50); *stx2* (0); *hlyA* (50); *hly* (0)	[43]
	Kofta ($n = 6$)	33.3	*eae* (0); *stx1* (50); *stx2* (50) *hlyA* (50); *hly* (0)	
	Luncheon ($n = 8$)	50	*eae* (0); *stx1* (0); *stx2* (0); *hlyA* (0); *hly* (0)	
	Chicken livers ($n = 6$)	50	*eae* (0); *stx1* (0); *stx2* (0); *hlyA* (0); *hly* (0)	
	Oysters ($n = 9$)	77.8	*eae* (0); *stx1* (0); *stx2* (0); *hlyA* (0); *hly* (0)	
	Calamari ($n = 7$)	57.1	*eae* (0); *stx1* (0); *stx2* (0); *hlyA* (0); *hly* (0)	
	Bivalves ($n = 7$)	100	*eae* (0); *stx1* (0); *stx2* (0); *hlyA* (0); *hly* (0)	
	Raw milk ($n = 6$)	66.7	*eae* (0); *stx1* (0); *stx2* (0); *hlyA* (0); *hly* (0)	
	Yogurt ($n = 4$)	100	*eae* (0); *stx1* (0); *stx2* (0); *hlyA* (0); *hly* (0)	
	Cheese ($n = 4$)	75	*eae* (0); *stx1* (0); *stx2* (0); *hlyA* (0); *hly* (0)	
	Cheese ($n = 4$)	75	*eae* (0); *stx1* (0); *stx2* (0); *hlyA* (0); *hly* (0)	

Table 1. *Cont.*

Country	Tested Food Samples (Total Number)	% of *E. coli*-Positive Samples or Isolates No.	Virulence Genes (%) (Out of Total Number) of *E. coli*-Positive Samples	References
Iraq	Beef carcasses (*n* = 120)	50 (41.6)	*eae* (30.7); *elt* (34.6); *esth* (34.6); *estp* (34.6); *stx*1 (53.8); *stx*2 (53.8); *aggR* (0)	[44]
	Imported chicken carcasses (*n* = 120)	52 (43.3)	*eae* (8.6); *elt* (91.3); *esth* (91.3); *estp* (91.3); *stx*1 (0); *stx*2 (0); *aggR* (8.6)	
	Fish surfaces (*n*= 120)	47 (39.1)	*eae* (0); *elt* (100); *esth* (100); *sstp* (100); *stx*1 (0); *stx*2 (0); *aggR* (0)	
	Imported and local raw burgers (*n* = 120)	45 (37.5)	*eae* (0); *elt* (71.4); *esth* (71.4); *estp* (71.4); *stx*1 (28.5); *stx*2 (28.5); *aggR* (0)	
	Local raw ground meat (*n* = 120)	46 (38.3)	*eae* (0); *slt* (100); *ssth* (100); *sstp* (100); *stx*1 (0); *stx*2 (0); *aggR* (0)	
	Local raw milk (*n* = 120)	43 (35.8)	*eae* (0); *elt* (0); *esth* (0); *estp* (0); *stx*1 (0); *stx*2 (0); *aggR* (0)	
	Fish (*n* = 78)	35.9	*stx1* (89.3); *stx1* (85.7); *rfb* (0)	[90]
	Camel meat (*n* = 50)	14	*sta* (100); *uspA* (42); *stb* (0); *stb* (0)	[72]
	Frozen burger (*n* = 50)	7	*sta* (100); *uspA* (42); *stb* (0); *lt* (0)	[91]
	Frozen chicken (*n* = 50)	8	*sta* (62.5); *uspA* (12.5); *stb* (0); *lt* (0)	
	Frozen fish (*n* = 50)	10	*sta* (40); *uspA* (10); *stb* (0); *lt* (0)	
Jordan	Drinking water (*n* = ND)	109 isolates	*aat* (12.8); *aaic* (2.8); *eae* (2.8); *ipaH* (1.8); *stx*1(0.9); *stx*2 (0)	[45]
Lebanon	Shankleesh (dairy products) (*n* = 340)	28.5	*eaeA* (13.5); *ehly* (8.1); *stx*1 (13.5); *stx*2 (13.5)	[47]
	Baladi (dairy products) (*n* = 340)	66.4	*eaeA* (2.7); *ehly* (5.4); *stx*1 (37.8); *stx*1 (37.8);	
	Kishk (dairy products) (*n* = 340)	7.2	*eaeA* (0); *ehly* (0); *stx*1 (10.8); *stx*1 (10.8);	
	Raw vegetables (*n* = ND)	60 isolates	*stx*1 (0); *stx*2 (0)	[92]
Morocco	Ground beef (*n* = 140)	45	*eaeA* (12.5); *aggA* (0); *stx*1 (4.7); *stx*2; (3.1); *lt* (0); *St* (0); *hlyA* (4.7); *Saa* (1.6); *astA* (4.7); *Ial* (0); *ipaH* (0); *iucD* (6.3); *cnf1* (0); *afa* (0); *sfa* (1.6)	[48]
	Sausages (*n* = 120)	30	*eaeA* (2.8); *aggA* (0); *stx*1 (2.8); *stx*2 (0); *Lt* (5.6); *St* (0); *hlyA* (0); *saa* (0); *astA* (27.8); *Ial* (0); *ipaH* (0); *iucD* (16.7); *cnf1* (0); *afa* (0); *pap* (5.6); *sfa* (0)	
	Turkey (*n* = 200)	35.5	*eaeA* (2.8); *aggA* (0); *stx*1 (0); *stx*2 (0); *lt* (0); *st* (1.4); *hlyA* (2.8); *saa* (0); *astA* (19.7); *ial* (0); *ipaH* (8.5); *iucD* (33.8); *cnf1* (0); *afa* (0); *pap* (2.8); *sfa* (0)	
	Well water (*n* = 50)	48	*eaeA* (0); *aggA* (0); *stx*1 (0); *stx*2 (0); *lt* (0); *st* (4.2); *hlyA* (4.4); *saa* (0); *astA* (0); *Ial* (0); *ipaH* (0); *iucD* (4.2); *cnf1* (0); *afa* (4.2); *pap* (0); *sfa* (0)	
	Shellfish (*n* = 82)	6.3	*eaeA* (0);*stx*1 (100); *stx*2 (60)	[93]
	Food products (*n* = 7200)	3.4	*hlyA* (4.3); *pap* (17.1); *sfa* (2.9); *stx*1 (10); *stx*2 (4.3); *eae* (4.3)	[94]
	Ground beef (*n* = 140)	2.1	*stx*1 (100); *stx*2 (66.7); *eaeA* (66.7); *hlyA* (100)	[95]
	Sausage (*n* = 120)	0.8	*stx*1 (100); *stx*2 (0); *eaeA* (0); *hlyA* (0)	

Table 1. *Cont.*

Country	Tested Food Samples (Total Number)	% of *E. coli*-Positive Samples or Isolates No.	Virulence Genes (%) (Out of Total Number) of *E. coli*-Positive Samples	References
Libya	Raw cow's milk (*n* = 28)	10.7	*vt* (33.3); *eaeA* (33.3)	[49]
	Raw camel's milk (*n* = 9)	33.3	*vt* (0); *eaeA* (0)	
	Raw goat's milk (*n* = 7)	28.6	*vt* (100); *eaeA* (100)	
	Fermented cow's milk (*n* = 28)	25	*vt* (75.7); *eaeA* (75.7)	
	Maasora cheese (*n* = 21)	42.9	*vt* (22.2); *eaeA* (22.2)	
	Ricotta cheese (*n* = 10)	30	*vt* (0); *eaeA* (0)	
Palestine	Raw beef (*n* = 300)	44 STEC isolates	*stx1* (68); *stx2* (63)	[96]
	Chicken meat (*n* = 15)	100	*vt* (0); *eaeA* (0); *bfpA* (0); *aggR* (6.6); *daaE* (0); *IT* (13.3); *sT* (46.6)	[50]
	Turkey (*n* = 10)	100	*vt* (0); *eaeA* (0); *bfpA* (0); *aggR* (0); *daaE* (0); *IT* (0); *sT* (20)	
Qatar	Chickens (*n* = 158)	65 APEC	*ompT* (69), *hlyF* (69%), *iroN* (68%); *tsh* (54%); *vat* (4%); *iss* (70%); *cvi/cva* (59%); *iucD* (65%)	[52]
Saudi Arabia	Raw beef (*n* = 100)	Two *E. coli* O157:H7 isolates	*stx1* (100); *stx2* (100); *eae* (50); *hlyA* (0)	[54]
	Raw mutton (*n* = 40)	One *E. coli* O157:H7 isolate	*stx1* (100); *stx2* (100); *eae* (100); *hlyA* (0)	
	Raw chicken (*n* = 40)	One *E. coli* O157:H7 isolate	*stx1* (100); *stx2* (100); *eae* (100); *hlyA* (0)	
	Ground beef (*n* = 80)	Four *E. coli* O157:H7 isolates	*stx1* (75); *stx2* (75); *eae* (0); *hlyA* (25)	
	Beef burger (*n* = 20)	Two *E. coli* O157:H7 isolates	*stx1* (50); *stx2* (0); *eae* (0); *hlyA* (50)	
	Ground chicken (*n* = 20)	One *E. coli* O157:H7 isolate	*stx1* (100); *stx2* (100); *eae* (0); *hlyA* (0)	
	Milk (*n* = 540)	15.93	*eaeA* (44.2); *stx2* (67.4)	[53]
	Raw meat (*n* = 150)	11.3	*eaeA* (58.8); *stx2* (94.1)	
	Fresh vegetables and fruits (*n* = ND)	16 *E. coli* isolates	*eae* (0); *stx1* (0); *stx2* (0)	[55]
Sudan	Drinking water (*n* = 184)	46	*IPaH* (12.7); *stx* (6.5); *AggR* (6.5); *eae* (0)	[97]
United Arab Emirates	Camel meat (*n* = 140)	4.3 (*E. coli* O157)	*rfbE* (100); *flicH7* (58.3); *hlyA* (75); *uidA* (0); *eaeA* (91.7); *stx2* (100); *stx1* (0)	[56]
	Goat (*n* = 150)	Two (*E. coli* O157)	*rfbE* (100); *flicH7* (0); *hlyA* (50); *uidA* (0); *eaeA* (100); *stx2* (100); *stx1* (0)	
	Cattle (*n* = 137)	1.5 (*E. coli* O157)	*rfbE* (100); *flicH7* (0); *hlyA* (60); *uidA* (0); *eaeA* (60); *stx2* (100); *stx1* (0)	

3. Pathogenic Antibiotic Resistance in *E. coli*

E. coli is the preferred organism when investigating bacterial resistance levels due to its ability to transfer genetic material not only among its own strains but also to other enteric pathogens [18]. In a study conducted in North Georgia, USA [98], 95 avian pathogenic *E. coli* (APEC) isolates were examined, revealing that 92% of them exhibited resistance to three or more antibiotics. A study by Yuan et al. [99] in China reported that 80% of 71 *E. coli* isolates from the livers of chickens that perished on 10 poultry farms displayed resistance to eight or more antibiotics. Similarly, between 2004 and 2005, Li et al. [100] identified high levels of antibiotic-resistant *E. coli* isolates from diseased chickens in China. These isolates demonstrated complete resistance to tetracycline and trimethoprim/sulfonamide, as well as resistance levels ranging from 79% to 83% to chloramphenicol, ampicillin, ciprofloxacin, and enrofloxacin. The presence of antibiotic resistance in commensal strains of *E. coli* could play a pivotal role in the dynamics of resistance and infectious diseases. European data from France, the UK, and the Netherlands indicate a moderate resistance pattern to ampicillin, streptomycin, tetracycline, and trimethoprim/sulfonamide; low resistance to gentamicin, chloramphenicol, and ciprofloxacin; and no resistance to cephalosporins [101].

Given the widespread detection of pathogenic *E. coli* in various food products (as shown in Table 1), there is a pressing need to enhance national and regional surveillance efforts aimed at monitoring antimicrobial resistance in *E. coli* within our food supply chain. In the primary production of animal-derived foods, there has been a noticeable increase in the use of traditional first-line antibiotics such as sulfonamides, chloramphenicol, ampicillin, tetracycline, and streptomycin. This surge in antibiotic use has led to the emergence and development of antibiotic-resistant *E. coli* due to the selective pressure it exerts [18].

As reported by Hemeg [102] in Saudi Arabia, all recovered pathogenic *E. coli* strains (carrying *stx2* and *eaeA* genes), including those from food samples (20 isolates) and individuals with colibacillosis (100 isolates), exhibited resistance to amoxicillin–clavulanic acid, penicillin, and erythromycin. Resistance rates among these *E. coli* strains included 83% for gentamicin, 75% for ampicillin, 65.3% for trimethoprim, 55.8% for oxytetracycline, 36.5% for chloramphenicol, 30.7% for norfloxacin, and 26.9% for nalidixic acid. Notably, 62.8% of the tested isolates remained sensitive to ciprofloxacin.

In Egypt, according to Elafify et al. [103], 36 STEC isolates recovered from milk and dairy products carried *stx1* and/or *stx2* genes, while 14 and 3 of those possessed the *eaeA* gene and the *rfbE* gene, respectively, exhibiting multidrug resistance. Approximately 86.11% of these isolates harbored extended-spectrum beta-lactamase encoding genes, specifically blaCTX-M-15, blaSHV-12, and blaCTX-M-14. Moreover, 33.33% of the isolates carried the plasmid-mediated quinolone resistance gene qnrS.

Given that pathogenic *E. coli* is associated with increased illness and mortality rates, assessing antimicrobial resistance profiles should be considered a crucial component of *E. coli* surveillance in food safety and public health laboratories across Arab countries. Studies in Arab countries, such as Algeria [104,105] and Iraq [106,107], have also shown antibiotic resistance in pathogenic E. coli isolated from food.

It is imperative to recognize that the extent of resistance serves as an informative indicator of the selection pressure resulting from antibiotic use and resistance issues in pathogens. Indiscriminate antibiotic use must be curtailed, as antibiotics may lose their effectiveness against pathogens, particularly since *E. coli* acquires antibiotic resistance at a faster rate than most other bacteria [18]. The escalating global prevalence of antibiotic resistance is a matter of significant concern. It is widely acknowledged that the primary driver of resistance development in pathogenic bacteria is the excessive use of antibiotics [108]. This prevailing situation has facilitated the emergence and dissemination of antibiotic-resistant bacteria and resistance genes. Antibiotic resistance can stem from antibiotic use for treatment in both humans and animals, as well as from prophylactic and growth promoting antibiotic use in animals [18].

4. Conclusions

Foodborne infections originating from bacterial pathogens like pathogenic *E. coli* are a prevalent cause of human illnesses in the Arab world, leading to significant economic losses and public health consequences. These *E. coli* pathogens' genetic material is frequently found in various food items across Arab countries. The existing evidence highlighted in this review emphasizes that the identification of these bacterial pathogens is common in animal-based food products. In contrast, when it comes to fruits and vegetables, the available data on these pathogens are limited compared to animal-derived foods. These bacteria can enter the human food supply chain from their initial production stages to the final consumption of products. The emergence of drug-resistant strains has raised serious concerns about public health regarding these bacterial pathogens. Despite some reports on the prevalence of foodborne bacteria in animal-based foods, livestock, and humans, the extent of these pathogens in animal-based foods within the Arab region remains insufficiently studied. The associated risk factors are not well defined, and there is a lack of comprehensive documentation on human infections resulting from foodborne exposure. This literature review underscores the persistent challenge posed by *E. coli* pathogens to food safety and public health in the Arab world. Consequently, we propose the following recommendations:

establish a coordinated surveillance and monitoring system for foodborne pathogens and their antimicrobial resistance at the national and regional levels across Arab countries to develop informed control and prevention strategies against these pathogens; generate epidemiological data on risk factors and the incidence of human infections linked to foodborne illnesses, focusing on national-level documentation; raise public awareness based on scientific risk analysis of bacterial pathogens responsible for foodborne infections; and employ advanced molecular-level characterization techniques, such as whole-genome sequencing, to guide the implementation of improved prevention and control strategies throughout Arab countries.

Author Contributions: Conceptualization, M.-Y.I.M.; formal analysis, M.-Y.I.M.; project administration, M.-Y.I.M. and I.H.; resources, I.H.; methodology, M.-Y.I.M.; supervision, I.H.; writing—original draft, M.-Y.I.M.; writing—review and editing, M.-Y.I.M. and I.H.; funding acquisition, I.H. All authors have read and agreed to the published version of the manuscript.

Funding: This research was funded by the ASPIRE Research Institute for Food Security in the Drylands (ARIFSID) project (Subtheme 4.1—One Health and Antimicrobial Resistance). ASPIRE is a Research and Development Funding Organization in the United Arab Emirates that works as the technology transition pillar of Abu Dhabi's Advanced Technology Research Council (ATRC).

Data Availability Statement: The data used to support the findings of this study can be made available by the corresponding author upon request.

Acknowledgments: I am grateful to my family for their support during this research work. I dedicate this work to my beloved mother soul (Siham Taha Salim Hamed), my beloved aunt soul (Somia Taha Salim Hamed), and my beloved uncle soul (Salim Taha Salim Hamed).

Conflicts of Interest: The author declares no conflict of interest.

References

1. EFSA. The European Union summary report on trends and sources of zoonoses, zoonotic agents and foodborne outbreaks in 2016. *EFSA J.* **2017**, *15*, e05077.
2. ECDC; EFSA. The European Union Summary Report on Antimicrobial Resistance in zoonotic and indicator bacteria from humans, animals and food in 2019–2020. *EFSA J.* **2022**, *20*, 7209. [CrossRef]
3. EFSA. The European Union summary report on trends and sources of zoonoses, zoonotic agents and foodborne outbreaks in 2017. *EFSA J.* **2018**, *16*, e05500.
4. Centers for Disease Control and Prevention (CDC). Preliminary FoodNet Data on the incidence of infection with pathogens transmitted commonly through food—10 states, 2009. *MMWR Morb. Mortal Wkly. Rep. J.* **2010**, *59*, 418–422.
5. World Health Organization (WHO). *WHO Estimates of the Global Burden of Foodborne Diseases: Foodborne Disease Burden Epidemiology Reference Group 2007–2015*; World Health Organization: Geneva, Switzerland, 2015.
6. Harb, A.; O'Dea, M.; Abraham, S.; Habib, I. Childhood diarrhoea in the Eastern Mediterranean region with special emphasis on non-typhoidal *Salmonella* at the human–food interface. *Pathogens* **2019**, *8*, 60. [CrossRef] [PubMed]
7. Rafei, R.; Hawli, M.; Osman, M.; Dabboussi, F.; Hamze, M. Distribution of emm types and macrolide resistance determinants among group A *streptococci* in the Middle East and North Africa. *J. Glob. Antimicrob. Resist.* **2020**, *22*, 334–348. [CrossRef] [PubMed]
8. Mohamed, M.-Y.I.; Abu, A.; Aziz, S.A.; Zakaria, Z.; Khan, A.R.; Habib, I. Public health significance of *Campylobacter jejuni*. *J. Biosci. Med.* **2021**, *9*, 100–112.
9. Paré, G.; Trudel, M.-C.; Jaana, M.; Kitsiou, S. Synthesizing information systems knowledge: A typology of literature reviews. *Inf. Manag.* **2015**, *52*, 183–199. [CrossRef]
10. Habib, I.; Mohamed, M.-Y.I.; Khan, M. Current State of *Salmonella*, *Campylobacter* and *Listeria* in the Food Chain across the Arab Countries: A Descriptive Review. *Foods* **2021**, *10*, 2369. [CrossRef]
11. Mohamed, M.-Y.I.; Saleha, A.A.; Jalila, A.; Khairani-Bejo, S.; Puan, C.L.; Bitrus, A.A.; Aliyu, A.B.; Awad, E.A. Occurrence of antibiotic resistant *Campylobacter* in wild birds and poultry. *Malays. J. Microbiol.* **2019**, *15*, 143–151. [CrossRef]
12. Nataro, J.P.; Kaper, J.B. Diarrheagenic *Escherichia coli*. *Clin. Microbiol. Rev.* **1998**, *11*, 142–201. [CrossRef] [PubMed]
13. Kaper, J.B.; Nataro, J.P.; Mobley, H.L.T. Pathogenic *Escherichia coli*. *Nat. Rev. Microbiol.* **2004**, *2*, 123. [CrossRef]
14. Smith, J.L.; Fratamico, P.M.; Gunther, N.W. Extraintestinal Pathogenic *Escherichia coli*. *Foodborne. Pathog. Dis.* **2007**, *4*, 134–163. [CrossRef]
15. Mohamed, M.-Y.I.; Jalila, A.A.A.; Saleha, A.A.; Zunita, Z.; Rashid, K.A. Occurrence of antibiotic resistant *C. jejuni* and *E. coli* in wild birds, chickens, humans, and the environment in Malay villages, Kedah, Malaysia. *Vet. Med-Czech.* **2022**, *67*, 298–308. [CrossRef]

16. Mohamed, M.-Y.I.; Jalila, A.; Saleha, A.A.; Zunita, Z.; Rashid, K.A.; Awad, E.A. Occurrence of antibiotic resistant *C. jejuni* and *E. coli* in wild birds, chickens, environment and humans from Orang Asli villages in Sungai Siput, Perak, Malaysia. *Am. J. Anim. Vet.* **2019**, *14*, 158–169. [CrossRef]
17. Winfield, M.D.; Groisman, E.A. Role of Nonhost Environments in the *Lifestyles* of *Salmonella* and *Escherichia coli*. *Appl. Environ. Microbiol.* **2003**, *69*, 3687–3694. [CrossRef]
18. Mohamed, M.-Y.I.; Jalila, A.; Zunita, Z.; Rashid, K.A.; Saleha, A.A.; Bitrus, A.A.; Habib, I. Multi-Drug Resistant Pathogenic *Escherichia coli* Isolated from Wild Birds, Chicken, and the Environment in Malaysia. *Antibiotics* **2022**, *11*, 1275. [CrossRef] [PubMed]
19. Habib, I.; Elbediwi, M.; Mohamed, M.-Y.I.; Ghazawi, A.; Abdalla, A.; Khalifa, H.O.; Khan, M. Enumeration, antimicrobial resistance and genomic characterization of extended-spectrum β-lactamases producing *Escherichia coli* from supermarket chicken meat in the United Arab Emirates. *Int. J. Food Microbiol.* **2023**, *398*, 110224. [CrossRef] [PubMed]
20. Habib, I.; Mohamed, M.-Y.I. Chapter 3—Foodborne infections in the Middle East. In *Food Safety in the Middle East*; Academic Press: Cambridge, MA, USA, 2022; pp. 71–107. [CrossRef]
21. Habib, I.; Al-Rifai, R.; Mohamed, M.-Y.I.; Ghazawi, A.; Abdalla, A.; Lakshmi, G.; Agamy, N.; Khan, M.A. Contamination Levels and Phenotypic and Genomic Characterization of Antimicrobial Resistance in *Escherichia coli* isolated from Fresh Salad Vegetables in the United Arab Emirates. Trop. *Med. Infect. Dis.* **2023**, *8*, 294. [CrossRef]
22. Russo, T.A.; Johnson, J.R. Proposal for a New Inclusive Designation for Extraintestinal Pathogenic Isolates of *Escherichia coli*: ExPEC. *J. Infect. Dis.* **2000**, *181*, 1753–1754. [CrossRef]
23. Madden, R.H.; Murray, K.A.; Gilmour, A. Carriage of Four Bacterial Pathogens by Beef Cattle in Northern Ireland at Time of Slaughter. *Lett. Appl. Microbiol.* **2007**, *44*, 115–119. [CrossRef] [PubMed]
24. Martinko, J.M.; Parker, J. *Brock Biology of Microorganisms*; Prentice-Hall: Upper Saffle River, NJ, USA, 2003.
25. Feng, P.C.H.; Jinneman, K.; Scheutz, F.; Monday, S.R. Specificity of PCR and Serological Assays in Detecting *Escherichia coli* Shiga Toxin Subtypes. *Appl. Environ. Microbiol.* **2011**, *77*. [CrossRef] [PubMed]
26. Habib, I.; Mohteshamuddin, K.; Mohamed, M.-Y.I.; Lakshmi, G.B.; Abdalla, A.; Alkaabi, A.B.A. Domestic Pets in the United Arab Emirates as Reservoirs for Antibiotic-Resistant Bacteria: A Comprehensive Analysis of Extended-Spectrum Beta-Lactamase Producing *Escherichia coli* Prevalence and Risk Factors. *Animals* **2023**, *13*, 1587. [CrossRef] [PubMed]
27. Baumann, D.; Salia, H.; Greune, L.; Norkowski, S.; Körner, B.; Uckeley, Z.M.; Schmidt, M.A. Multitalented EspB of Enteropathogenic *Escherichia coli* (EPEC) Enters Cells Autonomously and Induces Programmed Cell Death in human Monocytic THP-1 Cells. *Int. J. Med. Microbiol.* **2018**, *308*, 387–404. [CrossRef] [PubMed]
28. Nissim-Eliraz, E.; Nir, E.; Shoval, I.; Marsiano, N.; Nissan, I.; Shemesh, H.; Rosenshine, I. T3SS-dependent Microvascular Thrombosis and Ischemic Enteritis in Human Gut Xenografts Infected with Enteropathogenic *Escherichia coli*. *Infect. Immun.* **2017**, 00558-17.
29. Wang, L.; Zhang, S.; Zheng, D.; Fujihara, S.; Wakabayashi, A.; Okahata, K.; Hara-Kudo, Y. Prevalence of Diarrheagenic *Escherichia coli* in Foods and Fecal Specimens Obtained from Cattle, Pigs, Chickens, Asymptomatic Carriers, and Patients in Osaka and Hyogo, Japan. *Jpn. J. Infect. Dis.* **2017**, *70*, 464–469. [CrossRef]
30. Alonso, M.Z.; Sanz, M.E.; Irino, K.; Krüger, A.; Lucchesi, P.M.A.; Padola, N.L. Isolation of Atypical Enteropathogenic *Escherichia coli* from Chicken and Chicken-derived Products. *Br. Poult. Sci.* **2016**, *57*, 161–164. [CrossRef]
31. Ferhat, L.; Chahed, A.; Hamrouche, S.; Korichi-Ouar, M.; Hamdi, T.-M. Research and molecular characteristic of Shiga toxin producing *Escherichia coli* isolated from sheep carcasses. *Lett. Appl. Microbiol.* **2019**, *68*, 546–552. [CrossRef] [PubMed]
32. Ferhat, L.; Chahed, A.; China, B.; Assaous, F.; Daube, G.; Rahal, K. Research and characterization of *Escherichia coli* O157 strains isolated from ovine carcasses of two slaughterhouses of Algiers city. *HVM Bioflux* **2018**, *10*, 46–50.
33. Chahed, A.; China, B.; Mainil, J.; Daube, G. Prevalence of enterohaemorrhagic *Escherichia coli* from serotype O157 and other attaching and effacing *Escherichia coli* on bovine carcasses in Algeria. *J. Appl. Microbiol.* **2006**, *101*, 361–368. [CrossRef]
34. Mohamed, L.; Ge, Z.; Yuehua, L.; Yubin, G.; Rachid, K.; Mustapha, O.; Junwei, W.; Karine, O. Virulence traits of avian pathogenic (APEC) and fecal (AFEC) *E. coli* isolated from broiler chickens in Algeria. *Trop. Anim. Health Prod.* **2018**, *50*, 547–553. [CrossRef]
35. Dib, A.L.; Agabou, A.; Chahed, A.; Kurekci, C.; Moreno, E.; Espigares, M.; Espigares, E. Isolation, molecular characterization and antimicrobial resistance of enterobacteriaceae isolated from fish and seafood. *Food Control* **2018**, *88*, 54–60. [CrossRef]
36. Benameur, Q.; Gervasi, T.; Giarratana, F.; Vitale, M.; Anzà, D.; La Camera, E.; Nostro, A.; Cicero, N.; Marino, A. Virulence, antimicrobial resistance and biofilm production of *Escherichia coli* isolates from healthy broiler chickens in western algeria. *Antibiotics* **2021**, *10*, 1157. [CrossRef] [PubMed]
37. Laarem, M.; Barguigua, A.; Nayme, K.; Akila, A.; Zerouali, K.; El Mdaghri, N.; Timinouni, M. Occurrence of plasmid-mediated quinolone resistance and virulence genes in avian escherichia coli isolates from Algeria. *J. Infect. Dev. Ctries.* **2017**, *11*, 143–151. [CrossRef] [PubMed]
38. Galal, H.; Hakim, A.; Dorgham, S.M. Phenotypic and virulence genes screening of *Escherichia coli* strains isolated from different sources in delta Egypt. *Life Sci.* **2013**, *10*, 352–361. Available online: https://www.researchgate.net/publication/286572318 (accessed on 10 February 2013).
39. Saqr, S.; Khaliel, R.; Ibrahim, M. Antibiotic resistance and virulence genes of *E. coli* isolated from fresh Nile Tilapia (Oreochromis Niloticus) in El-Behera Governorate, Egypt. *Alex. J. Vet. Sci.* **2016**, *48*, 83. [CrossRef]

40. Merwad, A.; Gharieb, R.; Saber, T. Occurrence of shiga toxin-producing *Escherichia coli* in lactating cows and in contact workers in Egypt: Serotypes, virulence genes and zoonotic significance Emerging zoonoses View project Occurrence of shiga toxin producing *Escherichia coli* in lactating cows and in contact workers in Egypt: Serotypes, virulence genes and zoonotic significance. *Life Sci* **2014**, *11*, 563–571.
41. Mohammed, M.A.; Sallam, K.I.; Eldaly, E.A.Z.; Ahdy, A.M.; Tamura, T. Occurrence, serotypes and virulence genes of non-O157 Shiga toxin-producing *Escherichia coli* in fresh beef, ground beef, and beef burger. *Food Control* **2014**, *37*, 182–187. [CrossRef]
42. Hamed, O.M.; Sabry, M.A.; Hassanain, N.A.; Hamza, E.; Hegazi, A.G.; Salman, M.B. Occurrence of virulent and antibiotic resistant Shiga toxin-producing *Escherichia coli* in some food products and human stool in Egypt. *Vet. World.* **2017**, *10*, 1233–1240. [CrossRef]
43. Sahar, M.E.A.; Salwa, F.A.; Samy, A.S.; Mohamed, H.A.A.; Amira, M.Z.; John, D.K. Prevalence and characterization of Shiga toxin O157 and non-O157 enterohemorrhagic *Escherichia coli* isolated from different sources in Ismailia, Egypt. *Afr. J. Microbiol. Res* **2013**, *7*, 2637–2645. [CrossRef]
44. Taha, Z.M.; Yassin, N.A. Prevalence of diarrheagenic *Escherichia coli* in animal products in Duhok province, Iraq. *Iran. J. Vet. Res* **2019**, *20*, 255. [PubMed]
45. Swedan, S.; Alrub, H.A. Antimicrobial resistance, virulence factors, and pathotypes of *Escherichia coli* isolated from drinking water sources in Jordan. *Pathogens* **2019**, *8*, 86. [CrossRef] [PubMed]
46. Tarawneh, K.A.; Al-Tawarah, N.M.; Abdel-Ghani, A.H.; Al-Majali, A.M.; Khleifat, K.M. Characterization of verotoxigenic *Escherichia coli* (VTEC) isolates from faeces of small ruminants and environmental samples in Southern Jordan. *J. Basic Microbiol* **2009**, *49*, 310–317. [CrossRef] [PubMed]
47. Saleh, I.; Zouhairi, O.; Alwan, N.; Hawi, A.; Barbour, E.; Harakeh, S. Antimicrobial resistance and pathogenicity of *Escherichia coli* isolated from common dairy products in the Lebanon. *Ann. Trop. Med. Parasitol.* **2009**, *103*, 39–52. [CrossRef] [PubMed]
48. Badri, S.; Filliol, I.; Carle, I.; Hassar, M.; Fassouane, A.; Cohen, N. Prevalence of virulence genes in *Escherichia coli* isolated from food in Casablanca (Morocco). *Food Control* **2009**, *20*, 560–564. [CrossRef]
49. Garbaj, A.M.; Awad, E.M.; Azwai, S.M.; Abolghait, S.K.; Naas, H.T.; Moawad, A.A.; Gammoudi, F.T.; Barbieri, I.; Eldaghayes, I.M. Enterohemorrhagic *Escherichia coli* O157 in milk and dairy products from Libya: Isolation and molecular identification by partial sequencing of 16S rDNA. *Vet. World.* **2016**, *9*, 1184–1189. [CrossRef] [PubMed]
50. Adwan, G.M.; Alqarem, B.R.; Adwan, K.M. Prevalence of foodborne pathogens in meat samples in Palestine. *Int. Food Res. J.* **2015**, *22*, 1806.
51. Adam, M.A.; Wang, J.; Enan, K.A.; Shen, H.; Wang, H.; El Hussein, A.R.; Musa, A.B.; Khidir, I.M.; Ma, X. Molecular survey of viral and bacterial causes of childhood diarrhea in Khartoum State, Sudan. *Front. Microbiol.* **2018**, *9*, 112. [CrossRef]
52. Johar, A.; Al-Thani, N.; Al-Hadidi, S.H.; Dlissi, E.; Mahmoud, M.H.; Eltai, N.O. Antibiotic resistance and virulence gene patterns associated with avian pathogenic *Escherichia coli* (APEC) from broiler chickens in Qatar. *Antibiotics* **2021**, *10*, 564. [CrossRef] [PubMed]
53. Al-Zogibi, O.G.; Mohamed, M.I.; Hessain, A.M.; El-Jakee, J.K.; Kabli, S.A. Molecular and serotyping characterization of shiga toxogenic *Escherichia coli* associated with food collected from Saudi Arabia. *Saudi J. Biol. Sci.* **2015**, *22*, 438–442. [CrossRef]
54. Hessain, A.M.; Al-Arfaj, A.A.; Zakri, A.M.; El-Jakee, J.K.; Al-Zogibi, O.G.; Hemeg, H.A.; Ibrahim, I.M. Molecular characterization of *Escherichia coli* O157:H7 recovered from meat and meat products relevant to human health in Riyadh, Saudi Arabia. *Saudi J. Biol. Sci.* **2015**, *22*, 725–729. [CrossRef]
55. Abu-Duhier, F.M. *Escherichia coli* contamination of selected vegetables and fruits from markets of Tabuk city, Saudi Arabia: An anticipatory surveillance using real-time PCR for the presence of pathogenic strain *E. coli* O104:H4. *Int. J. Healthc. Biomed. Res.* **2015**, *4*, 126–134.
56. Al-Ajmi, D.; Rahman, S.; Banu, S. Occurrence, virulence genes, and antimicrobial profiles of *Escherichia coli* O157 isolated from ruminants slaughtered in Al Ain, United Arab Emirates. *BMC Microbiol.* **2020**, *20*, 210. [CrossRef]
57. Gomes, T.A.T.; Elias, W.P.; Scaletsky, I.C.A.; Guth, B.E.C.; Rodrigues, J.F.; Piazza, R.M.F.; Martinez, M.B. Diarrheagenic *Escherichia coli*. *Braz. J. Microbiol.* **2016**, *47*, 3–30. [CrossRef]
58. Pasqua, M.; Michelacci, V.; Di Martino, M.L.; Tozzoli, R.; Grossi, M.; Colonna, B.; Prosseda, G. The Intriguing Evolutionary Journey of Enteroinvasive, *E. coli* (EIEC) Toward Pathogenicity. *Front. Microbiol.* **2017**, *8*, 2390. [CrossRef]
59. Marier, R.; Wells, J.G.; Swanson, R.C.; Callahan, W.; Mehlman, I.J. An Outbreak of Enteropathogenic *Escherichia coli* Foodborne Disease Traced to Imported French Cheese. *Lancet* **1973**, *302*, 1376–1378. [CrossRef]
60. Escher, M.; Scavia, G.; Morabito, S.; Tozzoli, R.; Maugliani, A.; Cantoni, S.; Gesu, G.P. A Severe Foodborne Outbreak of Diarrhoea Linked to a Canteen in Italy Caused by Enteroinvasive *Escherichia coli*, an Uncommon Agent. *Epidemiol. Infect.* **2014**, *142*, 2559–2566. [CrossRef] [PubMed]
61. Pettengill, E.A.; Hoffmann, M.; Binet, R.; Roberts, R.J.; Payne, J.; Allard, M.; Morabito, S. Complete Genome Sequence of Enteroinvasive *Escherichia coli* O96: H19 Associated with a Severe Foodborne Outbreak. *Genome Announc.* **2015**, *3*, 10-1128. [CrossRef] [PubMed]
62. Newitt, S.; MacGregor, V.; Robbins, V.; Bayliss, L.; Chattaway, M.A.; Dallman, T.; Hawker, J. Two Linked Enteroinvasive *Escherichia coli* Outbreaks, Nottingham, UK, June 2014. *Emerg. Infect. Dis.* **2016**, *22*, 1178. [CrossRef]
63. Zhang, Y.; Tan, P.; Zhao, Y.; Ma, X. Enterotoxigenic *Escherichia coli*: Intestinal pathogenesis mechanisms and colonization resistance by gut microbiota. In *Gut Microbes*; Taylor and Francis: Abingdon, UK, 2022; Volume 14. [CrossRef]

64. Von Mentzer, A.; Connor, T.R.; Wieler, L.H.; Semmler, T.; Iguchi, A.; Thomson, N.R.; Pickard, D. Identification of Enterotoxigenic *Escherichia coli* (ETEC) Clades with Long-term Global Distribution. *Nat. Genet.* **2014**, *46*, 1321. [CrossRef]
65. Abdulrazzaq, K.M.; Oain, M.S.; Majeed, H.M.; Alhyani, O.H. Molecular detection of rfbO157, shiga toxins and hemolysin genes for *Escherichia coli* O157:H7 from canine feces in Tikrit and Mosul cities, Iraq. *Iraqi J. Vet. Sci.* **2021**, *35*, 325–329. [CrossRef]
66. Mühlen, S.; Ramming, I.; Pils, M.C.; Koeppel, M.; Glaser, J.; Leong, J.; Flieger, A.; Stecher, B.; Dersch, P. Identification of antibiotics that diminish disease in a murine model of enterohemorrhagic *Escherichia coli* infection. *Antimicrob. Agents Chemother.* **2020**, *64*, 10–1128. [CrossRef]
67. Amin, M.A.; Hashem, H.R.; El-Mahallawy, H.S.; Abdelrahman, A.A.; Zaki, H.M.; Azab, M.M. Characterization of enterohemorrhagic *Escherichia coli* from diarrhoeic patients with particular reference to production of Shiga-like toxin. *Microb. Pathog.* **2022**, *166*, 105538. [CrossRef] [PubMed]
68. Kehl, S.C. Role of the Laboratory in the Diagnosis of Enterohemorrhagic *Escherichia coli* Infections. *J. Clin. Microbiol.* **2002**, *40*, 2711–2715. [CrossRef]
69. Surendran-Nair, M.; Kollanoor-Johny, A.; Ananda-Baskaran, S.; Norris, C.; Lee, J.Y.; Venkitanarayanan, K. Selenium Reduces Enterohemorrhagic *Escherichia coli* O157:H7 Verotoxin Production and Globotriaosylceramide Receptor Expression on Host Cells. *Future Microbiol.* **2016**, *11*, 745–756. [CrossRef] [PubMed]
70. Yamasaki, E.; Watahiki, M.; Isobe, J.; Sata, T.; Nair, G.B.; Kurazono, H. Quantitative Detection of Shiga Toxins Directly from Stool Specimens of Patients Associated with an Outbreak of Enterohemorrhagic *Escherichia coli* in Japan—Quantitative Shiga Toxin Detection from Stool During EHEC Outbreak. *Toxins* **2015**, *7*, 4381–4389. [CrossRef] [PubMed]
71. Mohammed, M.A.M. Molecular characterization of diarrheagenic *Escherichia coli* isolated from meat products sold at Mansoura city, Egypt. *Food Control* **2012**, *25*, 159–164. [CrossRef]
72. Klaif, S.F.; Saleh, Z.F.; Hussein, M.T.; Jawad, A.A.; Jawad, M.S. Molecular characterization of enterohemorrhagic *E. coli* O157 and O153 isolated from tissue camel and human stool samples in Al-Diwaniyah, Iraq. *Iraqi J. Vet. Sci.* **2019**, *33*, 81–86. Available online: http://www.vetmedmosul.com (accessed on 8 September 2023). [CrossRef]
73. Nada, H.G.; El-Tahan, A.S.; El-Didamony, G.; Askora, A. Detection of multidrug-resistant Shiga toxin-producing *Escherichia coli* in some food products and cattle faeces in Al-Sharkia, Egypt: One health menace. *BMC Microbiol.* **2023**, *23*, 127. [CrossRef]
74. Auvray, F.; Bièche-Terrier, C.; Um, M.M.; Dupouy, V.; Nzuzi, N.; David, L.; Allais, L.; Drouet, M.; Oswald, E.; Bibbal, D.; et al. Prevalence and characterization of the seven major serotypes of Shiga toxin-producing *Escherichia coli* (STEC) in veal calves slaughtered in France. *Vet. Microbiol.* **2023**, *282*, 109754. [CrossRef]
75. Salih, M.H. Prevalence and characterization of virulence genes toxin-producing *Escherichia coli* enterohemorragic 0157:H7 Strain isolated from frozen imported bovine meat in Algeria. *Adv. Environ. Biol.* **2016**, *8*, 6–13. Available online: https://www.researchgate.net/publication/287314728 (accessed on 8 September 2023).
76. Rogawski, E.T.; Guerrant, R.L.; Havt, A.; Lima, I.F.N.; Medeiros, P.H.Q.S.; Seidman, J.C.; Bodhidatta, L. Epidemiology of Enteroaggregative *Escherichia coli* Infections and Associated Outcomes in the MAL-ED Birth Cohort. *PLoS Negl. Trop. Dis.* **2017**, *11*, e0005798. [CrossRef]
77. Rajan, A.; Robertson, M.J.; Carter, H.E.; Poole, N.M.; Clark, J.R. Enteroaggregative *E. coli* Adherence to Human Heparan Sulfate Proteoglycans Drives Segment and Host Specific Responses to Infection. *PLoS Pathog.* **2020**, *16*, e1008851. [CrossRef]
78. Petro, C.D.; Duncan, J.K.; Seldina, Y.I.; Allué-Guardia, A.; Eppinger, M.; Riddle, M.S.; Tribble, D.R.; Johnson, R.C.; Dalgard, C.L.; Sukumar, G. Genetic and Virulence Profiles of Enteroaggregative *Escherichia coli* (EAEC) Isolated from Deployed Military Personnel (DMP) With Travelers' Diarrhea. *Front. Cell. Infect. Microbiol.* **2020**, *10*, 200. [CrossRef]
79. Bejide, O.S.; Odebode, M.A.; Ogunbosi, B.O.; Adekanmbi, O.; Akande, K.O.; Ilori, T.; Ogunleye, V.O.; Nwachukwu, V.U.; Grey-Areben, A.; Akande, E.T.; et al. Diarrhoeal pathogens in the stools of children living with HIV in Ibadan, Nigeria. *Front. Cell. Infect. Microbiol.* **2023**, *13*, 1108923. [CrossRef]
80. Huang, D.B.; Nataro, J.P.; DuPont, H.L.; Kamat, P.P.; Mhatre, A.D.; Okhuysen, P.C.; Chiang, T. Enteroaggregative *Escherichia coli* is A Cause of Acute Diarrheal Illness: A Meta-analysis. *Clin. Infect. Dis.* **2006**, *43*, 556–563. [CrossRef] [PubMed]
81. Kotloff, K.L.; Nataro, J.P.; Blackwelder, W.C.; Nasrin, D.; Farag, T.H.; Panchalingam, S.; Breiman, R.F. Burden and Aetiology of Diarrhoeal Disease in Infants and Young Children in Developing Countries (The Global Enteric Multicenter Study, GEMS): A Prospective, Case-control Study. *Lancet* **2013**, *382*, 209–222. [CrossRef] [PubMed]
82. Mero, S.; Timonen, S.; Lääveri, T.; Løfberg, S.; Kirveskari, J.; Ursing, J.; Rombo, L.; Kofoed, P.E.; Kantele, A. Prevalence of diarrhoeal pathogens among children under five years of age with and without diarrhoea in guinea-bissau. *PLoS Negl. Trop. Dis.* **2021**, *15*, e0009709. [CrossRef]
83. Lounis, M.; Zhao, G.; Li, Y.; Gao, Y.; Wang, J.; Oumouna, M.; Oumouna, K. Molecular profile of avian pathogenic *Escherichia coli* (APEC) from poultry associated with colibacillosis in Algeria. *J. Hellenic. Vet. Med Soc.* **2020**, *71*, 2113. [CrossRef]
84. El-Jakee, J.K.; EL-Jakee, J.; Moussa, E.; Mohamed, K.F.; Mohamed, G. Using Molecular Techniques for Characterization of *Escherichia coli* Isolated from Water Sources in Egypt. *Glob. Vet.* **2009**, *3*, 354–362.
85. Ammar, A.M.; Abd El-Hamid, M.I.; Eid, S.E.A.; El Oksh, A.S. Insights into antimicrobial resistance and virulence genes of emergent multidrug resistant avian pathogenic *Escherichia coli* in Egypt: How closely related are they? Detection of Shiga-like Toxin Producing *Escherichia coli* in Food of Animal Origin by Street Vendors at Luxor City View project. *In. Revue. Méd. Vét.* **2015**, *166*, 304–314. Available online: https://www.researchgate.net/publication/285812668 (accessed on 8 September 2023).

86. Ombarak, R.A.; Hinenoya, A.; Awasthi, S.P.; Iguchi, A.; Shima, A.; Elbagory, A.R.M.; Yamasaki, S. Prevalence and pathogenic potential of *Escherichia coli* isolates from raw milk and raw milk cheese in Egypt. *Int. J. Food Microbiol.* **2016**, *221*, 69–76. [CrossRef] [PubMed]
87. Elmonir, W.; Abo-Remela, E.M.; Sobeih, A. Public health risks of *Escherichia coli* and *staphylococcus aureus* in raw bovine milk sold in informal markets in Egypt. *J. Infect. Dev. Ctries.* **2018**, *12*, 533–541. [CrossRef] [PubMed]
88. Fakhr, A.E.; Gohar, M.K.; Atta, A.H. Impact of Some Ecological Factors on Fecal Contamination of Drinking Water by Diarrheagenic Antibiotic-Resistant *Escherichia coli* in Zagazig City, Egypt. *Int. J. Microbiol.* **2016**, *2016*, 6240703. [CrossRef] [PubMed]
89. Elmonir, W.; Shalaan, S.; Tahoun, A.; Mahmoud, S.F.; Remela, E.M.A.; Eissa, R.; El-Sharkawy, H.; Shukry, M.; Zahran, R.N. Prevalence, antimicrobial resistance, and genotyping of Shiga toxin-producing *Escherichia coli* in foods of cattle origin, diarrheic cattle, and diarrheic humans in Egypt. *Gut Pathog.* **2021**, *13*, 1–11. [CrossRef] [PubMed]
90. Alttai, N.A.; Alsanjary, R.A.; Sheet, O.H. Detection of some virulence gene *stx1*, *stx2* and *rfb* of *Escherichia coli* isolated from fish in Nineveh governorate, Iraq. *Iraqi J. Vet. Sci.* **2023**, *37*, 453–457. [CrossRef]
91. Abbas, B.A.; Alghanim, A.M. 34 Detection of Virulence Genes in *Escherichia coli* Isolated from Frozen Meat in Basrah Market. *Basra J. Vet. Res.* **2016**, *15*, 134–138.
92. Faour-Klingbeil, D.; Kuri, V.; Fadlallah, S.; Matar, G.M. Prevalence of antimicrobial-resistant *Escherichia coli* from raw vegetables in Lebanon. *J. Infect. Dev. Ctries.* **2016**, *10*, 354–362. [CrossRef] [PubMed]
93. Bennani, M.; Badri, S.; Baibai, T.; Oubrim, N.; Hassar, M.; Cohen, N.; Amarouch, H. First detection of shiga toxin-producing *Escherichia coli* in shellfish and coastal environments of morocco. *Appl. Biochem. Biotechnol.* **2011**, *165*, 290–299. [CrossRef]
94. Nayme, K.; Barguigua, A.; Bouchrif, B.; Karraouan, B.; El Otmani, F.; Elmdaghri, N.; Zerouali, K.; Timinouni, M. Genotypic characterization of quinolone resistant-*Escherichia coli* isolates from retail food in Morocco. *J. Environ. Sci. Health—Part B Pestic. Food Contam. Agric. Wastes* **2017**, *52*, 107–114. [CrossRef]
95. Badri, S.; Fassouane, A.; Filliol, I.; Hassar, M.; Cohen, N. Detection of shiga toxin-producing *Escherichia coli* in meat marketed in casablanca (Morocco). *Cell. Mol. Biol.* **2011**, *57*, 1476–1479. [CrossRef]
96. Adwan, G.M.; Adwan, K.M. Isolation of shiga toxigenic *Escherichia coli* from raw beef in Palestine. *Int. J. Food Microbiol.* **2004**, *97*, 81–84. [CrossRef] [PubMed]
97. Adam, O.A.E.J. Detection of Virulence Genes of Diarrhea Genic *Escherichia coli* Strains, Isolated from Drinking Water in Khartoum State. Ph.D. Thesis, Sudan University of Science and Technology College of Graduate Studies, Khartoum, Sudan, 2017.
98. Zhao, S.; Maurer, J.J.; Hubert, S.; De Villena, J.F.; McDermott, P.F.; Meng, J.; White, D.G. Antimicrobial Susceptibility and Molecular Characterization of Avian Pathogenic *Escherichia coli* Isolates. *Vet. Microbiol.* **2005**, *107*, 215–224. [CrossRef]
99. Yuan, L.; Liu, J.H.; Hu, G.Z.; Pan, Y.S.; Liu, Z.M.; Mo, J.; Wei, Y.J. Molecular Characterization of Extended-spectrum β-lactamase-producing *Escherichia coli* Isolates from Chickens in Henan Province, China. *J. Med. Microbiol.* **2009**, *58*, 1449–1453. [CrossRef] [PubMed]
100. Li, L.; Jiang, Z.G.; Xia, L.N.; Shen, J.Z.; Dai, L.; Wang, Y.; Wu, C.M. Characterization of Antimicrobial Resistance and Molecular Determinants of Beta- lactamase in *Escherichia coli* Isolated from Chickens in China During 1970–2007. *Vet. Microbiol.* **2010**, *144*, 505–510. [CrossRef] [PubMed]
101. Bywater, R.; Deluyker, H.; Deroover, E.; De Jong, A.; Marion, H.; McConville, M.; Walters, J. A European Survey of Antimicrobial Susceptibility Among Zoonotic and Commensal Bacteria Isolated from Food-producing Animals. *J. Antimicrob. Chemother.* **2004**, *54*, 744–754. [CrossRef] [PubMed]
102. Hemeg, H.A. Molecular characterization of antibiotic resistant *Escherichia coli* isolates recovered from food samples and outpatient Clinics, KSA. *Saudi J. Biol. Sci.* **2018**, *25*, 928–931. [CrossRef] [PubMed]
103. Elafify, M.; Khalifa, H.O.; Al-Ashmawy, M.; Elsherbini, M.; El Latif, A.A.; Okanda, T.; Matsumoto, T.; Koseki, S.; Abdelkhalek, A. Prevalence and antimicrobial resistance of Shiga toxin-producing *Escherichia coli* in milk and dairy products in Egypt. *J. Environ. Sci. Health—Part B Pestic. Food Contam. Agric. Wastes* **2020**, *55*, 265–272. [CrossRef] [PubMed]
104. Messaili, C.; Messai, Y.; Bakour, R. Virulence gene profiles, antimicrobial resistance and phylogenetic groups of fecal *Escherichia coli* strains isolated from broiler chickens in Algeria. *Vet. Ital.* **2019**, *55*, 35–46. [CrossRef] [PubMed]
105. Meguenni, N.; Chanteloup, N.; Tourtereau, A.; Ahmed, C.A.; Bounar-Kechih, S.; Schouler, C. Virulence and antibiotic resistance profile of avian *Escherichia coli* strains isolated from colibacillosis lesions in central of Algeria. *Vet. World.* **2019**, *12*, 1840–1848. [CrossRef]
106. Allami, M.; Bahreini, M.; Sharifmoghadam, M.R. Antibiotic Resistance, Phylogenetic Typing and Virulence Genes Prole Analysis of Uropathogenic *Escherichia coli* Isolated from Patients in Southern Iraq. *J. Appl. Genet.* **2022**, *63*, 401–412. [CrossRef] [PubMed]
107. Al-Sa'ady, A.T.; Mohammad, G.J.; Hussen, B.M. Genetic relation and virulence factors of carbapenemase-producing Uropathogenic *Escherichia coli* from urinary tract infections in Iraq. *Gene Rep.* **2020**, *21*, 100911. [CrossRef]
108. Martinez, J.L. The Role of Natural Environments in the Evolution of Resistance Traits in Pathogenic Bacteria. *Proc. R. Soc. Lond. B Biol. Sci.* **2009**, *276*, 2521–2530. [CrossRef] [PubMed]

foods

MDPI

Article

Effect of Amino Acids on *Fusarium oxysporum* Growth and Pathogenicity Regulated by TORC1-*Tap42* Gene and Related Interaction Protein Analysis

Yijia Deng [1,2], Rundong Wang [1,2,*], Yuhao Zhang [1,3,*], Jianrong Li [2] and Ravi Gooneratne [4]

1 College of Food Science, Southwest University, Chongqing 400715, China; ikea7713@163.com
2 College of Food Science and Engineering, Bohai University, Jinzhou 121013, China; lijr6491@163.com
3 Chongqing Key Laboratory of Speciality Food Co-Built by Sichuan and Chongqing, Chongqing 400715, China
4 Department of Wine, Food and Molecular Biosciences, Faculty of Agriculture and Life Sciences, Lincoln University, Lincoln 7647, New Zealand
* Correspondence: wrd2011777@163.com (R.W.); zhy1203@163.com (Y.Z.); Tel.: +86-137-6309-4052 (R.W.); +86-159-2335-8038 (Y.Z.)

Abstract: Free amino acids (AAs) formed in fermented meat products are important nitrogen sources for the survival and metabolism of contaminating fungi. These AAs are mainly regulated by the TORC1-*Tap42* signaling pathway. *Fusarium* spp., a common contaminant of fermented products, is a potential threat to food safety. Therefore, there is an urgent need to clarify the effect of different AAs on *Fusarium* spp. growth and metabolism. This study investigated the effect of 18 AAs on *Fusarium oxysporum* (Fo17) growth, sporulation, T-2 toxin (T-2) synthesis and *Tri5* expression through *Tap42* gene regulation. Co-immunoprecipitation and Q Exactive LC-MS/MS methods were used to detect the interacting protein of Tap42 during specific AA treatment. *Tap42* positively regulated L-His, L-Ile and L-Tyr absorption for Fo17 colony growth. Acidic (L-Asp, L-Glu) and sulfur-containing (L-Cys, L-Met) AAs significantly inhibited the Fo17 growth which was not regulated by *Tap42*. The L-Ile and L-Pro addition significantly activated the sporulation of ΔFo*Tap42*. L-His and L-Ser inhibited the sporulation of ΔFo*Tap42*. In T-2 synthesis, ΔFo*Tap42* was increased in GYM medium, but was markedly inhibited in L-Asp and L-Glu addition groups. Dose–response experiments showed that 10–70 mg/mL of neutral AA (L-Thr) and alkaline AA (L-His) significantly increased the T-2 production and *Tri5* expression of Fo17, but *Tri5* expression was not activated in ΔFo*Tap42*. Inhibition of T-2 synthesis and *Tri5* expression were observed in Fo17 following the addition of 30–70 mg/mL L-Asp. KEGG enrichment pathway analysis demonstrated that interacting proteins of Tap42 were from glycerophospholipid metabolism, pentose phosphate pathway, glyoxylate and dicarboxylate metabolism, glycolysis and gluconeogenesis, and were related to the MAPK and Hippo signaling pathways. This study enhanced our understanding of AA regulation in fermented foods and its effect on *Fusarium* growth and metabolism, and provided insight into potential ways to control fungal contamination in high-protein fermented foods.

Keywords: amino acids; *Fusarium oxysporum*; T-2 toxin; *Tap42*; KEGG

Citation: Deng, Y.; Wang, R.; Zhang, Y.; Li, J.; Gooneratne, R. Effect of Amino Acids on *Fusarium oxysporum* Growth and Pathogenicity Regulated by TORC1-*Tap42* Gene and Related Interaction Protein Analysis. *Foods* **2023**, *12*, 1829. https://doi.org/10.3390/foods12091829

Academic Editors: Xinjun Du and Philip G. Crandall

Received: 9 March 2023
Revised: 26 April 2023
Accepted: 27 April 2023
Published: 28 April 2023

1. Introduction

Fermented meat (sausage, bacon, ham) and fermented dried aquatic (dried/smoked fish, dried shrimp) products are two types of flavor foods, processed under natural or control drying conditions. Flavor formation occurs through a series of biochemical and physical changes and depends on microorganism fermentation and/or endogenous enzyme action [1–3]. During fermentation, a number of aromatic substances, such as free amino acids (AAs), aldehydes, alcohols, acids, heterocyclic compounds and nucleotides, can be formed which impart a unique flavor to the food [4–7]. Due to lower water activity (aw) and higher salinity, fermented dry/semi-dry products can be stored for a long time with minimal microbial reproduction. However, unpackaged fermented products are at risk of contamination by environmental and suspended microorganisms in the air, including fungi, which adhere easily to the product surface. During long-term storage, protein breakdown in food forms free AAs to provide nutrients for fungal growth and metabolism. In addition, some fungal contaminants such as *Penicillium* spp., *Aspergillus* spp., *Fusarium* spp. and *Candida* spp. can utilize AAs to grow at aw < 0.9 [8,9] and contaminate fermented food [10,11]. Among the common contaminated fungi, *Fusarium* spp. are plant pathogenic fungi which generally infect high-carbon crops such as grain, corn, soybean and wheat to produce mycotoxins such as T-2 toxin (T-2), HT-2 toxin (HT-2), deoxynivalenol (DON and 15-acetyl deoxynivalenol (15Ac-DON) [12–14]. Among these, T-2 is highly toxic and causes weight loss [15], neurological disorders [16], immunosuppression [17], bone marrow damage [18] and cutaneous toxicity [19]. The *Tri5* gene plays an important role as the initiator of T-2 synthesis [20]. Several studies have reported that *Fusarium* contamination in dried/smoked fish and fermented meat products sold at markets poses a potential threat to consumer health [10,21–24]. Proteins in meat products are degraded to AAs during long-term storage due to microbial growth. Rabie et al. (2014) reported a reduction in free AA concentration in horse and beef sausage during 28-day storage period due to contaminating microorganism metabolism [25]. Although protein degradation into free AAs such as aspartic acid (L-Asp), glutamic acid (L-Glu) and alanine (L-Ala) can enhance the unique flavor of fermented products, long storage periods can lead to excessive proteolysis, resulting in the accumulation of free AAs, which provide conducive conditions for *Fusarium* growth and the production of spores and mycotoxins. Furthermore, fermented meat products can degrade into more than 22 AAs, including threonine, serine, glutamine, lysine, tyrosine and histidine. These AAs can be classified into acidic-, alkaline-, neutral and sulfur-containing AAs. Regulatory mechanism of each AA on fungal metabolism vary between species. At present, few studies have reported the effects of amino acid on *Fusarium* sp. growth and metabolism [26]. To understand the metabolic mechanism of *Fusarium* sp. on AAs, it is important to determine how fungi utilize different nutrients in high-protein fermented foods.

The target of rapamycin (TOR) protein is a key regulator of eukaryotic growth and a class of evolutionarily conserved serine/threonine (Ser/Thr) protein kinase [27]. The TORC1 signaling pathway includes two parts: (1) upstream activation pathway of TORC1, consisting of GTPases Gln and Gtr [28] regulated by *Gln3* [29] and *Gtr1/Gtr2* [30], respectively, allowing fungi to respond to AAs, stress and other environmental signals; (2) two branches of the downstream pathway, controlled by *Sch9* and *Tap42*, which regulate fungal growth and metabolism, respectively [31], including protein translation, ribosomal synthesis, gene transcription, protein degradation and autophagy [32]. Most studies on the regulation of AAs on fungal growth and metabolism have examined the TORC1 signaling pathway [33].

A complete response system was investigated in *Saccharomyces cerevisiae* under a limited nitrogen condition [34]. *Tap42*-dependent phosphorylation occurs when the TOR signal is activated, promoting the interaction between *Tap42* and phosphatase. When the external nutrient conditions are unfavorable, phosphatase isolated from Tap42 and many downstream targets of the TOR signaling pathway, such as *Npr1*, *Ure2* and *Gln3*, are dephosphorylated to regulate metabolism [35]. Therefore, *Tap42* is an important metabolic regulator used by fungi to respond to external AAs and control *Fusarium* metabolism.

The aim of this study was to elucidate the role of 18 AAs as nitrogen sources for *Fusarium* spp. growth (colony morphology) and metabolism (sporulation, toxin production, *Tr5* expression) regulated by the *Tap42* gene. Co-immunoprecipitation was used to capture the interacting proteins of Tap42 following exposure to specific AAs. This provides an experimental basis for controlling fungal contamination in fermented meat products from an AA perspective.

2. Materials and Methods

2.1. Chemicals and Experimental Strains

Biological reagents (≥99%) L-Aspartic acid (L-Asp), L-Glutamic acid (L-Glu), L-Serine (L-Ser), Glycine (Gly), L-Histidine (L-His), L-Threonine (L-Thr), L-Arginine (L-Arg), L-Proline (L-Pro), L-Alanine (L-Ala), L-Tyrosine (L-Tyr), L-Valine (L-Val), L-Methionine (L-Met), L-Cysteine (L-Cys), L-Isoleucine (L-Ile), L-Leucine (L-Leu), L-Tryptophan (L-Try), L-Phenylalanine (L-Phe) and L-Lysine (L-Lys) were purchased from Ruiyong Biological Technology (Shanghai, China). T-2 toxin standard was purchased from Enzo Life Science (Farmingdale, NY, USA). Chromatographic-grade reagents (≥99%) methanol, ethyl acetate, acetonitrile and formic acid were purchased from Sigma-Aldrich (Shanghai, China). Kanamycin, dithiothreitol (DTT), isopropyl-beta-D-thiogalactopyranoside (IPTG), and 1-(p-Toluenesulfonyl) imidazole ($C_{10}H_{10}N_2O_2$ S) were purchased from Xiya Chemical Technology (Qingdao, China). IP lysis, 1× loading buffer and trypsin were obtained from Sangon Biotech (Shanghai, China). Disodium hydrogen phosphate, sodium chloride, iodoacetamide, ammonium bicarbonate, ammonium dihydrogen phosphate, potassium chloride, magnesium sulfate heptahydrate, dipotassium hydrogen phosphate, ferrous sulfate heptahydrate and ammonium bicarbonate were purchased from Xilong Science (Shantou, China).

Wild-type *Fusarium oxysporum* (Fo17, GDMCC 60824, isolated from dried fish) has a strong T-2 toxin synthesis ability. Gene deletion strain ΔFo*Tap42* and complement strain ΔFo*Tap42*-C were obtained by the double-crossover homologous recombination method (in a preliminary experiment). Top10 *Escherichia coli* and Rossetta (DE3) competent cells were purchased from Kamede Biological (Tianjin, China).

2.2. Colony Growth Analysis

A dextrose agar medium (20 g glucose, 20 g agar 0.1 g chloramphenicol) containing 5 mg/mL each of 18 AAs was prepared. For the control group, PDA medium was used without AA addition. The wild-type Fo17, ΔFoT*ap42* and ΔFoT*ap42*-C were inoculated into a AA dextrose agar medium with a 5 mm mycelium disk, cultured for 7 d at 28 °C. The colony's morphology was observed and photographed.

2.3. Sporulation Capacity Analysis

A 1L Czapek dox agar (CDA) liquid medium (nitrogen (N) source removed, containing 1 g K_2HPO_4, 0.5 g KCl, 0.5 g $MgSO_4 \cdot 7H_2O$, 0.01 g $FeSO_4 \cdot 7H_2O$, 30 g saccharose) containing 10 mg/mL each of 18 AAs were prepared. The control group used a CDA liquid medium with an original nitrogen source and without AA addition. The wild-type Fo17, ΔFoT*ap42* and ΔFoT*ap42*-C were inoculated into the CDA medium with a 5 mm mycelium disk, followed by shaking at 120 rpm/min at 28 °C for 7 d. Next, spores in cultured solution were dispersed with a 84-1A magnetic agitator (Sile, Shanghai, China) at a speed of 500 rpm/min and filtered with three layers of gauze. The 50 μL filtrate was transferred to a blood cell-counting plate to calculate the number of spores using a CX23 optical microscope (Puch, Shanghai, China).

2.4. T-2 Toxin-Producing Ability Analysis

A 1L GYM liquid medium (N source removed, containing 0.2 g KCl, 0.2 g $MgSO_4 \cdot 7H_2O$, 10 g glucose, 5 g yeast extract, 1 mL 0.005 g/L $CuSO_4 \cdot 5H_2O$, 1 mL 0.01 g/L $ZnSO_4 \cdot 7H_2O$) containing 5 mg/mL each of 18 AAs were prepared. The control group used a GYM liquid medium with original N source and without AA addition. A 1 mL GYM solution was added into a 2 mL centrifuge tube. Next, wild-type Fo17, ΔFo*Tap42* and ΔFo*Tap42*-C were inoculated into the GYM medium with two 5 mm mycelium disks and cultured at 28 °C for 14 d. After culturing, the solution was centrifuged (Xiangyi, Changsha, China) at 5000 rpm for 10 min, the supernatant was collected, 1 mL ethyl acetate was added and the solution was mixed using a XW-80A vortex (Fudi, Fuzhou, China) for 5 min. The supernatant was collected and dried at 60 °C in N. Next, 1 mL 30% methanol was added to redissolve it; then, it was filtered with a 0.22 μm filter and T-2 was detected by LC-MS/MS.

Toxin analysis was performed on a Thermo Scientific Surveyor HPLC (Thermofisher, Waltham, MA, USA)system which comprised a Surveyor MS Pump Plus, an on-line degasser and a Surveyor Autosampler Plus coupled with a Thermo TSQ Quantum Access tandem mass spectrometer equipped with an electrospray ionization (ESI) source (Thermofisher, Waltham, MA, USA). The separation was performed at 35 °C using a Hypersil GOLD column (5 μm, 100 mm × 2.1 mm) (Thermofisher, Waltham, MA, USA) at a flow rate of 0.25 mL/min. The mobile phase consisted of methanol (A) and water containing 5 mM ammonium acetate 0.1% formic acid (B), with a gradient elution program as follows: 0 min 30% A, 3.0 min 90% A, 5 min 90% A and 3 min 30% MS/MS detection was carried out using a triple quadruple mass spectrometer coupled with an electrospray ionization source operating in positive (ESI+) mode (Shimadzu, Kyoto, Japan). The ionization source parameters were set as follows: spray voltage, 4500 V; sheath gas pressure, 35 au; ion sweep gas pressure, 0 au; auxiliary gas pressure, 15 au; capillary temperature, 350 °C; tube lens offset, 118 V; skimmer offset, 0; collision energy, 1.5 eV; and collision pressure, 1.5 mTorr.

2.5. *T-2 Synthesis and Tri5 Expression Dose-Response Relationship Analysis*

The 50 mL GYM liquid medium (N source removed) containing 10, 30, 50 or 70 mg/mL of L-Thr, L-His and L-Asp was prepared, and the *F. oxysporum* Fo17, ΔFo*Tap42* and ΔFo*Tap42*-C were inoculated into a 1 mL GYM medium with a 5 mm mycelium disk. Next, the samples were cultured at 28 °C and centrifuged at 120 rpm/min for 14 days. Then, the culture solution was treated as explained in Section 2.4, and T-2 was detected by LC-MS/MS.

The 25–50 mg (dry weight) cultured mycelia of Fo17, ΔFo*Tap42* and ΔFo*Tap42*-C were weighed and ground into powder with liquid N. Total RNA was extracted using a Spin Column Fungal Total RNA Purification Kit (Sangon Biotech, Shanghai, China). Purity and concentration were determined by a nucleic acid quantitative analyzer (Thermofisher, Waltham, MA, USA). A StarScript II First-strand cDNA Synthesis Kit (Genstar, Beijing, China) was used to transcribe RNA into cDNA. The HATri/F (5-CAGATGGAGAACtGGATGGT-3) and HATri/R (5-GCACAAGTGCCACGTGAC-3) were used as primer pairs, and the β-Tubulin/F (TTCCCCCGTCTCCACTTCTT) and β-Tubulin/R (GACGAGATCGTTCATGTTG) were used as the internal control gene primers. The reaction system of the qRT-PCR mixture was prepared as a 10 μL reaction (containing 5 μL 2× SYBR Green Supermix, 0.4 μL of each pair of primers (10 μM), 0.2 μL ROX Reference Dye, 1.0 μL cDNA and 3.0 μL PCR-certified water). The amplification reaction cycling procedure was as follows: 5 min at 95 °C, followed by 40 PCR cycles of 10 s at 95 °C for denaturation and 30 s at 60 °C for annealing and elongation. The melting curve analysis consisted of 3 s at 60 °C, followed by heating up to 95 °C with a ramp rate of 1 °C/3 s. If the melting curves showed a clear single peak, it meant the primers were specific. The experiment was conducted for each sample in triplicate. Based on the PCR reaction's CT value, *Tri5* expression was analyzed by the $2^{-\Delta\Delta Ct}$ method.

2.6. *TORC1-Tap42 Interacting Protein Analysis*

2.6.1. Construction of Tap42 Expression Vector

Based on the NCBI website (https://www.ncbi.nlm.nih.gov/, accessed on 27 May 2022) for the Tap42 sequence and related protein sequence, analyzed the protein signal peptide to obtain the corresponding base sequence. The Tap42 template was synthesized by chemical synthesis and cloned into a pET-28a(+) (3228 bp) expression vector by the BamHI-XhoI double enzyme digestion method. The constructed expression vector was transformed into a Top10 *E. coli* clone, coated on an agar plate (containing 30 μg/mL kanamycin) and cultured overnight at 37 °C. A single cloned strain with the correct sequence was selected. A 100 μL sample of Rossetta (DE3) competent cells was mixed with 10 μL pET28a(+)-TAP42 plasmid, bathed in ice for 30 min, quickly heated at 42 °C for 90 s and placed in an ice bath for 5 min. The solution was coated on LB medium (containing 50 μg/mL kanamycin) and cultured at 37 °C for 14 h. Monoclones were selected for the induction culture. The cloned strain was cultured at 37 °C for 200 mL until the OD 600 was 0.6–0.8. IPTG was added to the medium for induction at 16 °C for 24 h and centrifuged at 5000 rpm for 5 min. The supernatant was filtered by a 0.22 μm filter and then enriched by Ni column affinity chromatography.

2.6.2. Protein Purification by Ni Column Affinity Chromatography

A binding buffer (0.02 M Na_2HPO_4, 0.5 M NaCl, pH = 7.4) was used to balance the Ni column, with a flow rate of 3 mL/min. The supernatant, collected in Section 2.6.1, was purified using a AKTA purifier (Explorer 10, Cytiva, Sweden). After sample loading, in order to remove the impurities, the Ni column was washed with buffer (0.02 M NaH_2PO_4, pH = 7.4, 0.5 M NaCl, 20 mM 1-(p-Toluenesulfonyl) imidazole) until the UV signal returned to the baseline. Elusion buffer (0.02 M Na_2HPO_4, 0.5 M NaCl, 500 mM 1-(p-Toluenesulfonyl)imidazole, pH = 7.4) was used to wash the Ni column until the sample was eluted. The samples were collected for SDS-PAGE expression.

2.6.3. Co-Immunoprecipitation Analysis

(1) Sample treatment: The cultured mycelia were washed with precooled 0.01 M PBS 3 times and centrifuged at 12,000 rpm for 1 min each time to collect the mycelia. Liquid N was used to grind it into a powder; then, it was redissolved in 1 mL IP lysis, centrifuged at 12,000 rpm for 2 min to collect the supernatant and, finally, collected on ice. (2) Next, we prepared 100 μL Ni beads in triplicate, centrifuged them at 12,000 rpm and 4 °C for 1 min and removed the supernatant. Then, we added 800 μL IP lysate, centrifuged it at 12,000 rpm and 4 °C for 1 min and removed the supernatant. This step was repeated 3 times, and the final supernatant was resuspended in 200 μL IP lysate. (3) The 10 μg of Tap42 protein was prepared by adding it into the solution and incubating at 4 °C for 60 min, followed by dividing the solution into one tube. The sample supernatant prepared in step (1) was added to the tube, incubated at 4 °C for 1 h and centrifuged at 2000 rpm at 4 °C for 1 min to remove the supernatant. Next, 1.6 mL IP lysate was added and centrifuged at 2000 rpm 4 °C for 1 min to remove the supernatant. This step was repeated five times. (4) A 30 μL sample of 1× loading buffer was added into the sample tube and boiled at 100 °C for 10 min. Next, the sample was stored at −80 °C for further studies.

2.6.4. Sodium Dodecyl Sulfate-Polyacrylamide Gel Electrophoresis (SDS-PAGE)

The samples (from procedure Section 2.6.3) were defrosted and centrifuged at 12,000 rpm at 4 °C for 10 min. The supernatant was mixed with one volume of load buffer (187.5 mM Tris (pH 6.8), 6% SDS, 30% glycerol and 15% β-mercaptoethanol) and heated at 98 °C for 5 min to completely denature the proteins (as sample solutions for SDS-PAGE). Polyacrylamide gels (6%) were prepared in Bio-Rad chambers (separation gel: 2.5 μL (acrylamide 30%, bisacrylamide 0.8%), 1.5 μL Tris-HCl (1.5 M pH 8.8), 52.5 μL 10% SDS, 955 μL distilled water, 150 μL PSA, and 7.5 μL TEMED; stacking gel: 312 μL (acrylamide 30%, bisacrylamide 0.8%), 450 μL Tris-HCl (0.5 M pH 6.8), 1.0 μL distilled water, 50 μL PSA and 4.0 μL TEMED). The electrophoresis chamber was filled with running buffer (1.44% glycine, 0.3% Tris, and 0.1% SDS). Next, the sample solutions (10 μL) and 10 μL of protein marker were loaded onto the gel. Gel electrophoresis was performed at a constant voltage of 50 V until the sample reached the separation gel, at which point the voltage was increased to 100 V. The gels were stained with a Protein Stain Q Kit (Sangon Biotech, Shanghai, China). Gel images were processed using the Image lab Software (version 6.01, BioRad, Hercules, CA, USA), adjusting the gamma setting to improve the contrast.

2.6.5. LC-MS/MS Analysis of Tap42 Interacting Proteins

The electrophoretic target band was cut, transferred into a 50 mL centrifuge tube and rinsed twice with ultrapure water; then, a mixture of 25 mM NH_4HCO_3 and 50%

acetonitrile was added to decolorize for 30 min. To the extracted discolored solution, a dehydrated solution 1 of 50% acetonitrile was added and let stand for 30 min; then, it was sucked out and the dehydrated solution 2 of absolute ethyl alcohol was added and let stand for 30 min. The target band was freeze-dried in a vacuum (Xiangyi, Changsha China) at −20 °C for 2 h to obtain a lyophilized gel block, 50 μL reductive solution 1 was added (10 mM DTT, 25 mM NH_4HCO_3) and it was left to stand in a bath at 57 °C for 1 h. Next, the solution was removed, 50 μL reductive solution 2 was added (50 mm iodoacetamide, 25 mM NH_4HCO_3), it was placed at room temperature (RT) for 30 min and then the solution was removed. For 10 min, 10% ethyl alcohol was added and the solution was removed. Then, the dehydrated solution 1 was added for 30 min, it was removed and the dehydrated solution 2 was added for 30 min. The dehydrating solution 2 was removed and 10 μL enzymatic hydrolysate was added (25 mM NH_4HCO_3 containing 0.02 μg/μL trypsin) to hydrolyze overnight at 37 °C. Next, the solution was centrifuged at 12,000 rpm for 2 min to obtain the supernatant for protein analysis. A Thermofisher QE Orbitrap high field electrostatic field orbital trap mass spectrometer (Thermofisher, Waltham, MA USA) was used to detect the interacting proteins. The first-order spectrum scanning range was 350–1600 m/z; the second-order mass spectrometry mode was CID (Elite). A new protein library was constructed using the theoretical amylase sequence. The main search parameters were as follows: (1) immobilization modification: carbomimomethyl on Cys; (2) variable modification: oxidation on Met, acetylation on protein N-terminal; (3) allowable error of parent ion of peptide: 10 ppm; (4) allowable error of fragment ion: 0.6 Da.

2.7. Statistical Analysis

All experiments were conducted three times in parallel. The data were expressed as mean ± standard deviation (SD) and statistically analyzed by IBM SPSS statistics software (Version 26.0, BioRad, Hercules, CA, USA). Significant differences between the control and the treated fish were determined by one-way analysis of variance (ANOVA), followed by the Tukey's test to compare the control and treatment group values. A p-value of <0.05 was considered significant.

3. Results and Discussion

3.1. Effects of AAs on Colony Growth of F. oxysporum Regulated by Tap42

Rapamycin-sensitive TORC1 protein kinase is an important component of a conserved signal-cascading mechanism which controls cellular absorption and the response of AAs to regulate fungal growth, metabolism and pathogenicity [36]. The *Tap42* gene is the key regulator to absorb and metabolize AAs for fungi. The effects of AAs on the colony morphology of Fo17, ΔFo*Tap42* and ΔFo*Tap42*-C are shown in Figure 1. On PDA medium, Fo17 grew normally with several aerial hyphae, while ΔFo*Tap42* grew relatively slowly with fewer aerial hyphae. Colony morphology recovered to normal conditions following *Tap42* supplementation. The L-His, L-Leu, L-Pro, L-Thr and L-Tyr additions as nitrogen sources significantly activated the growth of Fo17, with larger hyphal diameters but fewer mycelia in the 7-d culture. The colony was significantly smaller in ΔFo*Tap42*, especially in the L-Try and L-Tyr media. This may be because the *Tap42* target-regulated the absorption of L-Try and L-Tyr, while a lack of *Tap42* decreased the response of the TORC1 pathway, which resulted in growth inhibition. In the L-Val medium, the colony of ΔFo*Tap42* was smaller, with radial mycelia, but recovered to the normal level in ΔFo*Tap42*-C. Sulfur-containing (L-Cys, L-Met) and acidic (L-Asp, L-Glu) AAs significantly inhibited the growth of Fo17, ΔFo*Tap42* and ΔFo*Tap42*-C, which indicated that the absorption of these two AAs was not regulated by *Tap42*. Supplementation with L-Cys and L-Met resulted in less mycelial growth, while L-Asp and L-Glu caused thicker hemispherical mycelia. These results are in accordance with Shiobara et al. [26], who reported an inhibitory effect of L-Met and L-Cys on *Fusarium grainei* growth. Hence, S-containing AAs can be used as mycotoxin antidotes [37] and fungal inhibitors [38] because the sulfhydryl groups prevent normal fungal growth and metabolism. The L-Asp and L-Glu were not conducive to Fo17 and ΔFo*Tap42* growth, possibly because acidic AAs significantly decreased the pH of the medium. *F. oxysporum* Fo17 prefers growth at pH > 6 (unpublished data).

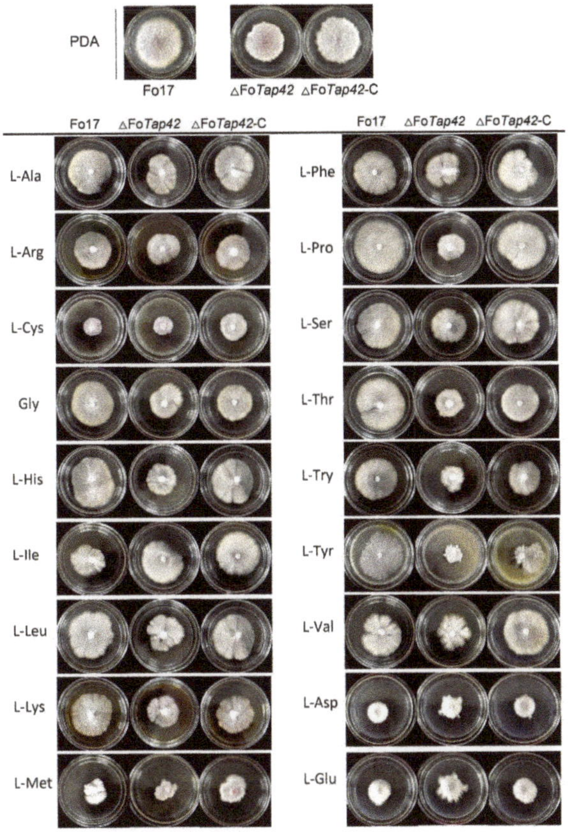

Figure 1. Colony morphology of wild-type *F. oxysporum* Fo17, ΔFo*Tap42* and ΔFo*Tap42*-C in dextrose agar medium, supplemented with 18 amino acids cultured for 7 d at 28 °C (n = 3). Control was cultured in the PDA medium.

3.2. Effect of AAs on Spore Production and T-2 Synthesis of F. oxysporum Regulated by Tap42

The spore production and T-2 synthesis of *F. oxysporum* Fo17, ΔFo*Tap42* and ΔFo*Tap42*-C were significantly different and regulated by AAs (Table 1). Compared to the control group (CDA medium), the spore production of Fo17 in the L-Arg, L-Lys, L-Met, L-Phe and L-Pro groups were significantly elevated ($p < 0.05$), especially in the L-Lys group (22.30×10^5 CFU/L, $p < 0.05$), but were decreased in the L-Asp, L-Cys, L-Gly, L-Leu, L-Thr, L-Try and L-Val groups, with values ranging from 3.14–7.50×10^5 CFU/L. Following *Tap42* deletion, the spore production of ΔFo*Tap42* significantly decreased ($p < 0.05$) in most groups with added AAs. In contrast, L-Arg, L-Ile, L-Lys, L-Met, L-Phe and L-Pro addition significantly increased ($p < 0.05$) the spore production of Fo17. This was most evident in the L-Ile- and L-Pro-added groups, especially the ΔFo*Tap42*, with values ranging from 35.80–36.26×10^5 CFU/L. Non-polar AAs such as L-Ile and L-Pro are generally hydrophobic, which is not conducive for the absorption of pathomycetes such as *Fusarium* sp.; thus, the speed of fungal growth and spore production is limited [39]. The spore production of *Fusarium* sp. Was positively regulated by *Tap42* in both the control and most AA-added groups, except L-Asp, Gly, L-Thr and L-Val. Alfatah et al. [40] reported that, in a nutrient-dependent developmental state, *Tap42-Sit4-Rrd1/Rrd2* signaled for the regulation of spore germination, while *Tap42* deletion inhibited the TORC1 pathway and blocked spore germination. Therefore, a lack of *Tap42* leads to inactivation of a part of TORC1 function, as well as an inability to respond to glucose and other exogenous AAs, resulting in reduced sporulation.

Table 1. Effects of 18 amino acids as nitrogen sources in T-2 toxin and spore production by *F. oxysporum* Fo17, ΔFo*Tap42* and ΔFo*Tap42*-C.

Group	Spore Concentration (10^5 CFU/L)			T-2 Toxin Concentration (ng/mL)		
	Fo17	**ΔFo*Tap42***	**ΔFo*Tap42*-C**	**Fo17**	**ΔFo*Tap42***	**ΔFo*Tap42*-C**
Control	12.35 ± 1.32^{c}	$3.51 \pm 1.12^{e*}$	$8.24 \pm 1.62^{bc*}$	22.86 ± 3.54^{bc}	$30.82 \pm 3.41^{b*}$	20.69 ± 1.82^{bc}
L-Ala	11.11 ± 2.25^{c}	$3.31 \pm 0.76^{e*}$	$7.61 \pm 1.24^{bc*}$	23.85 ± 4.34^{bc}	20.73 ± 4.61^{c}	25.63 ± 4.41^{b}
L-Asp	6.52 ± 0.64^{d}	8.75 ± 3.83^{de}	4.25 ± 0.71^{c}	14.87 ± 2.94^{c}	$6.68 \pm 2.65^{e*}$	14.43 ± 3.66^{d}
L-Arg	18.80 ± 1.35^{b}	$6.25 \pm 1.70^{de*}$	$12.50 \pm 1.13^{ab*}$	23.19 ± 4.01^{bc}	24.32 ± 3.83^{c}	23.39 ± 3.64^{b}
L-Cys	7.50 ± 1.32^{d}	$2.75 \pm 1.11^{e*}$	$4.62 \pm 0.92^{c*}$	33.64 ± 1.41^{a}	$22.23 \pm 2.08^{c*}$	28.21 ± 4.36^{ab}
L-Glu	10.52 ± 2.10^{c}	$14.80 \pm 4.62^{c*}$	$6.75 \pm 0.35^{c*}$	12.64 ± 1.40^{c}	$2.09 \pm 1.61^{e*}$	17.03 ± 5.27^{c}
Gly	5.56 ± 0.61^{de}	4.00 ± 1.23^{de}	5.64 ± 0.95^{c}	22.39 ± 0.36^{bc}	19.43 ± 4.05^{c}	19.08 ± 2.03^{c}
L-His	11.59 ± 2.46^{c}	$2.50 \pm 0.64^{e*}$	13.88 ± 1.66^{ab}	23.67 ± 3.72^{bc}	$36.38 \pm 0.64^{a*}$	23.87 ± 4.08^{b}
L-Ile	18.30 ± 3.65^{b}	$35.80 \pm 4.87^{a*}$	16.05 ± 1.45^{a}	28.49 ± 1.32^{ab}	28.03 ± 5.46^{b}	26.15 ± 4.47^{ab}
L-Leu	3.02 ± 0.57^{e}	$1.25 \pm 0.36^{e*}$	4.25 ± 1.04^{c}	27.07 ± 2.25^{ab}	$11.22 \pm 1.54^{d*}$	26.81 ± 1.44^{ab}
L-Lys	22.30 ± 3.55^{a}	$14.85 \pm 0.98^{c*}$	$16.32 \pm 3.52^{a*}$	21.51 ± 2.77^{bc}	27.78 ± 1.39^{bc}	18.71 ± 2.92^{c}
L-Met	17.20 ± 1.28^{b}	$24.21 \pm 1.16^{b*}$	$10.84 \pm 2.69^{b*}$	20.19 ± 1.54^{bc}	$29.25 \pm 2.02^{b*}$	19.43 ± 3.91^{c}
L-Phe	19.34 ± 13.35^{b}	$10.01 \pm 2.06^{d*}$	$14.30 \pm 0.98^{a*}$	30.49 ± 1.54^{a}	30.84 ± 3.05^{b}	26.95 ± 2.98^{ab}
L-Pro	18.05 ± 4.21^{b}	$36.26 \pm 4.53^{a*}$	15.80 ± 3.36^{a}	30.95 ± 5.17^{a}	24.29 ± 5.37^{c}	34.78 ± 4.19^{a}
L-Ser	12.24 ± 2.17^{c}	$2.56 \pm 0.53^{e*}$	14.32 ± 4.39^{a}	29.61 ± 2.97^{a}	22.02 ± 0.64^{c}	25.79 ± 4.33^{b}
L-Thr	4.01 ± 0.75^{e}	6.32 ± 1.27^{de}	5.11 ± 1.24^{cd}	25.96 ± 1.54^{b}	23.58 ± 1.36^{c}	24.85 ± 1.57^{b}
L-Try	7.25 ± 0.69^{d}	$3.21 \pm 0.28^{e*}$	6.25 ± 0.35^{c}	25.75 ± 1.61^{b}	26.41 ± 2.62^{bc}	26.56 ± 5.17^{ab}
L-Tyr	14.30 ± 3.22^{c}	$17.25 \pm 3.54^{c*}$	14.5 ± 2.12^{a}	25.26 ± 1.88^{b}	$31.22 \pm 4.92^{b*}$	28.03 ± 1.68^{ab}
L-Val	3.14 ± 0.53^{e}	5.25 ± 1.20^{e}	4.51 ± 1.31^{d}	25.54 ± 3.91^{b}	$38.15 \pm 0.64^{a*}$	23.93 ± 3.02^{b}

Three strains were cultured in CDA and GYM media (nitrogen removal sources) and supplemented with 18 amino acids for 14 d ($n = 3$). The original CDA and GYM media were the controls. Different letters (a–e) in the same column indicate significant differences ($p < 0.05$). Compared with Fo17, significant differences ($p < 0.05$) of ΔFo*Tap42* and ΔFo*Tap42*-C are marked with '*' in each row.

The T-2 toxin production of Fo17 was 22.86 ng/mL in the control group (GYM medium), which increased in ΔFo*Tap42* (30.82 ng/mL). When supplemented with *Tap42*, the T-2 production returned to normal. Contrary to the spore production capacity, this result indicated that a lack of the *Tap42* gene can induce T-2 synthesis of *F. oxysporum*. Targeted gene deletion has a significant effect on fungal metabolism, and sometimes will specifically activate other metabolism pathways. He et al. [1] reported that inactivation of TORC1-Tap42 leads to autophagy in yeast. The addition of neutral AAs, such as L-Ile, L-Leu and L-Ser, increased T-2 production of Fo17, with values reaching up to 28.49, 27.07 and 29.61 ng/mL, respectively. Similarly, L-Phe, L-Pro and L-Cys significantly activated ($p < 0.05$) the T-2 production in Fo17, with values up to 30.49, 30.95 and 33.64 ng/mL, respectively ($p < 0.05$). However, acidic AAs L-Asp and L-Glu, as N sources, significantly decreased ($p < 0.05$) the T-2 toxin-production in Fo17 to 14.87 and 12.64 ng/mL, respectively ($p < 0.05$). After *Tap42* deletion, the T-2 production was significantly decreased ($p < 0.05$) to 6.68 and 2.09 ng/mL. Thus, acidic AAs were not conducive to *F. oxysporum* growth and secondary metabolism, and sporulation was the main path to survival. However, the inhibition of toxin production does not mean toxicity reduction. When pathogenic fungi are exposed to pH stress, AAs are used for biotransformation and to drive environmental alkalinization for growth. *Candida albicans* uses AAs as the sole N source to raise the environmental pH from 4 to neutral within 6–8 h, which is another critical pathogenicity trait [41].

Compared with Fo17, the T-2-producing capacity of ΔFo*Tap42* was significantly increased ($p < 0.05$) by L-His and L-Val addition, which indicates that *Tap42* negatively regulates L-His and L-Val absorption. The L-Asp and L-Glu addition significantly inhibited the T-2 synthesis by ΔFo*Tap42* to only 6.68 and 2.09 ng T-2/mL, respectively ($p < 0.05$). L-Ala, L-Arg, L-Cys, Gly, L-Leu, L-Pro, L-Ser, L-Thr and L-Try addition as N sources also reduced the T-2 production by ΔFo*Tap42*, which may be due to a lack of AA absorption and metabolism by the corresponding pathway gene (*Tap42*), resulting in a decline in T-2 production. A similar result was reported by Phasha et al. [42], who similarly demonstrated that the *Ras2* gene can control growth and toxin synthesis of *F. circinatum* though encoding mitogen-activated protein kinase.

As shown in Figure 2, heat map analysis showed the effect patterns of AAs on sporulation and T-2 production by Fo17, ΔFo*Tap42* and ΔFo*Tap42*-C. During sporulation, L-Ile and L-Pro groups clustered closely together, showing significantly activated ($p < 0.05$) spore production of ΔFo*Tap42*, but a reduction in ΔFo*Tap42*-C, which indicates that *Tap42* negatively regulated L-Ile and L-Pro absorption and, thereby, affected the *F. oxysporum* spore formation. In contrast, in the L-His and L-Ser groups, the spore production of ΔFo*Tap42* was lower than in Fo17 and ΔFo*Tap42*-C, meaning that *Tap42* positively activated L-His and L-Ser absorption in *F. oxysporum*. When a lack of a regulated gene markedly reduces spore production, it indicates a decrease in reproductive dispersal capacity. The L-Ala group clustered with the control group means that L-Ala addition did not significantly affect the spore production of Fo17, ΔFo*Tap42* or ΔFo*Tap42*-C (Figure 2A). L-Asp and L-Glu addition inhibited T-2 production by Fo17, and a lack of *Tap42* significantly inhibited the T-2 production (<20 ng/mL). This is because acidic AAs negatively regulated the pathogenicity of *Fusarium* spp.. The L-His, L-Val, L-Lys and L-Met groups clustered together, and the T-2 production was significantly increased when *Tap42* was lacking (Figure 2B).

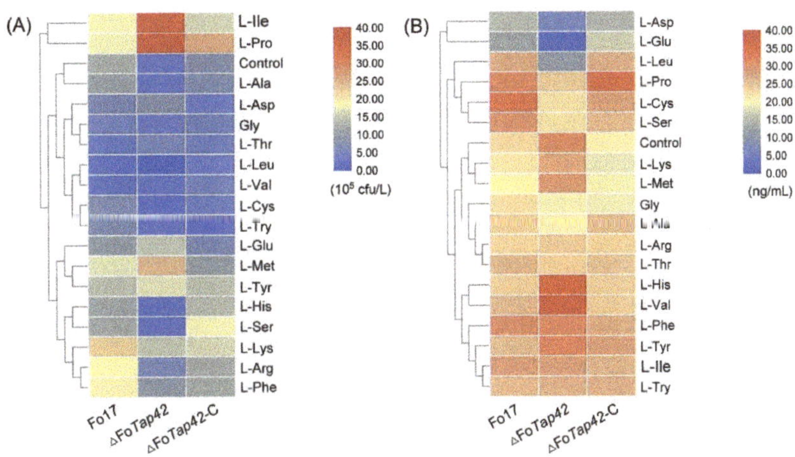

Figure 2. Changed patterns of sporation (**A**) and T-2 synthesis (**B**) by *F. oxysporum* Fo17, ΔFo*Tap42* and ΔFo*Tap42*-C exposed to 18 amino acids as nitrogen sources ($n = 3$). Control was cultured in CDA and GYM media.

3.3. Dose-Effect Relationship between L-Thr, L-His and L-Asp on T-2 Toxin Synthesis by F. oxysporum Regulated by Tap42

Tri5 is the initiating gene of trichothecene synthesis, and plays an important role in *Fusarium* spp. pathogenicity [43]. In this study, neutral (L-Thr), basic (L-His) and acidic (L-Asp) AAs were selected to assess the effect on T-2 synthesis and *Tri5* expression. Increasing L-Thr in the GYM medium significantly activated the T-2 production of Fo17, ΔFo*Tap42* and ΔFo*Tap42*-C ($p < 0.05$) with the concentrations of T-2 at 70 mg/mL L-Thr, 13.86, 16.87

and 13.85 ng/mL, respectively. Compared with the control group, a significant increase ($p < 0.05$) in *Tri5*'s expression of Fo17 was observed in the 50–70 mg/mL L-Thr groups. Δ*FoTap42* showed a marked activation of *Tri5* expression in the 10–70 mg/mL L-Thr groups. When complemented with the *Tap42* gene, the *Tri5* expression was significantly increased ($p < 0.05$), but only in the 70 mg/mL L-Thr group (Figure 3A). Similarly, 10–70 mg/mL L-His addition significantly increased ($p < 0.05$) the T-2 production of Fo17, ΔFo*Tap42* and ΔFo*Tap42*-C, with the T-2 concentrations reaching 33.03, 31.66 and 32.75 ng/mL, respectively, in the 70 mg/mL L-His added group. However, the *Tri5* expression of Fo17 increased, but the increase in amplitude of ΔFo*Tap42* and ΔFo*Tap42*-C were lower than in Fo17 (Figure 3B). In the L-His group, higher T-2 production and lower *Tri5* expression were observed in ΔFo*Tap42*, which may due to the lack of the TORC1-*Tap42* gene, thus reducing the sensitivity to L-His and, therefore, reducing *Tri5* signal transduction. However, T-2 synthesis is not only regulated by *Tri5*. Multiple trichothecene synthetic gene clusters, such as three P450 oxygenase genes, *Tri4*, *Tri11* and *Tri13* [44]; esterase gene *Tri8* [45]; and acetyltransferase genes *Tri3* and *Tri7* [46] also affect T-2 synthesis. Conversely, 10 mg/mL L-Asp addition significantly increased ($p < 0.05$) T-2 production and *Tri5* expression, while L-Asp addition gradually reduced T-2 production and *Tri5* expression in Fo17, ΔFo*Tap42* and ΔFo*Tap42*-C, which indicated that a low dosage of L-Asp led to a significant increase in the growth and T-2 production of *F. oxysporum*, whereas a high dosage reduced this effect. An inhibitory effect of *Tri5* expression was detected in ΔFo*Tap42* following the addition of 70 mg/mL L-Asp. Thus, acidic AA L-Asp showed a significant reduction in *Fusarium* sp. growth and metabolism at a relatively high dose (Figure 3C). The inhibitory effect of L-Asp has been well-studied and applied for fungal inhibition in food products. Bitu et al. [47] used L-Asp combined with Th (IV) and Zr (IV) ions to synthesize new peroxo complexes, and demonstrated high antibacterial activity in *Aspergillusflavus*, *Penicillium* sp., *Candida* sp. and *Aspergillus niger*. Thus, based on its excellent antifungal properties, L-Asp can be used as an additive in the development of natural preservatives.

3.4. L-Thr, L-His and L-Asp Activated Tap42 Interacting Proteins and Metabolic Pathways

A Tap42 expression vector was constructed using BamHi-XhoI restriction enzyme digestion. Two bands (3359 bp and 5192 bp) were observed in the pET-28a (+) plasmid vector after PCR amplification, indicating that the target gene, *Tap42*, was cloned into the expression vector (Figure 4A). Through comparative analysis, the pET28a(+)-Tap42 expression vector was found to be ~140 kDa. *E. oil* was transformed with the correctly identified positive clone plasmid, and single bacterial colonies were selected for induction expression. The results of SDS-PAGE electrophoresis are shown in Figure 4B. The recombinant strain induced by IPTG showed a fusion protein, pET28a(+)-Tap42, with a target band of ~140 kDa (electrophoretic band 2–4), but not the non-induced strain (electrophoretic band 1). A purified protein (~140 kDa) was obtained after washing and eluting the protein (electrophoretic band 3) (Figure 4C). Based on SDS-PAGE, no significant interacting proteins of Fo17 were detected in the GYM medium (electrophoretic band 1). In the L-His group, clear protein bands were detected at ~10 and 39–43 kDa (electrophoretic band 2). In the L-Thr group, as well, clear protein bands were evident at ~10, 27 and 39–52 kDa (electrophoretic band 3). In the L-Asp group, bands were found at ~10, 27 and 39–52 kDa (electrophoretic band 4) (Figure 4D).

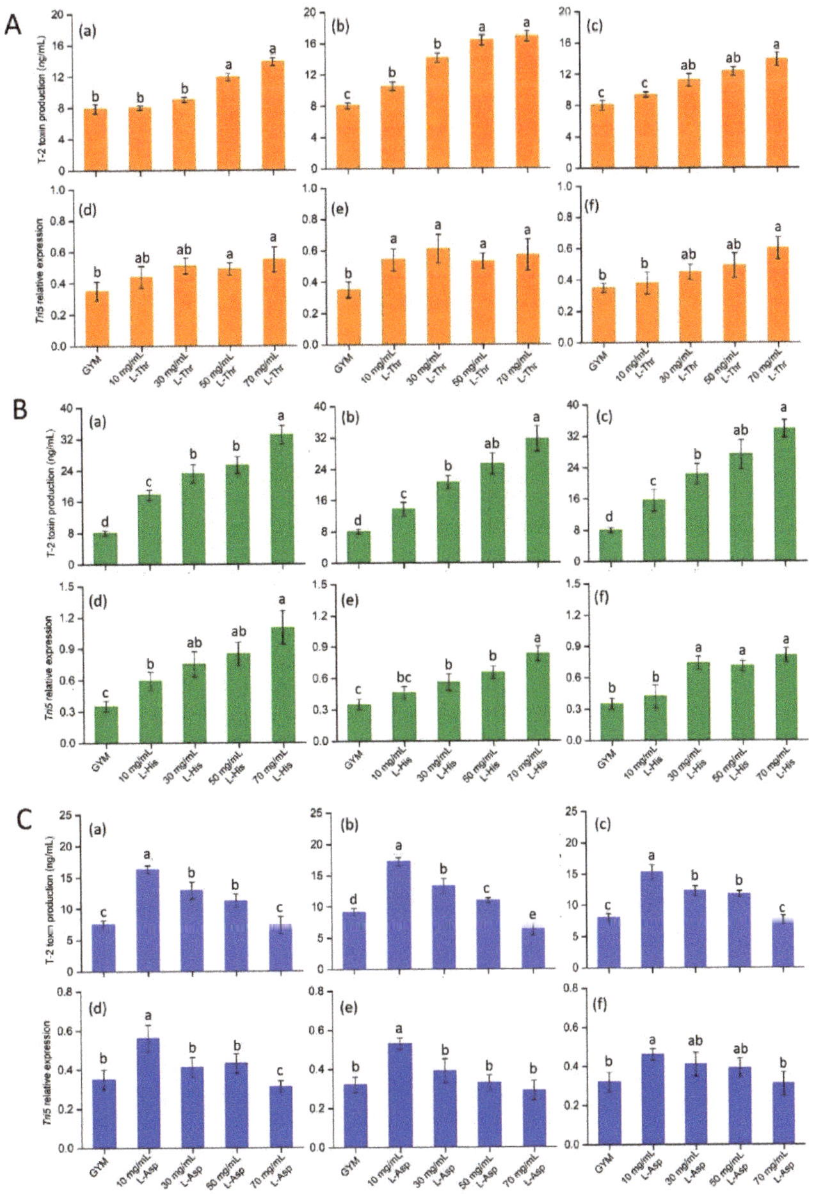

Figure 3. Effects of L-Threonine (L-Thr) (**A**), L-Histidine (L-His) (**B**) and L-Aspartic acid (L-Asp) (**C**) on T-2 toxin production and *Tri5* expression of different strains (n = 3), cultured at 28 °C for 14 d. The control was cultured in GYM medium. (**a–c**): T-2 toxin production of wild-type Fo17, ΔFo*Tap42* and ΔFo*Tap42*-C; (**d–f**): *Tri5* expression of wild-type Fo17, ΔFo*Tap42* and ΔFo*Tap42*-C. Different letters "a–d" in the same plot indicate significant differences (p < 0.05).

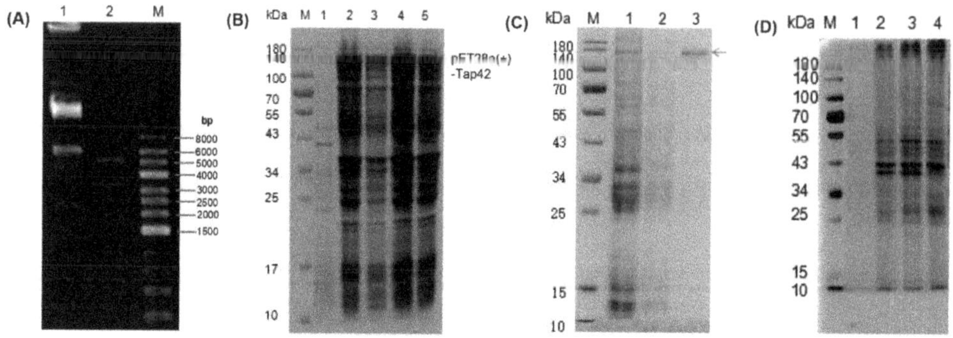

Figure 4. (**A**) PCR amplification analysis of a plasmid digested by the XbaI-XhoI enzyme. A1: plasmid; A2: XbaI-XhoI digestion plasmid; M: DNA marker. (**B**) Prokaryotic expression of pET28a(+)-Tap42 fusion protein. B1: protein not induced by IPTG, B2-B4: clone 1–3 protein induced by IPTG. (**C**) Purification of pET28a(+)-Tap42 fusion protein. M: marker; C1: unpurified protein; C2: washed protein; C3: eluted protein; M: molecular protein marker. (**D**) Co-immunoprecipitation products of Tap42 protein identified by SDS-PAGE. D1: GYM control group, D2: L-Histidine (L-His)-treated group, D3: L-Threonine (L-Thr)-treated group; D4: L-Aspartic acid (L-Asp)-treated group, M: molecular protein marker.

By Q Exactive LC-MS/MS analysis, 172, 201 and 89 potential interacting proteins of Tap42 were detected in the L-Thr, L-His and L-Asp treatment groups, respectively. A total of 86 interacting proteins co-existed in the three treatment groups. In the L-Thr and L-His groups, 84 similar interacting proteins were detected. However, only 1 and 2 interacting proteins from the L-Thr and L-His groups were similar to those observed in the L-Asp group (Figure 5). Based on mass analysis, the common potential interacting proteins with Tap42 treatment after L-Thr, L-His and L-Asp treatments were mostly DNA topoisomerase 2, glyceraldehyde-3-phosphate dehydrogenase 2, heat shock 70 kDa protein, elongation factor 1-alpha, actin, plasma membrane ATPase, enolase, transaldolase and ATP-dependent RNA helicase and citrate synthase (Table 2). Comparative analysis showed that the L-His and L-Asp groups specifically activated the ribosome-associated molecular chaperone SSB1, regulated by *SSB1*, and phosphoglycerate kinase regulated by *pgkA*, respectively. In the L-Thr and L-His groups, pyruvate decarboxylase regulated by *pdcA* was markedly activated. L-His specifically activated the Tubulin alpha-B chain regulated by *tba-2* (Table 3). DNA topoisomerases were the enzymes mostly responsible for the encapsulation of double-helix DNA and the transcription of proteins, and were shown to play an important role in the growth and metabolism of fungi [48]. *Candida albicans* and *Aspergillus niger* have high levels of both type I and type II topoisomerases, and increase pathogenicity by converting cellular proteins [49]. Therefore, the inhibition of DNA topoisomerases activity is also an important target of antifungal drugs [50]. Glyceraldehyde 3-phosphate dehydrogenase is an enzymatic protein highly related to the catalyzation of oxidation (dehydrogenation) and phosphorylation of glyceraldehyde 3-phosphate to generate 1, 3-diphosphate glyceric acid, which is the central link to the glycolytic pathway and, hence, plays an important role in glycometabolism [43]. In addition, Tap42-related enolase and pyruvate decarboxylase are important enzymes in the glycolytic pathway, and play important roles in maintaining cell wall protection and fungal pathogenicity [51]. The interaction of Tap42 with these enzymatic proteins cultured in GYM medium indicates that the growth and metabolism of *Fusarium* firstly utilize glucose through the glycolytic pathway to acquire energy.

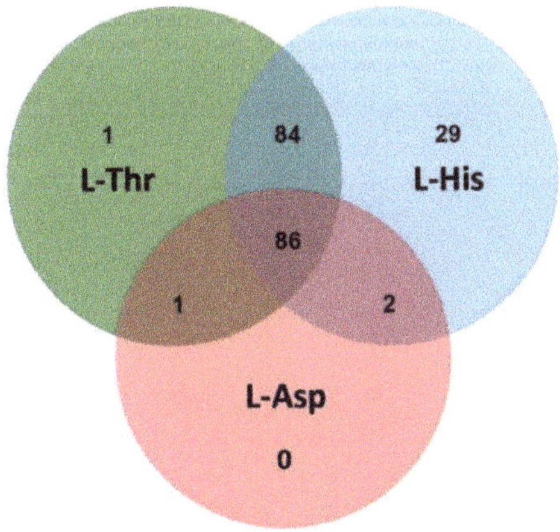

Figure 5. The common and specific numbers of potential Tap42-interacting proteins in *F. oxysporum* following L-threonine (L-Thr), L-Histidine (L-His) and L-Aspartic acid (L-Asp) treatments (n = 3).

Table 2. Common potential interacting proteins of the Tap42 immunoprecipitation products treated by L-Thr, L-His and L-Asp, identified by mass spectrometry.

Accession	Gene Name	Protein Name	Molecular Mass	Score
O93794	TOP2	DNA topoisomerase 2	161,402	1039
P17730	gpd2	Glyceraldehyde-3-phosphate dehydrogenase 2	36,198	904
P54117	GPDA	Glyceraldehyde-3-phosphate dehydrogenase	36,405	795
P06786	TOP2	DNA topoisomerase	164,626	784
P35143	GPDA	Glyceraldehyde-3-phosphate dehydrogenase	36,500	712
Q01233	hsps-1	Heat shock 70 kDa protein	70,738	683
Q01372	tef-1	Elongation factor 1-alpha	49,983	612
Q6TCF2	ACT	Actin	41,809	604
Q07421	PMA1	Plasma membrane ATPase	99,393	554
Q6RG04	ENO1	Enolase	47,388	553
Q76KF9	enoA	Enolase	47,264	545
Q01765	TEF	Elongation factor 1-alpha	50,215	500
P23704	atp-2	ATP synthase subunit beta	55,556	464
A0A075DVI9	FPRO05_10296	Transaldolase	35,571	389
P10592	SSA2	Heat shock protein SSA2	69,599	368
P37211	atp-1	ATP synthase subunit alpha	59,713	318
P07038	pma-1	Plasma membrane ATPase	100,280	299
Q96X45	cot-3	Elongation factor 2	93,545	270
A7EGL7	tif1	ATP-dependent RNA helicase eIF4A	45,088	237
Q8X097	B14D6.310	ATP-citrate synthase subunit 1	73,037	201
P28876	pma2	Plasma membrane ATPase 2	110,743	199
Q99002	BMH1	14-3-3 protein homolog	30,094	195
C7YTD6	RPS1	40S ribosomal protein S1	29,288	185
Q99170	KAR2	reticulum chaperone BiP	73,593	178
P34085	cit-1	Citrate synthase	52,241	166
C7C436	mcsA	2-methylcitrate synthase	52,184	159
P50142	HSP60	Heat shock protein 60	62,079	158
P0C016	ubi3	Ubiquitin-40S ribosomal protein S27a	17,475	142
Q9P3A7	cdc48	Cell division cycle protein 48	90,354	139
Q5I2J3	TUB1	Tubulin alpha chain	50,737	137
P51044	cit-1	Citrate synthase	52,406	135
P24634	tubB	Tubulin alpha-2 chain	50,541	126
Q8J0N6	FBA2	Fructose-bisphosphate aldolase 2	39,973	126
Q5AWS6	cdc48	Cell division control protein 48	90,769	124

Table 2. *Cont.*

Accession	Gene Name	Protein Name	Molecular Mass	Score
O13639	pi047	Adenosylhomocysteinase	47,866	121
Q7RVA8	ace-8	Pyruvate kinase	58,290	115
P49382	AAC	ADP, ATP carrier protein	33,187	111
Q12629	PDC1	Pyruvate decarboxylase	61,905	110
Q6C1F3	ENO	Enolase	47,277	108
Q9HES8	pyc	Pyruvate carboxylase	131,526	106
P87252	hex-1	Woronin body major protein	19,229	104
Q7RVI1	rps-5	40S ribosomal protein S5	23,836	103

The Tap42 potential interacting proteins of *F. oxysporum,* cultured in GYM medium, with L-threonine (L-Thr) L-Histidine (L-His) and L-Aspartic acid (L-Asp) (n = 3). A high score indicates that the protein is closely related to Tap42.

Table 3. Proteins of *F. oxysporum* potentially interacting with Tap42, activated by specific amino acids.

Amino Acid Treatment	Accession	Gene Name	Protein Name	Molecular Mass	Score
L-His group L-Asp group	P11484	*SSB1*	Ribosome-associated molecular chaperone SSB1	66,732	234
	P41756	*pgkA*	Phosphoglycerate kinase	44,413	209
L-Thr group L-His group	Q0CNV1	*pdcA*	Pyruvate decarboxylase	63,206	111
L-His group	P38669	*tba-2*	Tubulin alpha-B chain	50,675	72

The Tap42 potential interacting proteins of *F. oxysporum* cultured in GYM medium with L-threonine (L-Thr), L-Histidine (L-His) and L-Aspartic acid (L-Asp) addition (n = 3). A high score indicates that the protein is closely related to Tap42.

According to KEGG enrichment analysis, Tap42-interacting proteins were mostly derived from metabolism, genetic information processing, cellular processes and environmental information processing. Most of the interacting proteins came from the metabolic pathways, mainly from glycerophospholipid metabolism, pentose phosphate pathway, glyoxylate and dicarboxylate metabolism, glycolysis, gluconeogenesis, methane metabolism and glutathione metabolism. The related secondary metabolic pathways were galactose metabolism, fatty acid biosynthesis, fatty acid degradation, steroid biosynthesis, phenylalanine metabolism, one carbon pool by folate, citrate cycle, TCA cycle and oxidative phosphorylation (Figure 6).

KEGG enrichment analysis showed that the proteins potentially interacting with Tap42 were from the mitogen-activated protein kinase (MAPK) signaling pathway and Hippo signaling pathway. The MAPK signaling pathway is composed of mitogen-activated protein kinase, which is involved in the intracellular signal regulation system and can be activated by different extracellular stimuli, such as cytokine neurotransmitter hormones, cell stress and cell adhesion [52,53]. Many reports have shown that the MAPK pathway is significantly associated with the regulation of the TORC1 pathway [54]. The Hippo signaling pathway mainly inhibits cell growth and regulates cell proliferation, apoptosis and stem cell self-renewal [55]. Recent studies have shown that the Hippo signaling pathway is closely related to cancer genesis, tissue regeneration and stem cell function regulation [56]. Identification of the associated signaling pathway provides a theoretical basis for the further exploration of the interaction network of *Fusarium* pathogenicity when exposed to different amino acid conditions, and also provides important information for the further control of fungal contamination in high-protein fermented foods.

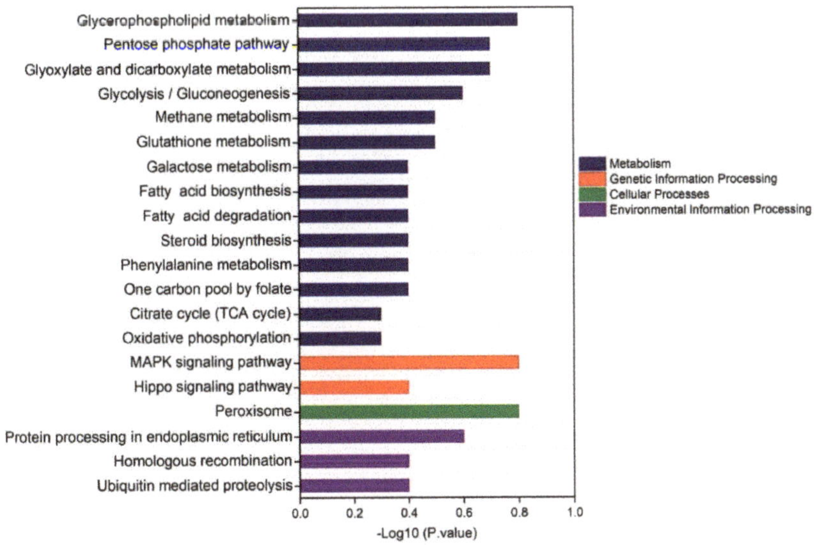

Figure 6. The KEGG enrichment analysis of the signaling pathway of Tap42, potentially interacting with proteins of *F. oxysporum* Fo17 cultured with L−threonine (L−Thr), L−Histidine (L−His) and L−Aspartic acid (L−Asp) ($n = 3$).

4. Conclusions

Exposure to free AAs (as N sources) showed a significant effect on the growth and metabolism of *F. oxysporum* regulated by *Tap42*. The absorption of L-Ile and L-Tyr was regulated by *Tap42*. Acidic (L-Asp, L-Glu) and S-containing (L-Cys, L-Met) AAs were not conducive to *F. oxysporum* growth, and were not regulated by *Tap42*. The addition of L-Ile and L-Pro activated sporulation in ΔFo*Tap42*, but this was negatively regulated by *Tap42*, while L-His and L-Ser inhibited sporulation in ΔFo*Tap42*, which was positively regulated by *Tap42*. Acidic AA showed a remarkable inhibitory effect on T-2 toxin production positively regulated by *Tap42*. Neutral (L-Thr) and alkaline (L-His) AAs significantly activated the T-2 synthesis and *Tri5* expression of *Fusarium*, while L-Asp showed an inhibitory effect at relatively high doses. The co-immunoprecipitation analysis showed that the interacting proteins of Tap42 were activated by L-The, L-His and L-Asp control metabolism; genetic information processing; cellular processes and environmental information processing. L-Asp inhibited the effects on *Fusarium* spp. growth and metabolism, and, thus, could be used as an inhibitor to further control fungal contamination in fermented foods.

Author Contributions: Conceptualization, formal analysis and writing—original draft, Y.D.; supervision, project administration and formal analysis, R.W.; validation, Y.Z.; funding acquisition, J.L.; editing and supervision, R.G. All authors have read and agreed to the published version of the manuscript.

Funding: This research was funded by the National Science Foundation of China (32202134; 32201929), China Postdoctoral Science Foundation (2022M712625) and 2021 and 2022 Chongqing Postdoctoral Special Funding Project.

Data Availability Statement: All related data and methods are presented in this paper. Additional inquiries should be addressed to the corresponding author.

Conflicts of Interest: The authors declare no conflict of interest.

References

1. He, C.; Yorimitsu, T.; Klionsky, D.J.; Wang, K. Tap42-associated protein phosphatase type 2A negatively regulates induction of autophagy. *Autophagy* **2009**, *5*, 616–624.
2. Wang, Y.B.; Li, F.; Chen, J.; Sun, Z.H.; Wang, F.F.; Wang, C.; Fu, L.L. High-throughput sequencing-based characterization of the predominant microbial community associated with characteristic flavor formation in Jinhua Ham. *Food Microbiol.* **2021**, *94*, 103643. [CrossRef] [PubMed]
3. Zhao, Y.; Wang, Y.Q.; Li, C.S.; Li, L.H.; Yang, X.Q.; Wu, Y.Y.; Chen, S.J.; Zhao, Y.Q. Novel insight into physicochemical and flavor formation in naturally fermented tilapia sausage based on microbial metabolic network. *Food Res. Int.* **2021**, *141*, 110122. [CrossRef] [PubMed]
4. Gallego, M.; Mora, L.; Escudero, E.; Toldr, F. Bioactive peptides and free amino acids profiles in different types of European dry-fermented sausages. *Int. J. Food Microbiol.* **2018**, *276*, 71–78. [CrossRef] [PubMed]
5. Shi, Y.A.; Li, X.; Huang, A.X. A metabolomics-based approach investigates volatile flavor formation and characteristic compounds of the dahe black pig dry-cured ham. *Meat Sci.* **2019**, *158*, 107904. [CrossRef] [PubMed]
6. Wu, K.; Xie, J.H.; Wang, Q.Q.; Ling, M.J.; Wu, J.Z. Effect of Monascus fermentation on aroma patterns of semi-dried grass carp. *Food Nutr. Sci.* **2019**, *10*, 923–936.
7. Yang, D.D.; He, Z.Y.; Gao, D.M.; Qin, F.; Deng, S.L.; Wang, P.; Xu, X.L.; Chen, J.; Zeng, M.M. Effects of smoking or baking procedures during sausage processing on the formation of heterocyclic amines measured using UPLC-MS/MS. *Food Chem.* **2019**, *276*, 195–201. [CrossRef] [PubMed]
8. Adekoya, I.; Obadina, A.; Phoku, J.; Nwinyi, O.; Njobeh, P. Contamination of fermented foods in Nigeria with fungi. *LWT-Food Sci. Technol.* **2017**, *86*, 76–84. [CrossRef]
9. Meftah, S.; Abid, S.; Dias, T.; Rodrigues, P. Effect of dry-sausage starter culture and endogenous yeasts on *Aspergillus westerdijkiae* and *Penicillium nordicum* growth and OTA production. *LWT-Food Sci. Technol.* **2018**, *87*, 250–258. [CrossRef]
10. Adeyeye, S.A.O.; Yildiz, F. Quality and safety assessment of sun dried meat product (kundi) from Ibadan, Oyo state, Nigeria. *Cogent Food Agr.* **2016**, *2*, 1209074. [CrossRef]
11. Wen, R.X.; Li, X.A.; Han, G.; Chen, Q.; Kong, B.H. Fungal community succession and volatile compound dynamics in Harbin dry sausage during fermentation. *Food Microbiol.* **2021**, *99*, 103764. [CrossRef] [PubMed]
12. Cendoya, C.; Chiotta, M.L.; Zachetti, V.; Chulze, S.N.; Ramirez, M.L. Fumonisins and fumonisin-producing *Fusarium* occurrence in wheat and wheat by products: A review. *J. Cereal Sci.* **2018**, *80*, 158–166. [CrossRef]
13. Nathanail, A.V.; Varga, E.; Meng-Reiterer, J.; Bueschl, C.; Michlmayr, H.; Malachova, A.; Fruhmann, P.; Jestoi, M.; Peltonen, K.; Adam, G.; et al. Metabolism of the *Fusarium* mycotoxins T-2 toxin and HT-2 toxin in wheat. *J. Agric. Food Chem.* **2015**, *63*, 7862–7872. [CrossRef]
14. Gilbert, J.; Clear, R.M.; Ward, T.J.; Gaba, D.; Tekauz, A.; Turkington, T.K.; Woods, S.M.; Nowicki, T.; O'Donnell, K. Relative aggressiveness and production of 3- or 15-acetyl deoxynivalenol and deoxynivalenol by *Fusarium graminearum* in spring wheat. *Can. J. Plant Pathol.* **2010**, *32*, 146–152. [CrossRef]
15. Chaudhari, M.; Jayaraj, R.; Santhosh, S.R.; Rao, P.V.L. Oxidative damage and gene expression profile of antioxidant enzymes after T-2 toxin exposure in mice. *J. Biochem. Mol. Toxic.* **2009**, *23*, 212–221. [CrossRef]
16. Guo, P.; Liu, A.M.; Huang, D.Y.; Wu, Q.H.; Fatima, Z.; Tao, Y.F.; Cheng, G.Y.; Wang, X.; Yuan, Z.H. Brain damage and neurological symptoms induced by T-2 toxin in rat brain. *Toxicol. Lett.* **2018**, *286*, 96–107. [CrossRef]
17. Agrawal, M.; Bhaskar, A.S.B.; Rao, P.V.L. Involvement of mitogen-activated protein kinase pathway in T-2 toxin-induced cell cycle alteration and apoptosis in human neuroblastoma cells. *Mol. Neurobiol.* **2015**, *51*, 1379–1394. [CrossRef] [PubMed]
18. Velazco, V.; Faifer, G.C.; Godoy, H.M. Differential effects of T-2 toxin on bone marrow and spleen erythropoiesis in mice. *Food Chem. Toxicol.* **1996**, *34*, 371–375. [CrossRef]
19. Agrawal, M.; Yadav, P.; Lomash, V.; Bhaskar, A.S.B.; Rao, L.P.V. T-2 toxin induced skin inflammation and cutaneous injury in mice. *Toxicology* **2012**, *302*, 255–265. [CrossRef] [PubMed]
20. Alexander, N.J.; Proctora, R.H.; McCormicka, S.P. Genes, gene clusters, and biosynthesis of trichothecenes and fumonisins in *Fusarium*. *Toxin Rev.* **2009**, *28*, 198–215. [CrossRef]
21. Adebayo-Tayo, B.C.; Onilude, A.A.; Ukpe, G.P. Mycoflora of smoke-dried fishes sold in Uyo, Eastern Nigeria. *World J. Agric. Sci.* **2008**, *4*, 346–350.
22. Ayeloja, A.A.; George, F.O.A.; Jimoh, W.A.; Shittu, M.O. Microbial load on smoked fish commonly traded in Ibadan, Oyo State, Nigeria. *J. Appl. Sci. Environ. Manag.* **2018**, *22*, 493–497. [CrossRef]
23. Fafioye, O.O.; Efuntoye, M.O.; Osho, A. Studies on the fungal infestation of five traditionally smoke-driedfreshwater fish in Ago-Iwoye, Nigeria. *Mycopathologia* **2001**, *154*, 177–179. [CrossRef] [PubMed]
24. Fagbohun, E.D.; Lawal, O.U. Aflatoxins investigation and mycobiota of selected marketed smoked - dried fish samples in Ado-Ekiti, Nigeria and their environmental Health implications. *Brit. Microbiol. Res. J.* **2015**, *7*, 126–132. [CrossRef]
25. Rabie, M.A.; Peres, C.; Malcata, F.X. Evolution of amino acids and biogenic amines throughout storage in sausages made of horse, beef and turkey meats. *Meat Sci.* **2014**, *96*, 82–87. [CrossRef]
26. Shiobara, T.; Nakajima, Y.; Maeda, K.; Akasaka, M.; Kitou, Y.; Kanamaru, K.; Ohsato, S.; Kobayashi, T.; Nishiuchi, T.; Kimura, M. Identification of amino acids negatively affecting *Fusarium* trichothecene biosynthesis. *Antonie Leeuwenhoek* **2019**, *111*, 471–478. [CrossRef]

27. Lushchak, O.; Strilbytska, O.; Piskovatska, V.; Storey, K.B.; Koliada, A.; Vaiserman, A. The role of the TOR pathway in mediating the link between nutrition and longevity. *Mech. Ageing Dev.* **2017**, *164*, 127–138. [CrossRef]
28. Tanigawa, M.; Maeda, T. An in vitro TORC1 kinase assay that recapitulates the Gtr-independent glutamine-responsive TORC1 activation mechanism on yeast vacuoles. *Mol. Cell. Biol.* **2017**, *37*, e00075-17. [CrossRef]
29. Rai, R.; Tate, J.J.; Cooper, T.G. gln3 mutations dissociate responses to nitrogen limitation (nitrogen catabolite repression) and rapamycin inhibition of TorC1. *J. Biol. Chem.* **2013**, *288*, 2789–2804. [CrossRef]
30. Valbuena, N.; Guan, K.L.; Moreno, S. The Vam6 and Gtr1-Gtr2 pathway activates TORC1 in response to amino acids in fission yeast. *J. Cell Sci.* **2012**, *125*, 1920–1928. [CrossRef]
31. Ajeet, S.; Daipayan, C.; Avinash, G.; Ramesh, C.M.; Amitabha, C. TORC1-signalling is down-regulated in *Saccharomyces cerevisiae* hsp30Δ cells by SNF1-dependent mechanisms. *Yeast* **2018**, *35*, 653–667.
32. Molinet, J.; Salinas, F.; Guillamón, J.M.; Martínez, C. *GTR1* affects bitrogen consumption and TORC1 activity in *Saccharomyces cerevisiae* under fermentation conditions. *Front Genet.* **2020**, *11*, 519. [CrossRef] [PubMed]
33. Chen, X.H.; Wang, G.Q.; Zhang, Y.; Dayhoff-Brannigan, M.; Diny, N.L.; Zhao, M.J.; He, G.; Sing, C.N.; Metz, K.A.; Stolp, Z.D.; et al. Whi2 is a conserved negative regulator of TORC1 in response to low amino acids. *PLoS Genet.* **2018**, *14*, e1007592. [CrossRef]
34. Li, J.M.; Yan, G.H.; Liu, S.C.; Jiang, T.; Zhong, M.M.; Yuan, W.J.; Chen, S.X.; Zheng, Y.; Jiang, Y.; Jiang, Y. Target of rapamycin complex 1 and Tap42-associated phosphatases are required for sensing changes in nitrogen conditions in the yeast *Saccharomyces cerevisiae. Mol. Microbiol.* **2017**, *106*, 938–948. [CrossRef] [PubMed]
35. Morrissette, V.A.; Rolfes, R.J. The intersection between stress responses and inositol pyrophosphates in *Saccharomyces cerevisiae. Currr. Genet.* **2020**, *66*, 901–910. [CrossRef]
36. Ma, N.; Ma, Y.; Nakashima, A.; Kikkawa, U.; Furuyashiki, T. The loss of Lam2 and Npr2-Npr3 diminishes the vacuolar localization of Gtr1-Gtr2 and disinhibits TORC1 activity in fission yeast. *PLoS ONE* **2016**, *11*, e0156239. [CrossRef]
37. Paimard, G.; Mohammadi, R.; Bahrami, R.; Khosravi -Darani, K.; Sarlak, Z.; Rouhi, M. Detoxification of patulin from juice simulator and apple juice via cross-linked Se-chitosan/L-cysteine nanoparticles. *LWT-Food Sci. Technol.* **2021**, *143*, 111146. [CrossRef]
38. Galgóczy, L.; Yap, A.; Marx, F. Cysteine-rich antifungal proteins from filamentous fungi are promising bioactive natural compounds in anti-candida therapy. *Isr. J. Chem.* **2019**, *59*, 360–370. [CrossRef]
39. Garbe, E.; Vylkova, S. Role of amino acid metabolism in the virulence of human pathogenic fungi. *Curr. Clin. Microbiol.* **2019**, *6*, 108–119. [CrossRef]
40. Alfatah, M.; Wong, J.H.; Krishnan, V.G.; Lee, Y.C.; Sin, Q.F.; Goh, C.J.H.; Kong, K.W.; Hoon, S.; Arumugam, P. TORC1 regulates the transcriptional response to glucose and developmental cycle via the Tap42-Sit4-Rrd1/2 pathway in *Saccharomyces cerevisiae. BioRxiv* **2019**, *62*, 115–120. [CrossRef]
41. Vylkova, S.; Carman, A.J.; Danhof, H.A.; Collette, J.R.; Zhou, H.; Lorenz, M.C. The fungal pathogen *Candida albicans* autoinduces hyphal morphogenesis by raising extracellular pH. *MBio* **2011**, *2*, e00055-11. [CrossRef] [PubMed]
42. Phasha, M.M.; Wingfield, M.J.; Wingfield, B.D.; Coetzee, M.P.A.; Hallen-Adams, H.; Fru, F.; Swalarsk-Parry, B.S.; Yilmaz, N.; Duong, T.A.; Steenkamp, E.T. Ras2 is important for growth and pathogenicity in *Fusarium circinatum. Fungal Genet. Biol.* **2021**, *150*, 103541. [CrossRef] [PubMed]
43. Tästensen, J.B.; Schönheit, P. Two distinct glyceraldehyde -3-phosphate dehydrogenases in glycolysis and gluconeogenesis in the archaeon Haloferax volcanii. *FEBS Lett.* **2018**, *592*, 1524–1534. [CrossRef]
44. Hohn, T.M.; Desjardins, A.E.; McCormick, S.P. The *Tri4* gene of *Fusarium sporotrichioides* encodes a cytochrome P450 monooxygenase involved in trichothecene biosynthesis. *Mol. Gen. Genet.* **1995**, *248*, 95–102. [CrossRef] [PubMed]
45. McCormick, S.P.; Alexander, N.J. *Fusarium Tri8* encodes a trichothecene C-3 esterase. *Appl. Environ. Microbiol.* **2002**, *68*, 2959–2964. [CrossRef]
46. McCormick, S.P.; Hohn, T.M.; Desjardins, A.E. Isolation and characterization of Tri3, a gene encoding 15-O-acetyltransferase from *Fusarium sporotrichioides. Appl. Environ. Microbiol.* **1996**, *62*, 353–359. [CrossRef]
47. Bitu, M.N.A.; Hossain, M.S.; Islam, A.N.; Islam, M.S.; Kudrat-E.-Zahan, M. Peroxo complexes of Th(IV) and Zr(IV) ions containing aspartic acid and amine bases as potential biological agents. *Russ. J. Gen. Chem.* **2020**, *90*, 1553–1557. [CrossRef]
48. Riccio, A.A.; Schellenberg, M.J.; Williams, R.S. Molecular mechanisms of topoisomerase 2 DNA-protein crosslink resolution. *Cell. Mol. Life Sci.* **2020**, *77*, 81–91. [CrossRef]
49. Shen, L.L.; Baranowski, J.; Fostel, J.; Montgomery, D.A.; Lartey, P.A. DNA topoisomerases from pathogenic fungi: Targets for the discovery of antifungal drugs. *Antimicrob. Agents Chemother.* **1992**, *36*, 2778–2784. [CrossRef] [PubMed]
50. Fostel, J.; Montgomery, D. Identification of the aminocatechol A-3253 as an in vitro poison of DNA topoisomerase I from *Candida albicans. Antimicrob. Agents Chemother.* **1995**, *39*, 586–592. [CrossRef]
51. Karkowska-Kuleta, J.; Satala, D.; Bochenska, O.; Rapala-Kozik, M.; Kozik, A. Moonlighting proteins are variably exposed at the cell surfaces of *Candida glabrata, Candida parapsilosis* and *Candida tropicalis* under certain growth conditions. *BMC Microbiol.* **2019**, *19*, 149. [CrossRef] [PubMed]
52. Hu, Q.L.; Wang, H.; He, C.; Jin, Y.X.; Fu, Z.W. Polystyrene nanoparticles trigger the activation of p38 MAPK and apoptosis via inducing oxidative stress in zebrafish and macrophage cells. *Environ. Pollut.* **2021**, *269*, 116075. [CrossRef] [PubMed]

53. Wu, J.; Ibtisham, F.; Niu, Y.F.; Wang, Z.; Li, G.H.; Zhao, Y.; Nawab, A.; Xiao, M.; An, L.L. Curcumin inhibits heat-induced oxidative stress by activating the MAPK-Nrf2 / ARE signaling pathway in chicken fibroblasts cells. *J. Therm. Biol.* **2019**, *79*, 112–119. [CrossRef] [PubMed]
54. Madrid, M.; Vzquez-Marn, B.; Franco, A.; Soto, T.; Vicente-Soler, J.; Gacto, M.; Cansado, J. Multiple crosstalk between TOR and the cell integrity MAPK signaling pathway in fission yeast. *Sci. Rep.* **2016**, *6*, 37515. [CrossRef] [PubMed]
55. Li, N.; Tong, X.; Zeng, J.; Meng, G.; Sun, F.Z.; Hu, H.; Song, J.B.; Lu, C.; Dai, F.Y. Hippo pathway regulates somatic development and cell proliferation of silkworm. *Genomics* **2019**, *111*, 391–397. [CrossRef] [PubMed]
56. Park, J.H.; Shin, J.E.; Park, H.W. The role of Hippo pathway in cancer stem cell biology. *Mol. Cells* **2018**, *41*, 83–89.

foods

MDPI

Article

Growth and Non-Thermal Inactivation of *Staphylococcus aureus* in Sliced Dry-Cured Ham in Relation to Water Activity, Packaging Type and Storage Temperature

Anna Austrich-Comas [1], Cristina Serra-Castelló [1], Maria Viella [1], Pere Gou [2], Anna Jofré [1] and Sara Bover-Cid [1,*]

[1] Food Safety and Functionality Program, IRTA, Finca Camps i Armet, E-17121 Monells, Spain; anna.jofre@irta.cat (A.J.)
[2] Food Quality and Technology Program, IRTA, Finca Camps i Armet, E-17121 Monells, Spain; pere.gou@irta.cat
* Correspondence: sara.bovercid@irta.cat

Abstract: Dry-cured ham (DCH) could support the growth of *Staphylococcus aureus* as a halotolerant bacterium, which may compromise the shelf-stability of the product according to the growth/no growth boundary models and the physicochemical parameters of commercial DCH. In the present study, the behavior of *S. aureus* is evaluated in sliced DCH with different water activity (a_w 0.861–0.925), packaged under air, vacuum, or modified atmosphere (MAP), and stored at different temperatures (2–25 °C) for up to 1 year. The Logistic and the Weibull models were fitted to data to estimate the primary kinetic parameters for the pathogen Log_{10} increase and Log_{10} reduction, respectively. Then, polynomial models were developed as secondary models following their integration into the primary Weibull model to obtain a global model for each packaging. Growth was observed for samples with the highest a_w stored at 20 and 25 °C in air-packaged DCH. For lower a_w, progressive inactivation of *S. aureus* was observed, being faster at the lowest temperature (15 °C) for air-packaged DCH. In contrast, for vacuum and MAP-packaged DCH, a higher storage temperature resulted in faster inactivation without a significant effect of the product a_w. The results of this study clearly indicate that the behavior of *S. aureus* is highly dependent on factors such as storage temperature, packaging conditions and product a_w. The developed models provide a management tool for evaluating the risk associated with DCH and for preventing the development of *S. aureus* by selecting the most appropriate packaging according to a_w range and storage temperature.

Keywords: predictive microbiology; ready-to-eat meat products; shelf-stable food; food safety

check for updates

Citation: Austrich-Comas, A.; Serra-Castelló, C.; Viella, M.; Gou, P.; Jofré, A.; Bover-Cid, S. Growth and Non-Thermal Inactivation of *Staphylococcus aureus* in Sliced Dry-Cured Ham in Relation to Water Activity, Packaging Type and Storage Temperature. *Foods* **2023**, *12*, 2199. https://doi.org/10.3390/foods12112199

Academic Editor: Xinjun Du

Received: 27 April 2023
Revised: 19 May 2023
Accepted: 20 May 2023
Published: 30 May 2023

1. Introduction

Dry-cured ham (DCH) has traditionally been considered a safe and microbiologically shelf-stable product because of the combination of hurdles (e.g., low moisture, high salt content and the presence of curing agents) that contribute to inhibiting pathogen growth and/or even promote pathogen inactivation [1–3]. However, DCH with high a_w has been reported to be associated particularly with commercial pre-packaged sliced products, which may compromise food safety [4]. For instance, according to the survey performed by Hereu [5], 50% of the DCH products sampled from retail showed an a_w equal to or higher than 0.92.

Serra-Castelló et al. [6] reported a progressive inactivation of *Listeria monocytogenes* in vacuum-packaged Serrano and Iberian DCH (a_w = 0.85–0.91) stored at different temperatures (4 to 25 °C). *Salmonella* viability also decreased on vacuum-packaged DCH stored at 1 to 25 °C [7]. However, compared with other pathogens, *Staphylococcus aureus* is a pathogen of concern for DCH due to its halotolerant nature, which enables it to grow over many adverse conditions, including at low a_w (≥0.83) and with high salt concentrations (up to 20%) [8–10]. Enterotoxigenic *S. aureus* strains are able to produce staphylococcal enterotoxins (SEs) when the concentration exceeds 5 Log_{10} CFU/g, although SE production requires higher a_w (i.e., 0.86) than growth [10–12].

The behavior of *S. aureus* has been quite widely studied through challenge tests under laboratory conditions in which food characteristics are mimicked [13–15]. However, only a few studies have evaluated the behavior of *S. aureus* on DCHs through challenge testing. Christieans et al. [16] observed no growth at 8 °C for any of the a_w studied (0.89–0.96). Conversely, growth was found on DCH samples stored at 20 °C regardless of the a_w. In another study, *S. aureus* growth on slices of vacuum-packaged DCH was reported at the highest temperature (25 °C); however, no SE was produced after storage for 28 days at 2 and 25 °C [1]. Márta et al. [17] detected SE in Serrano ham with low a_w and high salt and fat levels after 5 days when stored aerobically at 23 °C. Unterman and Müller [18] showed that in minced DCH with a_w of 0.89, enterotoxin was produced when it was stored at 35 °C for 7 days. These studies tested specific experimental conditions, but none of them were designed to simultaneously cover a wide range of a_w, storage temperature (from refrigeration to room temperature) and atmosphere compositions (such as air, vacuum and modified atmosphere packaging (MAP) with CO_2) usually found in commercial DCH. Accordingly, additional studies are needed to be able to draw conclusions regarding the conditions that pose a risk. In this respect, the development and application of predictive models, if available, represent a valuable complementary approach to challenge testing to quantitatively characterize the behavior of pathogens in food [19], to identify either the growth/no growth boundaries, or the growth or the inactivation (survival) kinetics throughout storage [20], which are used to assess the impact of relevant intrinsic and extrinsic factors taking into account the DCH variability [21–23].

The overall aim of the present study was to evaluate the behavior of *S. aureus* in Spanish dry-cured ham considering intrinsic (a_w and pH) and extrinsic factors (storage temperature and packaging conditions) through predictive modeling and challenge testing approaches. First, the physicochemical characteristics of pre-packaged sliced DCH were used as inputs of selected growth/no-growth (G/NG) models to assess the growth probability of *S. aureus* at different temperatures (Study 1). Afterwards, the growth of *S. aureus* was evaluated through challenge testing in DCH slices packaged using different packaging methods (air, vacuum and MAP) and stored at different temperatures (2 to 25 °C) with the subsequent development of three predictive models (Study 2).

2. Materials and Methods

2.1. Dry-Cured Ham (DCH) Samples

For Study 1, a total of 20 pH and a_w historical data provided by a food producer of Spanish DCH, corresponding to different batches and representative of their products, were used. Data representing the physicochemical characteristics of the sliced product (before the final packaging) was used.

For Study 2, blocks of deboned DCH (pH 5.80 ± 0.06) showing three different levels of a_w—low, medium and high—were provided by the same food producer and were selected to cover the range of a_w variability usually found (ca. 0.860, 0.901, 0.925, respectively). To equalize the value of a_w throughout the matrix, DCH blocks were vacuum packaged and stored at 4 °C for 15 days. In this way, the differences in a_w value within different sections of a DCH block were lower than 0.028.

Figure S1 shows a graphical summary of the experimental design of Study 1 and Study 2.

2.2. Challenge Test

2.2.1. Inoculum Preparation

A cocktail of three strains of *S. aureus* was used: CECT976 (SEA producer) and CECT4466 (SED producer), from the Spanish Type Culture Collection, and CTC1008, as a meat isolate from the IRTA culture collection. Each strain was independently grown in Brain Heart Infusion (BHI) broth (Becton Dickinson, Sparks, MD, USA) at 37 °C for 24 h.

The cultures were cryopreserved at −80 °C with 20% glycerol until use. Thawed cultures of each strain were mixed at equal concentrations before being inoculated on DCH.

2.2.2. DCH Inoculation

DCH was aseptically sliced (ca. 20 g/slice) and inoculated in a laminar flow cabinet. The cocktail of the *S. aureus* strains was inoculated on the surface of DCH slices at 0.5% (v/w) to reach a different final concentration, i.e., from 5×10^2 (to characterize growth) to 2.5×10^6 CFU/g (to characterize inactivation). For air- and vacuum-packaged samples, the inoculum was spread on the surface with a single-use Digralsky spreader. The DCH was packaged in PA/PE bags (oxygen permeability of 50 $cm^3/m^2/24$ h and a low water vapor permeability of 2.8 $g/m^2/24$ h; Sistemvac, Estudi Graf SA, Girona, Spain) thermosealed or vacuum packaged (EV-15-2-CD; Tecnotrip, Terrassa, Spain), respectively. Meanwhile, MAP samples were inoculated after packaging (80% N_2 and 20% CO_2) with a sterile syringe through a septum to avoid gas leakage.

2.2.3. DCH Storage and Sampling

DCH samples were stored at different temperatures depending on the packaging type: air-packaged samples were kept at 15, 20 and 25 °C; vacuum-packaged samples at 2, 8, 15, 20 and 25 °C; and MAP samples at 2, 8, 15 and 25 °C. Storage time ranged from 1 month for the DCH with the highest a_w at the highest temperature up to 1 year for the DCH with the lowest a_w at the lower temperature. Sampling points were distributed throughout the storage time. A total of 36 experimental conditions combining a_w, packaging format and storage temperature were assayed (resulting in 615 data points).

2.3. Microbiological and Physicochemical Determinations

For microbiological analysis, 10 g of sample were transferred into a bag blender Smasher® (bioMérieux, Marcy-l'Étoile, France) and 10-fold diluted and homogenized in physiological saline (0.85% NaCl and 0.1 % Bacto Peptone (Becton Dickinson, Sparks, MD, USA)) for 60 s with a Smasher™ device (bioMérieux Espãna S.A, Madrid, Spain). Serial decimal dilutions were prepared in physiological saline. Enumeration of *S. aureus* was performed on selective and differential chromogenic agar (CHROMagar Staphylococcus, CHROMagar, Paris, France) incubated at 37 °C for 48 h. LAB levels were determined in Man–Rogosa–Sharpe (MRS) agar (Merck, Darmstadt, Germany), incubated at 30 °C for 72 h anaerobically in sealed jars with an AnaeroGen sachet (Oxoid Ltd.)

The a_w was measured with an AquaLab™ instrument (Series 3; Decagon Devices Inc., Pullman, WA, USA). The pH was measured with a penetration probe (52-32; Crison Instrument SA, Alella, Spain) connected to a portable pH meter (PH25; Crison Instruments). The detection of SEs was determined according to ISO 19020 [24] by automated immunofluorescence.

The gas concentration of MAP-packaged samples was measured with the gas analyzer PBI Dansensor CheckMate II (Ametek Instrumentos, S.L.U., Barcelona).

2.4. Predictive Microbiology Approaches

2.4.1. Growth/No Growth Prediction

For Study 1, predictive models about G/NG boundaries for *S. aureus* were used to identify the pH and a_w combinations defining the 10% probability (as a moderately conservative threshold) and predict the growth probability of *S. aureus* associated with the physicochemical characteristics of commercial DCH (Section 2.1). The main features of the selected predictive models used are summarized in Table S1. The G/NG model of Borneman et al. [25] is a logistic regression-based polynomial that determines the probability of *S. aureus* growth on vacuum-packaged RTE meat products at 21 °C with pH and a_w as input factors. Polese et al. [14] used a gamma-concept model with pH, a_w and temperature as input factors, and the model was tested for a variety of foods stored between 2 to 30 °C. Finally, the model available in the *Sym'Previus* [26] portal predicts the G/NG interface for

S. aureus depending on a_w, pH and temperature using a gamma-concept approach and the mean cardinal parameters for the growth of eight *S. aureus* strains.

2.4.2. Primary Model Fitting

For Study 2, challenge test data were used to estimate kinetic parameters of growth or inactivation by fitting a primary model. For each data point, the Log_{10} change in the concentration in relation to the initial inoculum concentration (e.g., Log_{10} increase or Log_{10} reduction) was calculated as Log_{10} (N/N_0), N is the concentration (CFU/g) at the sampling time and N_0 is the initial concentration (CFU/g) after inoculation of DCH samples.

For conditions supporting the growth of *S. aureus*, the Logistic model (Equation (1)) [27] was used to estimate the growth kinetic parameters.

$$\begin{aligned} &\text{For } t < \lambda,\ \mathrm{Log}\left(\frac{N_t}{N_0}\right) = 0 \\ &\text{For } t \geq \lambda,\ \mathrm{Log}\left(\frac{N_t}{N_0}\right) = \mathrm{Log}\left(\frac{MGP}{1+(MGP-1)*(\exp(-\mu_{max}*(t-\lambda)))}\right) \end{aligned} \quad (1)$$

where N_0 is the concentration of the pathogen (CFU/g) at time zero; N_t is the concentration of the pathogen (CFU/g) at time t; MGP is the maximum growth potential as the ratio N_{max}/N_0 (N_{max} is the maximum population density, CFU/g); λ is the lag time (h); μ_{max} is the maximum specific growth rate (ln/h); and t is the storage time (h).

For conditions causing the inactivation of *S. aureus*, the Weibull model (Equation (2)) [6] was used to estimate the inactivation kinetic parameters.

$$\mathrm{Log}\left(\frac{N}{N_0}\right) = -\left(\frac{t}{\delta}\right)^p \quad (2)$$

where Log_{10} (N/N_0) is the inactivation in Log_{10} reduction (Log_{10} units) at a given time (t) of the storage, being equal to 0 at storage time 0; t is the storage time (h); δ is the time (h) for the first Log_{10} reduction and p is the shape of the inactivation curve.

Model fitting was performed with the *nls2* package of R software [28]. In addition to the standard error of the estimates, to evaluate the goodness of fit, the Root Mean Square Error (RMSE) values were calculated.

2.4.3. Secondary and Global Model Fitting

To evaluate the effect of storage temperature and DCH's a_w on the inactivation kinetics parameters (δ, p), a secondary model was developed based on a second-order polynomial equation (Equation (3)) for each packaging condition.

$$\mathrm{y} = a + b \cdot \mathrm{X} + c \cdot \mathrm{X}^2 \quad (3)$$

where y is the dependent variable, i.e., the primary kinetic parameter (e.g., δ), X is the independent variable, i.e., the environmental factor (e.g., temperature), and a, b and c are the model coefficients to be estimated.

Different parameter transformations (including square root and Log_{10}) were assessed. The stepwise linear regression was applied throughout the *step* function of the R software [28] to obtain the polynomial models with only significant parameters according to the parsimony principle. In addition to the standard error of the estimates, the goodness of fit was assessed in terms of RMSE and the adjusted coefficient of determination (R^2_{adj}).

In addition to the classical two-step modeling (primary and secondary model fitting), the one-step modeling approach was applied by integrating the secondary polynomial model for δ into the primary Weibull model. The global model was fitted to the whole dataset to obtain a global model with re-adjusted coefficients for each type of packaging [29,30]. Estimation of the model parameters with the standard error was carried out using *nls2* package of R software [28]. The goodness of fit of the global model was assessed on the basis of RMSE. The F-test (Equation (4)) was used to evaluate the statistical differences ($p < 0.05$) between the models developed to characterize the behavior of *S. aureus* in air, vacuum, and MAP conditions [31].

$$F = \frac{(RSS_{NH} - RSS_{AH})/(df_{NH} - df_{AH})}{RSS_{AH} - df_{AH}} \tag{4}$$

where RSS_{NH} and df_{NH} are the Residual Sum of Squares and the degrees of freedom (number of points minus the number of parameters of the model), respectively, of the global model common to all types of packages (null hypothesis), and RSS_{AH} and df_{AH} are the Residual Sum of Squares and the number of degrees of freedom, respectively, of the global model with specific parameter coefficients for each type of package (alternative hypothesis).

Moreover, due to the statistical correlation between δ and p parameters [32], the F-test was applied to test the statistical significance of the effect of storage temperature and the DCH's a_w on the shape of the inactivation curve of *S. aureus*. The global model with a fixed p value independent of environmental conditions (null hypothesis) was compared with the global model with a polynomial model describing the effect of environmental conditions on the p parameter (alternative hypothesis).

2.4.4. Model Validation

In order to evaluate the predictive performance of the developed model, the Acceptable Simulated Zone (ASZ) was applied [33]. Independent data were obtained from three published articles dealing with the behavior of *S. aureus* in dry-cured ham, with a total of 80 sampling points: 7 for air-packaged DCH, 24 for vacuum-packaged DCH, and 49 for MAP-packaged DCH. Log_{10} count data over time were extracted from published scientific literature using WebPlotDigitizer v.4.4 software. The observed and predicted Log_{10} reduction data during the storage time were compared. The predictive performance of the model was considered acceptable when at least 70% of the independent data were inside the ASZ $\pm$ 0.5 Log_{10}.

3. Results

3.1. Characteristics of Commercial DCH and the Associated Probability of S. aureus Growth (Study 1)

The distribution of physicochemical characteristics (a_w and pH) of commercial vacuum-packaged DCH is shown in Figure 1, with the prediction of the growth boundaries according to the predictive models available for *S. aureus*. Despite a 50% probability of growth being frequently used to assess the G/NG boundary, in the present study, a growth probability of 10% was used as a conservative reference boundary. Although the variability of the pH was rather limited (within 5.5. to 6.0), the values of a_w were scattered within a range from 0.85 to 0.92, with a considerable proportion (82%) of samples at above 0.88, the minimum a_w for growth reported for anaerobic conditions when the other factors (pH and temperature) were optimal for growth [34]. In fact, according to the models of Borneman et al. [25] and Polese et al. [14], for all the observed DCH characteristics, a growth probability higher than 10% was predicted at temperatures above 15 °C, while only 15% of samples would not support growth (probability below 10%) at these temperatures according to the model of the Sym'Previus portal [26]. Only when storage temperature decreased to below 8 °C for Polese et al. model [14] and to below 5 °C for Sym'Previus model [26] did the growth probability fall below 10% for almost all samples, indicating that refrigeration storage would be needed to control the growth of *S. aureus* in DCH during the shelf life.

However, these predictive models were not specifically developed for DCH, and do not take into consideration the effect of relevant factors related to the specific characteristics of the product (i.e., lactic acid concentration, lactic acid bacteria) and packaging (e.g., oxygen reduction of vacuum packaging and CO_2 concentration of MAP), which may contribute to further inhibiting the growth of *S. aureus*. Therefore, product-specific studies were required.

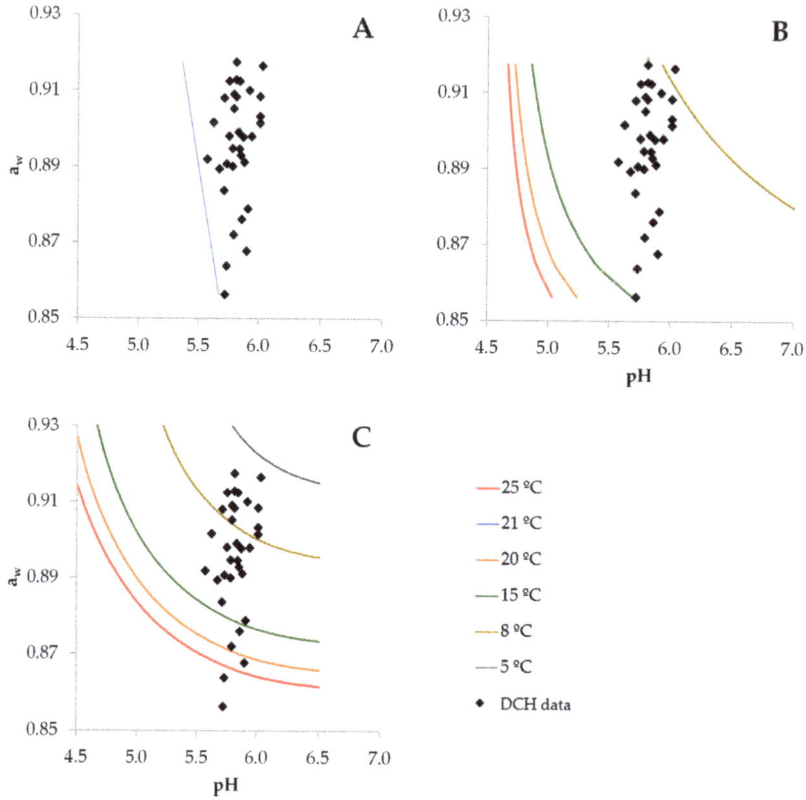

Figure 1. Distribution of pH and a_w values of commercial dry-cured ham (DCH, diamond dots) and pH–a_w boundaries for the growth probability of 10% for *S. aureus* at different temperatures according to the predictive models (lines) available in Borneman et al. [25] (**A**), Polese et al. [14] (**B**) and the Sym'Previus portal [26] (**C**).

3.2. Behavior of S. aureus *on Sliced DCH Stored under Different Conditions (Challenge Test Experiment, Study 2)*

Growth of *S. aureus* was observed in three out of the 36 trials, which corresponded to those where DCH had the highest a_w (0.925) stored in air at temperatures ≥20 °C and under vacuum at 25 °C (Figures 2 and 3). The growth kinetic parameters estimated for each trial are shown in Table 1, including the growth rate (μ_{max}) and the maximum growth potential (MGP). No lag time was observed. In air-packaged DCH, *S. aureus* increased by up to 2.7 and 4.54 Log_{10} units after 1.7 and 4.7 days of storage at 20 and 25 °C, respectively. At 25 °C, the growth rate was slightly higher compared to the growth at 20 °C. However, due to the high variability in Log_{10} increase data (especially at 20 °C), growth rates were not statistically different. Under vacuum, a slight increase in *S. aureus* (1.62 Log_{10} units in 22 days) was observed during the early stages of storage at 25 °C in the DCH with the highest a_w. Afterwards, the pathogen started to die off, and growth kinetic parameters could not be estimated.

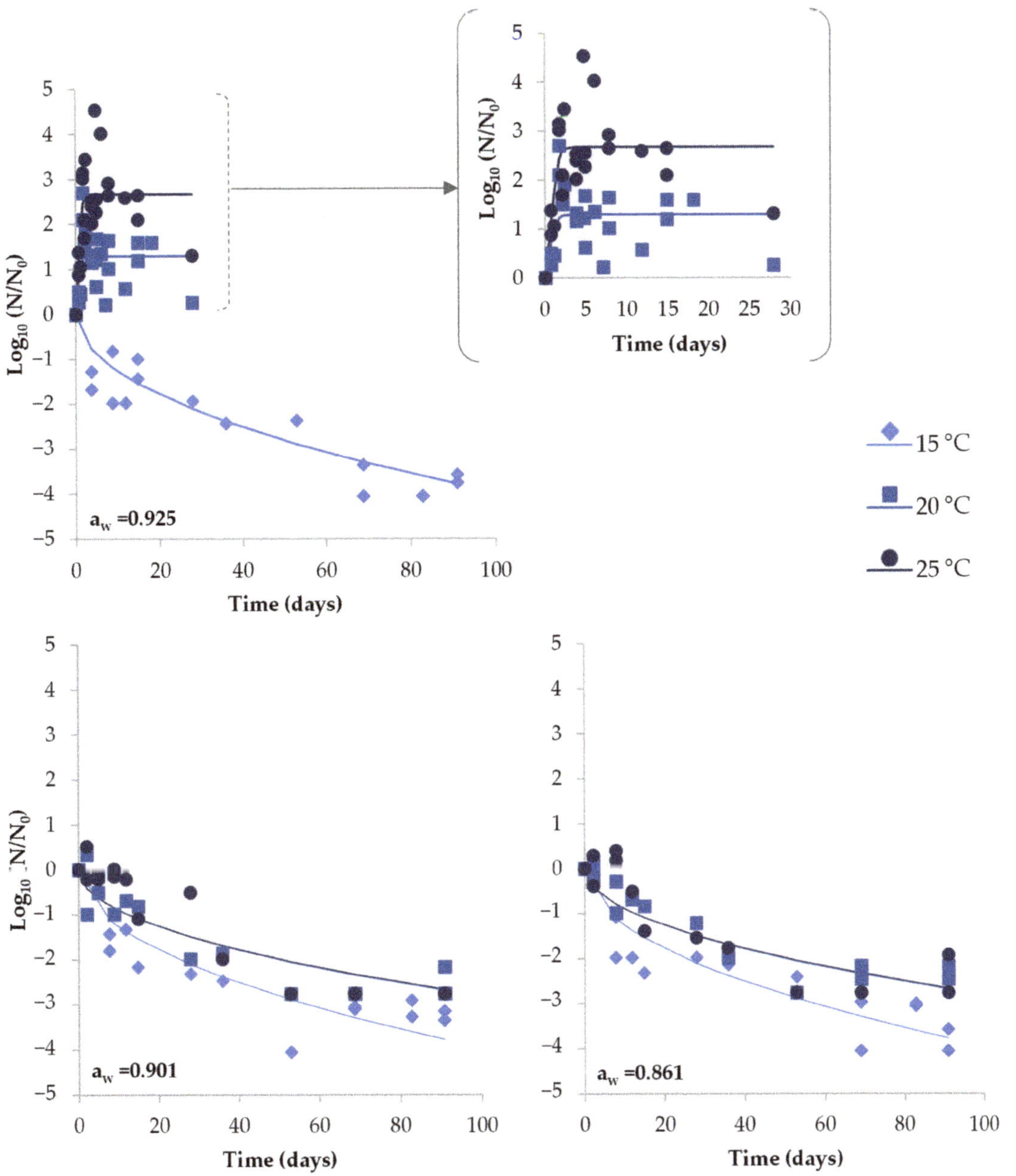

Figure 2. Behavior of *S. aureus* in sliced DCH with different a_w (0.861, 0.901 and 0.925) when air packaged and stored at different temperatures (15, 20 and 25 °C). Dots represent the observed *S. aureus* values ($Log_{10}\ N/N_0$). Lines show the fit of the global model.

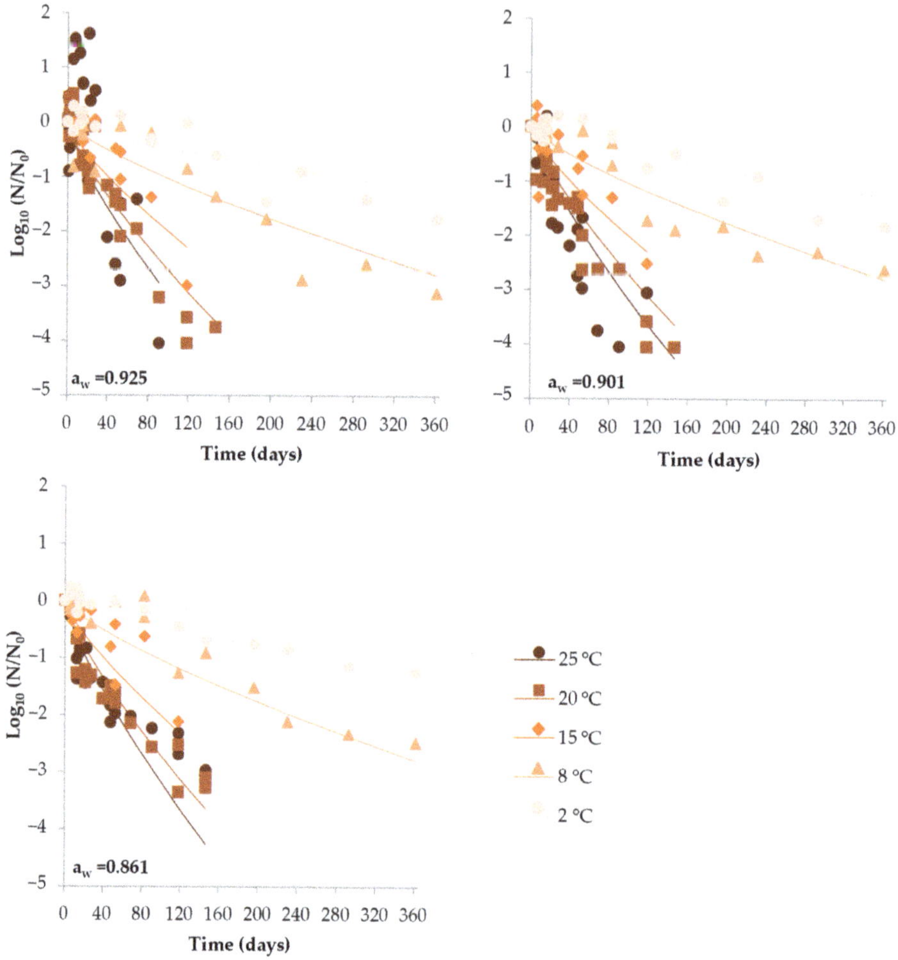

Figure 3. Behavior of *S. aureus* in DCH sliced with different a_w (0.861, 0.901 and 0.925) when vacuum packaged and stored under different storage temperatures (2, 8, 15, 20 and 25 °C). Dots represent the observed *S. aureus* values (Log_{10} N/N_0). Lines show the fit of the global model.

Under the rest of assessed conditions, inactivation of *S. aureus* was observed, and the kinetic parameters estimated with the Weibull model fit, i.e., the time for the first Log_{10} reduction (δ) and the shape of the inactivation curve (p) were obtained (Table 1).

In air-packaged DCH with medium a_w (0.902) and low a_w (0.861), *S. aureus* concentration had decreased by 2.5 Log_{10} units after 91 days at 20 and 25 °C, while higher inactivation occurred at 15 °C, with reductions of 3.66, 3.25 and 3.81 Log_{10} in DCH with high, medium and low a_w after 91 days, respectively (Figure 2). Higher δ values were found with increasing storage temperature (Table 1), although the kinetic curve at 20 °C was very similar to that at 25 °C (Figure 2). Conversely, similar δ values were observed for DCH with different a_w values for each storage temperature. Regarding the shape of the inactivation curve, the fit of the Weibull model resulted in p values below 1 for all the studied conditions, regardless of the storage temperature and DCH a_w, indicating a more pronounced inactivation of *S. aureus* at the early stages of the storage followed by a sort of a resistance tail showing a slower inactivation.

Table 1. Estimated kinetic parameters (for the inactivation or growth) resulting from fitting the primary models to data obtained for each challenge test of *S. aureus* on DCH with different a_w contents and types of packaging when stored at various storage temperatures.

Experimental Conditions			Kinetic Parameters [a]				Goodness of Fit [b]	
			Inactivation Weibull Model		Growth Logistic Model			
Packaging	a_w	**Temperature (°C)**	δ **(Days)**	p	μ_{max} **(ln/h)**	**MGP (Log_{10})**	**n**	**RMSE**
Air	0.861	15	2.26 ± 1.43	0.34 ± 0.07	-	-	15	0.446
		20	12.38 ± 3.64	0.48 ± 0.09	-	-	15	0.369
		25	19.01 ± 6.15	0.68 ± 0.17	-	-	15	0.553
	0.901	15	1.53 ± 1.03	0.30 ± 0.06	-	-	15	0.389
		20	14.21 ± 3.88	0.58 ± 0.10	-	-	17	0.453
		25	19.11 ± 4.30	0.80 ± 0.13	-	-	17	0.435
	0.925	15	4.29 ± 1.56	0.44 ± 0.06	-	-	18	0.465
		20	-	-	0.12 ± 0.06	1.29 ± 0.14	25	0.584
		25	-	-	0.17 ± 0.04	2.67 ± 0.19	24	0.737
Vacuum	0.861	2	271.65 ± 15.04	1.11 ± 0.15	-	-	18	0.135
		8	143.59 ± 14.01	1.11 ± 0.15	-	-	18	0.271
		15	66.56 ± 8.23	1.07 ± 0.25	-	-	16	0.325
		20	18.79 ± 2.25	0.58 ± 0.04	-	-	24	0.264
		25	16.81 ± 2.29	0.50 ± 0.04	-	-	24	0.255
	0.901	2	210.43 ± 14.81	1.27 ± 0.20	-	-	18	0.219
		8	97.69 ± 13.21	0.82 ± 0.11	-	-	18	0.305
		15	60.01 ± 7.60	1.24 ± 0.29	-	-	17	0.395
		20	20.57 ± 2.28	0.73 ± 0.05	-	-	24	0.316
		25	14.06 ± 3.10	0.64 ± 0.08	-	-	23	0.551
	0.925	2	223.31 ± 16.41	1.30 ± 0.23	-	-	18	0.232
		8	126.57 ± 18.30	1.11 ± 0.19	-	-	18	0.393
		15	64.04 ± 3.77	1.74 ± 0.19	-	-	17	0.209
		20	31.25 ± 2.98	0.95 ± 0.07	-	-	34	0.347
		25	38.55 ± 6.83	1.70 ± 0.45	-	-	33	0.936
MAP	0.861	2	324.91 ± 68.51	2.66 ± 2.12	-	-	14	0.332
		8	199.73 ± 18.99	1.66 ± 0.47	-	-	15	0.343
		15	104.76 ± 80.30	0.42 ± 0.28	-	-	14	0.516
		25	27.45 ± 5.12	0.67 ± 0.08	-	-	17	0.386
	0.901	2	506.40 ± 252.13	1.80 ± 1.56	-	-	16	0.301
		8	202.56 ± 14.40	2.04 ± 0.30	-	-	16	0.292
		15	77.35 ± 2.66	2.61 ± 0.22	-	-	16	0.163
		25	29.39 ± 3.32	0.77 ± 0.04	-	-	18	0.310
	0.925	2	478.37 ± 126.42	1.16 ± 0.44	-	-	16	0.193
		8	112.95 ± 18.42	0.47 ± 0.09	-	-	16	0.247
		15	73.63 ± 14.38	0.88 ± 0.32	-	-	19	0.411
		25	36.73 ± 7.60	0.90 ± 0.13	-	-	20	0.622

[a]: Parameter estimate ± standard error: "-" not applicable [b]: n number of data points used for model fitting; RMSE: root mean squared error (Log_{10} units).

Under vacuum conditions, the progressive inactivation of *S. aureus* after a slight increase during the first 22 days in DCH with the highest a_w stored at 25 °C resulted in an overall 4.05 Log_{10} reduction after 91 days. For the rest of the temperatures studied, *S. aureus* was unable to grow at all (Figure 3). Instead, a progressive reduction was observed, which was dependent on the storage temperature but not on the product a_w. Contrary to air-packaged DCH, for vacuum-packaged the time for the first Log_{10} reduction (δ value) decreased as the temperature increased from 2 to 25 °C (Table 1).

Storage under MAP promoted the loss of viability of *S. aureus* in all combinations of a_w and temperature tested from the beginning of the storage. As in other packaging types, the a_w of DCH had no relevant effect on inactivation (Figure 4), with similar δ values for DCH with different a_w for each storage temperature. The extent of the inactivation in MAP tended to be lower than under vacuum-packaged conditions. Moreover, at the lowest temperature (2 °C), no microbiologically relevant inactivation (less than 1 Log_{10}) occurred in DCH with medium and high a_w during the 365 days of storage, making the estimates of δ values higher than the studied storage time.

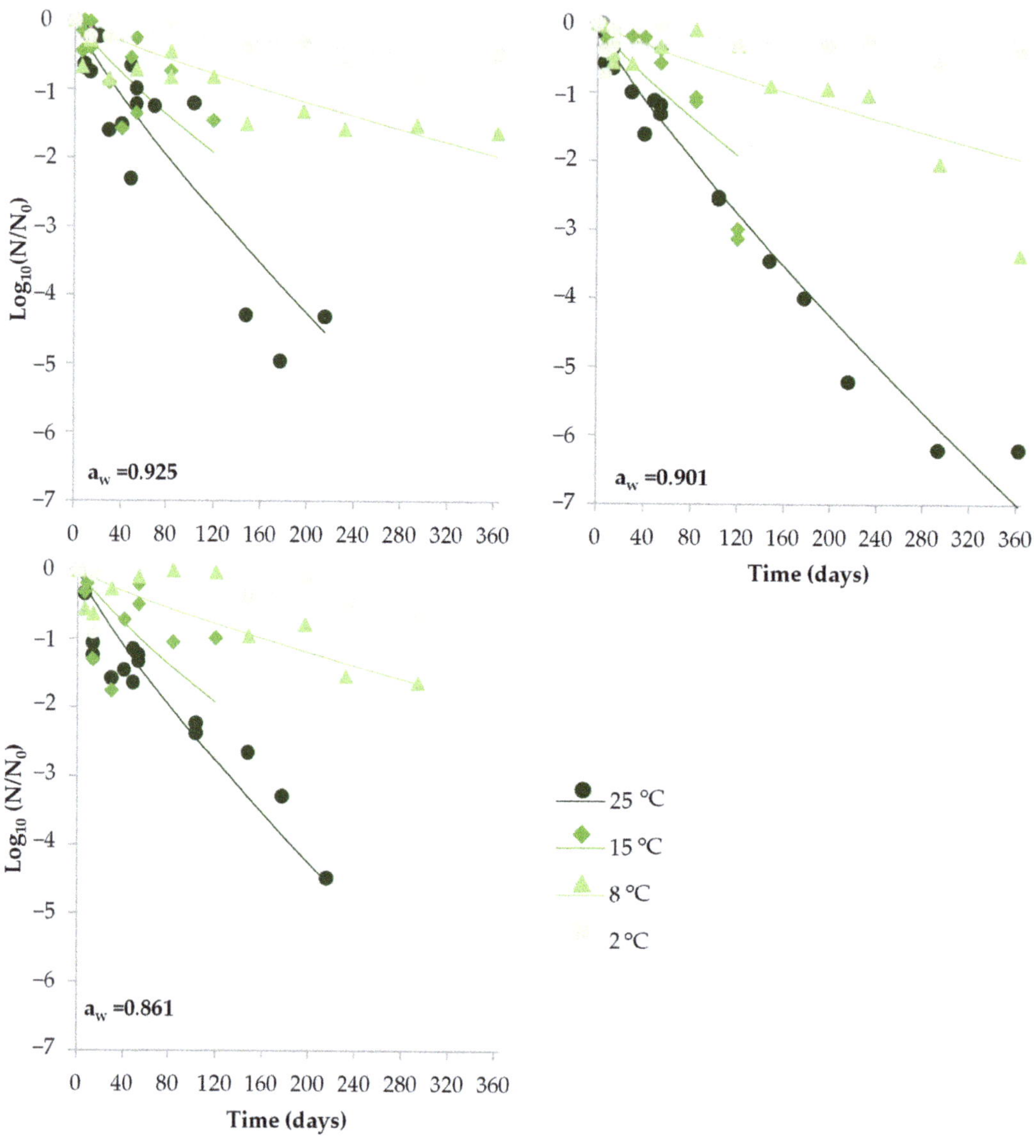

Figure 4. Behavior of *S. aureus* in sliced DCH with different a_w (0.861, 0.901 and 0.925) when MAP packaged and stored under different storage temperatures (2, 8, 15 and 25 °C). Symbols represent the observed *S. aureus* inactivation (Log_{10} N/N_0). Lines show the fit of the global model.

3.3. Physicochemical Determinations and Lactic Acid Bacteria Counts

The physicochemical characteristics were measured throughout the study. As all DCH samples were packaged with impermeable bags, values of a_w of DCH did not change during the storage at any temperature. The values of pH changed slightly depending on the packaging type and temperature or the a_w values (Table S2). The behavior of LAB levels during the storage time depended on the a_w of the DCH, the packaging type and the storage temperature (Table S3). Under air packaging, LAB was not able to grow at lower a_w, irrespective of the temperature; LAB was able to grow only in the DCH with medium and high a_w at all three temperatures tested, with a slight pH reduction (i.e., a 4.39 Log_{10} increase in LAB was associated with a 0.17 pH decrease in DCH with high a_w stored at 25 °C). Under vacuum packaging gand MAP, a similar trend was observed. At lower temperatures, no LAB growth was observed. Additionally, at higher temperatures (>15 °C), in general, LAB was able to grow, and a reduction in pH was observed, especially at 40 days of storage.

No differences in the gas composition on MAP-packaged DCH were detected over the course of the storage time.

The potential occurrence of staphylococcus enterotoxin (SE) was analyzed in DCH samples where the *S. aureus* grew at the maximum concentration, which corresponded to air-packaged DCH with higher a_w stored at 25 °C (6.38 and 5.87 Log_{10} CFU/g). No SEs were detected.

3.4. Secondary and Global Modeling

To describe the effect of storage temperature and a_w on the inactivation kinetics of *S. aureus* through a polynomial model, different transformations of the parameters (δ and p) were assessed, and the Log_{10} transformation of δ values provided the best results. Table 2 gathers the coefficients of the polynomial model and Figure 5 shows the Log_{10} δ values as a function of temperature for the three different packaging conditions. Only the temperature, and not the a_w, was found to significantly affect δ values in each packaging type. The effect of storage temperature on the time for the first Log_{10} reduction (δ values) in air-packaged DCH (δ decreased as temperature decreased) was the opposite of that in packaging systems without oxygen, i.e., vacuum packaging and MAP (δ decreased as temperature increased).

Table 2. Coefficients of polynomial models describing the effect of storage temperature (T, °C) on Log_{10} δ.

Packaging	Coefficients [a]			Goodness of Fit [b]		
	a (Intercept)	*b* (T)	*c* (T^2)	n	RMSE	R^2_{adj}
Air	−5.254 ± 1.191	0.549 ± 0.125	−0.0115 ± 0.0031	7	0.261	0.675
Vacuum	2.493 ± 0.031	−0.055 ± 0.006	0.0002 ± 0.0002	15	0.141	0.895
MAP	2.752 ± 0.028	−0.068 ± 0.005	0.0007 ± 0.0002	12	0.103	0.947

[a] Parameter estimate ± standard error. [b] n: number of data points; RMSE: root mean squared error (Log_{10} units).

Regarding the p parameter, no clear relationship could be established with either storage temperature or the product a_w, as indicated by the lack of fit of the polynomial model, not even with the different parameter transformations. Therefore, a fixed parameter corresponding to the mean p value for all the tested temperatures for each packaging condition (air, vacuum and MAP) was assumed.

A global (one-step) model integrating the secondary model for δ values into the Weibull primary model with a fixed p value for each packaging type was fitted to Log_{10} reduction data. The re-adjusted parameter estimates are shown in Table 3 (see the final equation in Table S4). All were statistically significant ($p < 0.05$). Despite the similar trend, the global model for vacuum and MAP were significantly different according to the F-test ($F = 16.61$, $p > 0.05$); therefore, specific model coefficients for each type of packaging are needed to predict the inactivation of *S. aureus* during the storage of DCH.

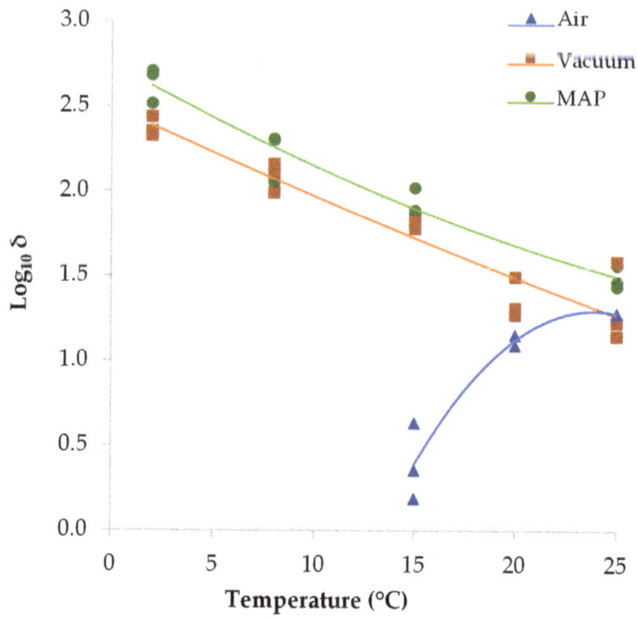

Figure 5. Time for the first Log_{10} reduction (δ value) of *S. aureus* in dry-cured ham (DCH) as a function of storage temperature for different packaging conditions (air, vacuum and MAP). Dots are the values estimated with the primary inactivation model (Weibull) and lines correspond the fit of the second-order polynomial model to Log_{10} transformed δ values.

Table 3. Coefficients of the global (one-step) model about the effect of storage temperature (T, °C) on *S. aureus* inactivation in DCH, integrating the secondary polynomial model into the primary Weibull model for each packaging type.

Packaging	Coefficients of the Polynomial Models [a]				Goodness of Fit [b]	
	δ			p	n	RMSE
	a (Intercept)	*b* (T)	*c* (T^2)			
Air	−1.848 ± 0.844	0.263 ± 0.087	−0.006 ± 0.002	0.495 ± 0.034	112	0.498
Vacuum	2.597 ± 0.086	−0.090 ± 0.013	0.002 ± 0.000	0.768 ± 0.050	311	0.441
MAP	3.088 ± 0.135	−0.133 ± 0.019	0.003 ± 0.001	0.838 ± 0.048	193	0.436

[a] Parameter estimate ± standard error. [b] n: number of data; RMSE: root mean squared error (Log_{10} units).

3.5. Predictive Performance of the Model

The predictive performance of the obtained global models (Table S5) was assessed using independent data obtained from three scientific articles dealing with *S. aureus* behavior during the storage of DCH for each packaging condition (i.e., Christieans et al. [16]; Untermann and Müller [18] and Iacumin et al. [35]), with a total of 80 data points. Considering an Acceptable Simulation Zone (ASZ) of ±0.5 Log_{10} units around the model predictions, 85.7% (6/7), 75% (18/24) and 83.67% (41/49) of agreement between the observed values and the model predictions were found for air, vacuum and MAP packaging, respectively. These results indicate a good predictive performance of the developed model in the wide range of a_w and temperature conditions assayed (Figure 6, Table S4).

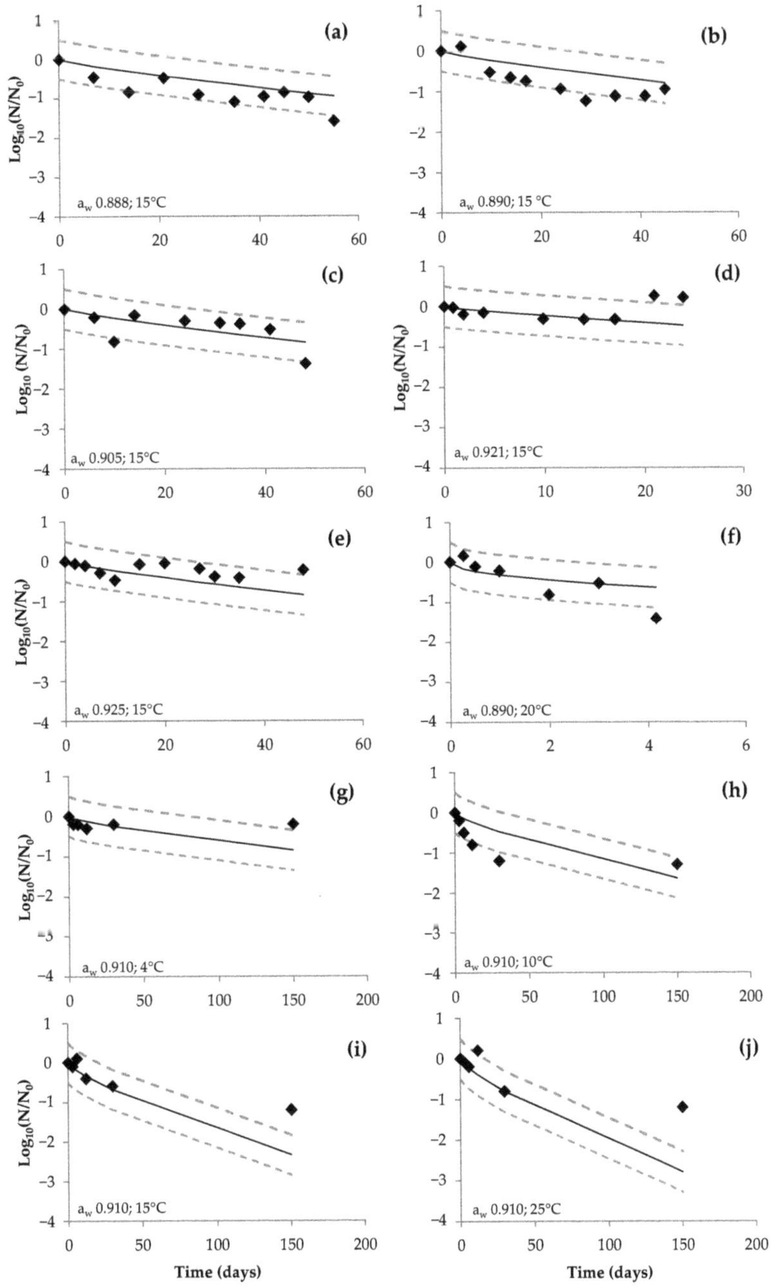

Figure 6. Observed Log_{10} reduction values (dot circles) and acceptable prediction zone −0.5 (fail—safe) to +0.5 (fail—dangerous) from different studies with respect to time (days); (**a**–**e**) data from Christieans et al. [16] in DCH with a_w values of 0.888 (**a**), 0.890 (**b**), 0.905 (**c**), 0.921 (**d**), 0.925 (**e**); (**f**) data from Untermann and Müller [18]; (**g**–**j**) data from Iacumin et al. [35] stored at 4 °C (**g**), 10 °C (**h**), 15 °C (**i**) and 25 °C (**j**).

4. Discussion

According to available predictive models, the physicochemical characteristics of most commercial DCH would support the growth of *S. aureus* with a probability higher than 10% when stored at >15 °C. The challenge test results confirmed the ability of *S. aureus* to grow on DCH, with the highest a_w tested (0.925) being obtained when stored at ≥20 °C under air conditions. These results are in agreement with those reported in Untermann and Müller [18] regarding the growth of *S. aureus* in DCH stored aerobically at temperature ≥20 °C, with an a_w of 0.918 and a pH of 5.60–6.07. However, staphylococcal enterotoxin (SE) was not detected in any of the DCH in which the maximum growth was observed (up to 10^5 CFU/g). Several studies have reported the production of SEs in different food matrixes when *S. aureus* reached 10^5–10^6 CFU/g [36–38], while others have reported no detection of SEs even at pathogen levels of up to 10^9–10^{10} CFU/g [39].

In DCH of medium and low a_w, under storage temperatures equal to or below 15 °C or when DCH was packaged without oxygen (i.e., vacuum and MAP), the viability of the pathogen was compromised. It is known that *S. aureus* grows better under air conditions, as it is a poorly competitive pathogen compared with other microorganisms, in particular LAB, which usually exerts a growth-inhibitory effect on *S. aureus* during meat fermentation processes associated with acidification and the production of antimicrobial substances [40,41]. In general, greater *S. aureus* inhibition is observed with higher LAB concentration and lower pH [42]. Although DCH does not go through a fermentation process, the DCHs studied in this work (i.e., medium- and high-a_w products) supported the growth of LAB, which grew faster and reached higher concentrations in DCH with oxygen-reduced packaging (i.e., vacuum packaging and, particularly, MAP) compared with air-packaged DCH. This LAB growth explains, at least partially, the small amount and lack of growth of *S. aureus* observed on DCH when vacuum packaged, as has been reported for raw beef [37]. Moreover, the addition of CO_2 in MAP-packaged DCH may have favored the selective growth of LAB that, in addition to the antimicrobial effect of CO_2, could promote the greater inactivation of *S. aureus* behavior.

The progressive loss of viability of microorganisms in harsh conditions occurring in shelf-stable foods such as DCH has been related to the metabolic exhaustion phenomenon associated with antimicrobial hurdles [6,43]. Due to this phenomenon, microorganisms tend to die, and their rate of death is faster when shelf-stability conditions approach the limits of growth [3], which in the present study would be the storage of DCH with the highest a_w at the highest temperature when vacuum or MAP packaged. Similar behavior has been reported for *L. monocytogenes* in vacuum-packaged DCH (a_w = 0.85–0.91) stored at different temperatures (4 to 25 °C) [6]. On the contrary, in DCH stored under air conditions, the inactivation of *S. aureus* at 15 °C was significantly enhanced compared to that observed at 20 and 25 °C. To the best of the authors' knowledge, there are no previous studies dealing with the effect of temperature on the non-thermal inactivation of *S. aureus* in DCH that compare aerobic and anaerobic environments. However, in the study of Ha et al. [44], the inactivation of *S. aureus* inoculated on beef jerky (a_w = 0.81) followed a similar trend during aerobic storage, and δ values at 20 and 25 °C were very similar, and were longer than that at 10 °C, confirming the higher inactivation at lower temperature. Therefore, the non-thermal inactivation seems to be affected by different mechanisms when oxygen is present compared with the anaerobic conditions occurring in vacuum and MAP. In any case, the a_w of the DCH did not have a significant effect on the *S. aureus* behavior for any of packaging types, in contrast to the reported behavior of *L. monocytogenes* in DCH [6] and *Salmonella* in dry-fermented sausages [43]. The halotolerance of *S. aureus* may explain the lack of the effect of decreasing the a_w of the DCH, at least within the range studied in the present study.

The modeling approach provided predictive models for the three packaging types with a satisfactory performance when assessed with independent data, which supports their suitability for predictive and simulation purposes. This fact provides a management tool for evaluating the risk associated with DCH and to prevent the development of *S. aureus* by selecting the most appropriate packaging according to a_w range and storage temperature.

5. Conclusions

DCH can support the growth of *S. aureus* at the a_w values found in commercial products (ca. 0.92) when stored at room temperature under aerobic conditions, although no staphylococcal enterotoxin was detected. Storage temperature ≤15 °C and reduced-oxygen packaging (i.e., vacuum packaging and, particularly, MAP) inhibits *S. aureus* growth and promotes its inactivation. The product a_w does not affect the survival of *S. aureus* on DCH, while the storage temperature has contrary effects in aerobic (higher inactivation at lower storage temperature) and anaerobic packaging (higher inactivation at higher storage temperature), suggesting the involvement of different mechanisms depending on the presence of oxygen in the environment. The predictive models developed are useful tools for stakeholders (e.g., risk assessors, food business operators, competent authority, etc.) for assessing and quantifying the behavior of *S. aureus* on sliced DCH commercialized in different packaging types as a function of storage temperature.

Supplementary Materials: The following supporting information can be downloaded at: https://www.mdpi.com/article/10.3390/foods12112199/s1, Figure S1: Graphical scheme of the experimental design of Study 1 and Study 2. Table S1: Main features of predictive models for growth/no growth (G/NG) boundaries of *S. aureus* used in Study 1; Table S2: pH values of DCH with different a_w content and type of packaging stored at various storage temperatures; Table S3: Lactic acid bacteria counts (Log_{10} CFU/g) measured in DCH with different a_w content and type of packaging stored at various storage temperatures; Table S4. Main features of predictive models for growth/no growth (G/NG) boundaries of S. aureus used in Study 1. Table S5: Comparison of observed and predicted *S. aureus* inactivation in sliced DCH.

Author Contributions: Data curation, C.S.-C., P.G. and S.B.-C.; Formal analysis, A.A.-C. and C.S.-C.; Funding acquisition, A.J. and S.B.-C.; Investigation, A.A.-C., C.S.-C. and M.V.; Methodology, A.A.-C., P.G., A.J. and S.B.-C.; Project administration, S.B.-C.; Resources, A.A.-C.; Supervision, P.G., A.J. and S.B.-C.; Visualization, A.J.; Writing—original draft, A.A.-C. and C.S.-C.; Writing—review and editing, P.G., A.J. and S.B.-C. All authors have read and agreed to the published version of the manuscript.

Funding: Pla de doctorats industrials de la Secretaria d'Universitats i Recerca del Departament d'empresa i coneixement (2018 DI 94), the Consolidated Research Group (2021 SGR00468) and CERCA Program of the Generalitat de Catalunya.

Data Availability Statement: Data available on request.

Conflicts of Interest: The authors declare no conflict of interest.

References

1. Ng, W.F.; Langlois, B.E.; Moody, W.G. Fate of Selected Pathogens in Vacuum-Packaged Dry-Cured (Country-Style) Ham Slices Stored at 2 and 25 °C. *J. Food Prot.* **1997**, *60*, 1541–1547. [CrossRef] [PubMed]
2. Menéndez, R.A.; Rendueles, E.; Sanz, J.J.; Santos, J.A.; García-Fernández, M.C. Physicochemical and Microbiological Characteristics of Diverse Spanish Cured Meat Products. *CYTA J. Food* **2018**, *16*, 199–204. [CrossRef]
3. Leistner, L. Basic Aspects of Food Preservation by Hurdle Technology. *Int. J. Food Microbiol.* **2000**, *55*, 181–186. [CrossRef] [PubMed]
4. Chitrakar, B.; Zhang, M.; Adhikari, B. Dehydrated Foods: Are They Microbiologically Safe? *Crit. Rev. Food Sci. Nutr.* **2019**, *59*, 2734–2745. [CrossRef]
5. Hereu, A. High Pressures and Biopreservation as Control Strategies for *Listeria monocytogenes* in Ready-to-Eat Meat Products. Inoculation Tests and Mathematical Modeling. Ph.D. Thesis, University of Girona, Girona, Spain, 2014.
6. Serra-Castelló, C.; Jofré, A.; Garriga, M.; Bover-Cid, S. Modeling and Designing a *Listeria monocytogenes* Control Strategy for Dry-Cured Ham Taking Advantage of Water Activity and Storage Temperature. *Meat Sci.* **2020**, *165*, 108131. [CrossRef] [PubMed]
7. Bover-Cid, S.; Jofré, A.; Garriga, M. Inactivation Kinetics of *Salmonella* and *L. monocytogenes* in Dry-Cured Ham Stored at Different Temperatures. In Proceedings of the 25th International ICFMH Conference—FoodMicro 2016. One Health Meets Food Microbiology, Dublin, Ireland, 19–22 July 2016; p. 472.
8. *FDA Bad Bug Book, Foodborne Pathogenic Microorganisms and Natural Toxins*, 2nd ed.; 2012; pp. 87–91. Available online: https://www.fda.gov/files/food/published/Bad-Bug-Book-2nd-Edition-(PDF).pdf (accessed on 26 April 2023).

9. Troller, J. Staphylococcal Growth and Enterotoxin Production. Factors and Control. *J. Milk Food Technol.* **1976**, *39*, 499–502. [CrossRef]
10. ANSES. *Staphylococcus aureus* and Staphylococcal Enterotoxins. Available online: https://www.anses.fr/en/system/files/MIC2011sa0117FiEN_0.pdf (accessed on 26 April 2023).
11. Busta, F.F.; Bernard, D.T.; Gravani, R.B.; Hall, P.; Pierson, M.D.; Prince, G.; Schaffner, D.W.; Swanson, K.M.J. Factors That Influence Microbial Growth. *Compr. Rev. Food Sci. Food Saf.* **2003**, *2*, 21–32. [CrossRef]
12. Gunvig, A.; Andresen, M.S.; Jacobsen, T.; Borggaard, C. Staphtox Predictor—A Dynamic Mathematical Model to Predict Formation of Staphylococcus Enterotoxin during Heating and Fermentation of Meat Products. *Int. J. Food Microbiol.* **2018**, *285*, 81–91. [CrossRef]
13. Jamshidi, A.; Kazerani, H.R.; Seifi, H.A.; Moghaddas, E. Growth Limits of *Staphylococcus aureus* as a Function of Temperature, Acetic Acid, NaCl Concentration, and Inoculum Level. *Iran. J. Vet. Res.* **2008**, *9*, 353–359. [CrossRef]
14. Polese, P.; Del Torre, M.; Spaziani, M.; Stecchini, M.L. A Simplified Approach for Modelling the Bacterial Growth/No Growth Boundary. *Food Microbiol.* **2011**, *28*, 384–391. [CrossRef]
15. Medved'ová, A.; Havlíková, A.; Valík, L'. Growth of *Staphylococcus aureus* 2064 Described by Predictive Microbiology: From Primary to Secondary Models. *Acta Chim. Slovaca* **2019**, *12*, 175–181. [CrossRef]
16. Christieans, S.; Denis, C.; Hanin, A.; Picgirard, L. Incidence of Storage Temperature and Water Activity in the Growth of *Staphylococcus aureus* in Sliced Dry Cured Ham Packed under Modified Atmosphere. *Viandes Prod. Carnés* **2018**, 1–9. Available online: https://www.viandesetproduitscarnes.fr/index.php/en/hygiene2/porc-charcuterie-salaison?download=749:risque-lie-a-staphylococcus-aureus-dans-le-jambon-sec-tranche (accessed on 26 April 2023).
17. Márta, D.; Wallin-Carlquist, N.; Schelin, J.; Borch, E.; Rådström, P. Extended Staphylococcal Enterotoxin D Expression in Ham Products. *Food Microbiol.* **2011**, *28*, 617–620. [CrossRef] [PubMed]
18. Untermann, F.; Müller, C. Influence of a_w Value and Storage Temperature on the Multiplication and Enterotoxin Formation of Staphylococci in Dry-Cured Raw Hams. *Int. J. Food Microbiol.* **1992**, *16*, 109–115. [CrossRef]
19. CAC—Guidelines for the Validation of Food Safety Control Measures. CAC/GL 69. Available online: http://www.fao.org/input/download/standards/11022/CXG_069e.pdf (accessed on 26 April 2023).
20. Stewart, C.M.; Cole, M.B.; Legan, J.D.; Slade, L.; Vandeven, M.H.; Schaffner, D.W. Modeling the Growth Boundary of *Staphylococcus aureus* for Risk Assessment Purposes. *J. Food Prot.* **2001**, *64*, 51–57. [CrossRef] [PubMed]
21. Bover-Cid, S.; Garriga, M. *Microbiología Predictiva: Herramienta de Soporte Para La Gestión de la Seguridad y la Calidad Alimentaria*; Eurocarne: Suzhou, China, 2008; Volume 166, pp. 1–8.
22. Bonilauri, P.; Grisenti, M.S.; Daminelli, P.; Merialdi, G.; Ramini, M.; Bardasi, L.; Taddei, R.; Cosciani-Cunico, E.; Dalzini, E.; Frustoli, M.A.; et al. Reduction of *Salmonella* spp. Populations in Italian Salami during Production Process and High Pressure Processing Treatment: Validation of Processes to Export to the U.S. *Meat Sci.* **2019**, *157*, 107869. [CrossRef]
23. Bover-Cid, S.; Serra-Castelló, C.; Dalgaard, P.; Garriga, M.; Jofré, A. New Insights on *Listeria monocytogenes* Growth in Pressurised Cooked Ham: A Piezo-Stimulation Effect Enhanced by Organic Acids during Storage. *Int. J. Food Microbiol.* **2019**, *290*, 150–158. [CrossRef]
24. *ISO 19020*; Microbiology of the Food Chain—Horizontal Method for the Immunoenzymatic Detection of Staphylococcal Enterotoxins in Foodstuffs. International Organization for Standardization: Geneva, Switzerland, 2017; 22p.
25. Borneman, D.L.; Ingham, S.C.; Ane, C. Predicting Growth-No Growth of *Staphylococcus aureus* on Vacuum-Packaged Ready-to-Eat Meats. *J. Food Prot.* **2009**, *72*, 539–548. [CrossRef]
26. Leporq, B.; Membré, J.-M.; Dervin, C.; Buche, P.; Guyonnet, J.P. The "Sym'Previus" Software, a Tool to Support Decisions to the Foodstuff Safety. *Int. J. Food Microbiol.* **2005**, *100*, 231–237. [CrossRef]
27. Rosso, L.; Bajard, S.; Flandrois, J.P.; Lahellec, C.; Fournaud, J.; Veit, P. Differential Growth of *Listeria monocytogenes* at 4 and 8 °C: Consequences for the Shelf Life of Chilled Products. *J. Food Prot.* **1996**, *59*, 944–949. [CrossRef]
28. Core Team, R. R: A Language and Environment for Statistical Computing. Available online: https://www.r-project.org/ (accessed on 26 April 2023).
29. Jewell, K. Comparison of 1-Step and 2-Step Methods of Fitting Microbiological Models. *Int. J. Food Microbiol.* **2012**, *160*, 145–161. [CrossRef]
30. Martino, K.; Marks, B. Comparing Uncertaintly Resulting from Two-Step and Global Regression Procedures Applied to Microbial Growth Models. *J. Food Prot.* **2007**, *70*, 2811–2818. [CrossRef] [PubMed]
31. Zwietering, M.H.; Jongenburger, I.; Rombouts, F.M.; van't Riet, K. Modeling of the Bacterial Growth Curve. *Appl. Environ. Microbiol.* **1990**, *56*, 1871–1875. [CrossRef]
32. Couvert, O.; Gaillard, S.; Savy, N.; Mafart, P.; Leguérinel, I. Survival Curves of Heated Bacterial Spores: Effect of Environmental Factors on Weibull Parameters. *Int. J. Food Microbiol.* **2005**, *101*, 73–81. [CrossRef]
33. Møller, C.O.A.; Ilg, Y.; Aabo, S.; Christensen, B.B.; Dalgaard, P.; Hansen, T.B. Effect of Natural Microbiota on Growth of *Salmonella* spp. in Fresh Pork—A Predictive Microbiology Approach. *Food Microbiol.* **2013**, *34*, 284–295. [CrossRef] [PubMed]
34. ICMSF (International Commission of Microbial Specifications of Food). *Microorganisms in Foods 6: Microbial Ecology of Food Commodities*, 2nd ed.; Roberts, T.A., Cordier, J.-L., Gram, L., Tompkin, R., Pitt, J.I., Gorris, L.G.M., Swanson, K.M.J., Eds.; Springer: New York, NY, USA, 2005; ISBN 978-0-306-48675-3.

35. Iacumin, L.; Zuccolo, C.; Comi, G. Fate of *Staphylococcus aureus* in Dry Cured Ham Packaged under Vacuum and Stored at Different Temperatures. *Ind. Aliment.* **2019**, *58*, 24–30.
36. Lindqvist, R.; Sylve, S.; Vagsholm, I. Quantitative Microbial Risk Assessment Exemplified by *Staphylococcus aureus* in Unripened Cheese Made from Raw Milk. *Int. J. Food Microbiol.* **2002**, *78*, 155–170. [CrossRef] [PubMed]
37. Yu, H.H.; Song, Y.J.; Kim, Y.J.; Lee, H.Y.; Choi, Y.S.; Lee, N.K.; Paik, H.D. Predictive Model of Growth Kinetics for *Staphylococcus aureus* in Raw Beef under Various Packaging Systems. *Meat Sci.* **2020**, *165*, 108108. [CrossRef]
38. Medina, M. Caracterización de *Staphylococcus aureus* Procedentes de Indústrias Cárnicas. In *Innovación en Productos Cárnicos Seguros y Saludables*; Córdoba, J.J., Medina, M., Carballo, J., Eds.; Agència Catalana de Seguretat Alimentària: Barcelona, Spain, 2021; p. 73.
39. Notermans, S.; van Otterdijk, R.L.M. Production of Enterotoxin A by *Staphylococcus aureus* in Food. *Int. J. Food Microbiol.* **1985**, *2*, 145–149. [CrossRef]
40. Kaban, G.; Kaya, M. Effect of Starter Culture on Growth of *Staphylococcus aureus* in Sucuk. *Food Control* **2006**, *17*, 797–801. [CrossRef]
41. Charlier, C.; Cretenet, M.; Even, S.; Le Loir, Y. Interactions between *Staphylococcus aureus* and Lactic Acid Bacteria: An Old Story with New Perspectives. *Int. J. Food Microbiol.* **2009**, *131*, 30–39. [CrossRef] [PubMed]
42. Metaxopoulos, J.; Genigeorgis, C.; Fanelli, M.J.; Franti, C.; Cosma, E. Production of Italian Dry Salami. I. Initiation of Staphylococcal Growth in Salami Under Commercial Manufacturing Conditions. *J. Food Prot.* **1981**, *44*, 347–352. [CrossRef] [PubMed]
43. Serra-Castelló, C.; Bover-Cid, S.; Garriga, M.; Beck Hansen, T.; Gunvig, A.; Jofré, A. Risk Management Tool to Define a Corrective Storage to Enhance *Salmonella* Inactivation in Dry Fermented Sausages. *Int. J. Food Microbiol.* **2021**, *346*, 109160. [CrossRef] [PubMed]
44. Ha, J.; Lee, J.; Lee, S.; Kim, S.; Choi, Y.; Oh, H.; Kim, Y.; Lee, Y.; Seo, Y.; Yoon, Y. Mathematical Models to Describe the Kinetic Behavior of *Staphylococcus Aureus* in Jerky. *Food Sci. Anim. Resour.* **2019**, *39*, 371–378. [CrossRef] [PubMed]

foods

Article

Effects of *ESA_00986* Gene on Adhesion/Invasion and Virulence of *Cronobacter sakazakii* and Its Molecular Mechanism

Yufei Fan [1], Ping Li [1], Dongdong Zhu [1], Chumin Zhao [1], Jingbo Jiao [1], Xuemeng Ji [2] and Xinjun Du [1,*]

[1] State Key Laboratory of Food Nutrition and Safety, College of Food Science and Engineering, Tianjin University of Science and Technology, Tianjin 300457, China; 17908002@mail.tust.edu.cn (Y.F.); zoelxx@126.com (P.L.), duiduiz@163.com (D.Z.); zcm3278@163.com (C.Z.); 18753391899@163.com (J.J.)

[2] Tianjin Key Laboratory of Food Science and Health, School of Medicine, Nankai University, Tianjin 300071, China; jixuemeng@nankai.edu.cn

* Correspondence: xjdu@tust.edu.cn; Tel.: +86-22-60912484; Fax: +86-22-60912484

Abstract: *Cronobacter sakazakii* is an opportunistic Gram-negative pathogen that has been identified as a causative agent of severe foodborne infections with a higher risk of mortality in neonates, premature infants, the elderly, and immunocompromised populations. The specific pathogenesis mechanisms of *C. sakazakii*, such as adhesion and colonization, remain unclear. Previously, we conducted comparative proteomic studies on the two strains with the stronger and weaker infection ability, respectively, and found an interesting protein, ESA_00986, which was more highly expressed in the strain with the stronger ability. This unknown protein, predicted to be a type of invasitin related to invasion, may be a critical factor contributing to its virulence. This study aimed to elucidate the precise roles of the *ESA_00986* gene in *C. sakazakii* by generating gene knockout mutants and complementary strains. The mutant and complementary strains were assessed for their biofilm formation, mobility, cell adhesion and invasion, and virulence in a rat model. Compared with the wild-type strain, the mutant strain exhibited a decrease in motility, whereas the complementary strain showed comparable motility to the wild-type. The biofilm-forming ability of the mutant was weakened, and the mutant also exhibited attenuated adhesion to/invasion of intestinal epithelial cells (HCT-8, HICE-6) and virulence in a rat model. This indicated that *ESA_00986* plays a positive role in adhesion/invasion and virulence. This study proves that the *ESA_00986* gene encodes a novel virulence factor and advances our understanding of the pathogenic mechanism of *C. sakazakii*.

Keywords: *Cronobacter sakazakii*; *ESA_00986*; adhesion/invasion; virulence

Citation: Fan, Y.; Li, P.; Zhu, D.; Zhao, C.; Jiao, J.; Ji, X.; Du, X. Effects of *ESA_00986* Gene on Adhesion/Invasion and Virulence of *Cronobacter sakazakii* and Its Molecular Mechanism. *Foods* **2023**, *12*, 2572. https://doi.org/10.3390/foods12132572

Academic Editor: Antonio Bevilacqua

Received: 31 May 2023
Revised: 28 June 2023
Accepted: 29 June 2023
Published: 30 June 2023

1. Introduction

Cronobacter spp. is a motile Gram-negative bacillus with the ability to form biofilm. It is a facultative anaerobic bacterium [1]. Among the seven species of this genus, *Cronobacter sakazakii* exhibits the highest prevalence and clinical relevance in the human population [2,3]. It is considered a causative agent of neonatal meningitis and necrotizing colitis [4], which can result in a high mortality rate of up to 80% [5,6]. In addition to its impact on neonates, *C. sakazakii* has the ability to induce severe infections in elderly individuals and people with compromised immune systems [7]. Numerous studies have suggested that *C. sakazakii* is often present in different food types and most clinical cases are associated with powdered infant formula (PIF) and other infant food products [8].

In recent years, significant advancements have been made in the study of the pathogenic mechanism of *C. sakazakii*. Numerous studies have provided evidence that *C. sakazakii* is capable of adhering to and invading human intestinal epithelial cells. Additionally, it has the ability to cross the blood–brain barrier and replicate within macrophages [9,10]. As a peroral pathogen that can cause systemic infections, *C. sakazakii* must have the necessary virulence factors to invade various epithelial and endothelial cells in human and animal hosts, evade host defense mechanisms, and traverse the blood–brain barrier. Several virulence determinants have been identified and validated, such as lipopolysaccharides (LPSs), outer membrane protein A (Omp A), and outer membrane protein X [11]. LPSs play a crucial role in the invasion of intestinal epithelial cells by *C. sakazakii* via the disruption of tight junctions [12–14]. Omp A can interact with glycoproteins on the surface of human brain microvascular endothelial cells (HBMECs) to mediate adhesion infection [15,16]. Omp X has been confirmed to play a critical role in adhesion to and invasion of Caco-2 cells [17,18]. The flagella of *C. sakazakii* serve as immune stimuli and can trigger the production of pro-inflammatory cytokines in human-derived monocytes [19]. LysR plays a critical role in the regulation of factors involved in adhesion to and invasion of human intestinal cells [20]. However, the details of the adhesion and infection mechanisms of *C. sakazakii* are still unclear.

Adhesins are one of the most important factors in bacterial adhesion. Some adhesins exist on fimbriae, which are conducive to the initial adhesion of bacteria to cells. Some adhesins can recognize and interact with receptors on the cell surface, triggering a series of signals that rearrange the cytoskeleton of the host cell and induce bacterial uptake. The Intimin/Invasin (Int/Inv) family encompasses a wide range of proteins that facilitate bacterial attachment to and/or invasion of host cells. These proteins play a crucial role in mediating the interaction between the bacterium and its target cells [21]. The Int/Inv family's prototype members originate from pathogenic strains of *Escherichia coli* (Int) [22] and *Yersinia* (Inv) [23]. These strains have been extensively studied and serve as important models for understanding the mechanisms of bacterial attachment and invasion in various host cell types. Jerse et al. first describe intimin in enteropathogenic and *E. coli* strains in [24]; it facilitates bacterial entry into eukaryotic cells by engaging in high-affinity binding with members of the β1 integrin family [25]. Some intimin/invasin proteins contain multiple bacterial Ig-like domains that are classified as belonging to the Big_1 superfamily [26]. Ig-like domains are present in the surface proteins of bacteria and have been implicated in bacterial pathogenesis [27]. Previous studies have reported that *Yersinia* or *Salmonella* invasins, which possess Ig domains, are involved in the invasion of different epithelial host cells such as M and Hep-2 cells [28,29].

In our previous study, we conducted comparative proteomic studies on the two strains with the stronger and weaker infection ability, respectively. We found an adherence-related protein, ESA_00986, which was highly expressed in the strain with the stronger ability. It has typical Ig-like domains typical of intimin/invasin and is presumed to be a protein in the intimin/invasin family. However, the homolog which is most similar to protein ESA_00986 is the intimin in *E. coli* [24], which is only 32% similar. Therefore, the detailed roles of this gene were unknown. However, the detailed roles of this gene have not been studied. In this study, the functions of the *ESA_00986* gene in *C. sakazakii* were explored via gene knockout and complementation. This study plays a significant part in understanding the detailed role of the ESA_00986 gene in the interactions between *C. sakazakii* and host cells, which is greatly helpful in controlling the foodborne pathogen.

2. Experimental Materials and Procedures

2.1. Strains and Plasmids

The strains and plasmids employed in this investigation are presented in Table 1. Bacteria were stored in LB broth containing 15% glycerol at −80 °C and grown on LB agar medium and in LB broth at 37 °C while being continuously shaken at 200 rpm. The broth contained chloramphenicol and ampicillin at a concentration of 100 μg/mL each.

Table 1. Bacterial strains and plasmids used in this study.

Strain or Plasmid	Genotype or Characteristics	Source
	Cronobacter sakazakii	
ATCC BAA-894	WT	ATCC
Δ*ESA_00986*	Δ*ESA_00986*::amp^r	This study
cpESA_00986	Δ*ESA_00986* with pACYC184-*ESA_00986*	This study
	E. coli	
S17 lambda pir	Strain for construction harboring lambda pir	29
	Plasmids	
pCVD442	Suicide plasmid for deletion: amp^r	29
pCVD442-Q-H	pCVD442 with homologous arms	This study
pACYC184	p15A ori Cm^r Tet^r	
pACYC184-*ESA_00986*	pACYC184 with *ESA_00986*	This study

2.2. Construction of ESA_00986 Mutant

The generation of a mutant with a deletion in the *ESA_00986* gene was carried out using a method previously established in [30]. To linearize pCVD442, PCR was performed using primers pCVD442F and pCVD442R (Table 2). The up and down homologous arms of *ESA_00986* were amplified via PCR using the two pairs of primers (*ESA_00986* QF/*ESA_00986* QR and *ESA_00986* HF/*ESA_00986* HR) listed in Table 2. To construct the pCVD442-QH vector, the fragments located upstream and downstream of the target region were inserted into the linearized pCVD442 suicide vector using a seamless cloning kit (VAZYME, China). Subsequently, the constructed vector was chemically transformed into *E. coli* S17 cells. In the next stage, the vector containing the targeted fragments was transformed into the wild-type (WT) strain of *C. sakazakii* ATCC BAA-894 via chemical transformation. All primer sequences used in these procedures can be found in Table 2.

Table 2. Primers used in this study.

Primer	Gene Amplified	Primer Sequences (5′-3′)	Amplification Size (bp)
pCVD442 F	Suicide plasmid for	CAATAACCCTGATAAATGCTTCAA	6345
pCVD442 R	markerless deletion	CTCATGAGCGGATACATATTTG	
ESA_00986 QF	Upstream of	TGATAAATGCTTCAACGTCAGCGTCACCTGGAACG	828
ESA_00986 QR	*ESA_00986* gene	GTGGTAGTTCAGGCGACTATTTTCCTGACGGAAACAGACG	
ESA_00986 HF	Downstream of	ATAGTCGCCTGAACTACCACAAATGG	780
ESA_00986 HR	*ESA_00986* gene	GCGGATACATATTTGCCAGCCCGTCAGTCGTAATG	
ESA_00986 1F	*ESA_00986* gene sequence	TACATGGCAACTTATTCTGTATTTAC	1166
ESA_00986 1R		TGACCATTTGTGGTAGTTCAGG	
ESA_00986 2F	Both ends of the *ESA_00986*	GCTCTGTCTCGGTGATGTAG	1620(WT)/600(Δ*ESA_00986*)
ESA_00986 2R	gene sequence	TGATCCTTCAATCCCAGTAA	

2.3. Complementation

The low-copy vector pACYC184 was used to complement the deletion mutants. To construct the complementary plasmid pACYC184-*ESA_00986*, the primers *ESA_00986* 1F and *ESA_00986* 1R (Table 2) were employed. To construct the Δ*ESA_00986* complementation, the pACYC184 plasmid was used, which carried an *ESA_00986* gene fragment. The plasmid was introduced into the Δ*ESA_00986* mutant strain, resulting in the complementary strain *cpESA_00986* harboring the *ESA_00986* gene.

2.4. Growth Curve Analysis

The bacterial growth curve was determined using established protocols as described previously [31]. In brief, the WT strain, Δ*ESA_00986*, and *cpESA_00986* were cultured overnight in LB medium at 37 °C, whereupon the cultures were transferred into 100 mL of fresh culture medium (1:100). Bacterial growth was monitored by measuring the optical density (OD) at 600 nm at hourly intervals for a total of 14 h using a UV-Vis spectrophotometer. Subsequently, the collected data points were used to plot the bacterial growth curves. Each sample was replicated three times to ensure the accuracy and reliability of the obtained results.

2.5. Motility Analysis

The motility of bacteria was evaluated by measuring the migration radius in LB containing 0.3% agar [32]. The WT, ΔESA_00986, and the cpESA_00986 strain were incubated at 37 °C overnight for cultivation. Subsequently, the bacterial cultures were transferred onto soft agar plates (LB medium containing 0.3% agar) and incubated at 30 °C for 16 h. Afterwards, colony size was observed, and the average migration radius was calculated.

2.6. Hydrophobicity Analysis

To assess hydrophobicity, a xylene contact angle assay was performed following previously established methods [33]. Bacterial cells were cultured in LB broth overnight and then subcultured in fresh LB medium at a 1% inoculum. The cultures were grown until they reached an optical density of 600 nm (OD600) within a range from 0.6–0.8, indicating mid-log phase. Next, the bacteria were collected via centrifugation and subsequently washed with phosphate-buffered saline (PBS), then adjusted to OD600 of 0.5. 2 mL bacterial suspension with a xylene mixture, trained for 3 h at room temperature, and then the OD600 of the bacterial cultures was measured using a spectrophotometer. The hydrophobicity of the bacterial cells was calculated using the following formula: $[(H0 - H)/H0] \times 100\%$, where H0 and H represent the optical density at 600 nm of the bacterial suspension before and after the addition of xylene.

2.7. Outer Membrane Permeability Analysis

Permeability was assessed via N-Phenyl-1-naphthylamine (NPN) assay. Briefly, bacterial cells were initially cultured in LB broth overnight. The culture was then transferred to fresh LB medium at a 1% inoculum and incubated until OD600 reached a range of 0.6-0.8. Cells were harvested and suspended in PBS, then adjusted to OD600 of 0.5. A final concentration of 1 μM of NPN was added to the sample. The fluorescence of the sample was then continuously measured using an excitation wavelength of 380 nm and an emission wavelength of 430 nm.

2.8. Biofilm Formation Ability Analysis

Biofilm formation capacity was assessed via crystal violet (CV) staining. In summary, suspensions of the WT, ΔESA_00986, and cpESA_00986 strains were prepared at an OD600 value of 0.7. These suspensions were then added to individual wells and incubated at 37 °C for a duration of 48 h. Following incubation, the biofilm was fixed with methanol after 15 min, and the supernatant was removed. The samples were air-dried at 25 °C. Subsequently, each well was treated with 1% CV solution and incubated for 30 min. After staining, the wells were washed three times, and 200 μL ethanol was added for decolorization purposes. The optical density was measured at a wavelength of 570 nm using a spectrophotometer. Each experiment was performed with three repetitions to ensure the accuracy and reliability of the results.

2.9. Adhesion/Invasion Capacity Analysis

The adhesion/invasion assay was conducted following the previously described method with slight modifications [34]. HCT-8 cells and HIEC-6 cells were used for this assay. The cell monolayer was cultured using RPMI-1640 medium and 10% fetal bovine serum. The WT, ΔESA_00986 and cpESA_00986 cells were collected at an OD600 of 0.6, then were washed and suspended in RPMI-1640 medium. Next, 1 mL of *C. sakazakii* bacteria was added to the HIEC-6 cells or HCT-8 cells at a Multiplicity of Infection (MOI) of 100. After incubation for 3 h, the cells were washed with PBS and subsequently lysed using 1% Triton X-100. The resulting cell suspensions were then serially diluted with PBS and plated on LB agar plates for colony counting analysis.

The invasion assay was conducted following the previously described method with slight modifications [35]. A 1 mL quantity of *C. sakazakii* bacteria was added to the HIEC-6 cells or HCT-8 cells at a Multiplicity of Infection (MOI) of 100. After incubation for 1 h, the cells were washed with PBS and covered with 2 mL RPMI-1640 medium containing gentamicin at bactericidal concentration of 100 μg/mL to kill extracellular bacteria. After 2 h, the cells were washed with PBS and subsequently lysed using 1% Triton X-100. The resulting cell suspensions were then serially diluted with PBS and plated onto LB agar plates for colony-counting analysis.

2.10. Rat Virulence Assay

Neonatal rats (48 h old) were sourced from SPF (Beijing) Biotechnology Co., Ltd. The rats were housed in separate clean cages that had been disinfected prior to use. A total of 12 neonatal rats were assigned to each experimental group. To administer the bacteria, each pup received an oral dose of 0.2 mL of infant formula with sheep's milk powder containing 1 × 109 bacterial cells. This was done using an animal gavage needle. The control group received an oral dose of 0.2 mL of sterile infant formula with sheep's milk powder only. During the post-challenge period, the pups were closely monitored for physiological symptoms and mortality rates every 2 h. After 48 h, all the pups were euthanized for further analysis. Blood samples were collected and then subjected to centrifugation at 2000× *g* for 15 min at 4 °C. This process allowed for the separation of serum, which was subsequently stored at −80 °C for future analysis. Next, liver, spleen, and brain samples were excised, washed with PBS, and dried with filter papers during necropsy. All tissues were rapidly frozen via immersion in liquid nitrogen and then stored at −80 °C for future utilization. Animal experiments were performed according to the guidelines of the institutional animal ethics committee and supported by the Institutional Animal Committee of Tianjin University of Science and Technology (2022117).

2.11. Quantification of Bacteria in Tissues

The tissues intended for *C. sakazakii* enumeration were washed three times with PBS (pH 7.4). Subsequently, they were homogenized in cold PBS and subjected to 10-fold serial dilutions until the desired concentrations were achieved. The dilutions were plated onto modified lauryl sulfate broth (mLST) plates and incubated at 37 °C for 24 h. After the incubation period, bacterial colonies were counted. The absolute quantities of *C. sakazakii* were determined based on their corresponding dilutions.

2.12. Serum Inflammatory Levels Analysis

The levels of inflammatory cytokines in the serum were analyzed via an ELISA assay kit (Nanjing Jiancheng, China). The inflammatory cytokines that were analyzed include IFN-γ, IL-6, IL-1β, IL-8, and TNF-α. The procedures were performed according to the instructions provided in the assay kit manual.

2.13. Gene Expression Analysis by qRT-PCR

RNA was extracted from tissues (n = 5) using an RNA extraction kit (ACCURATE, Changsha, China) according to the manufacturer's instructions. Subsequently, the RNA was reverse-converted into complementary DNA (cDNA) via reverse transcription using a PrimeScript™ RT reagent Kit (ACCURATE, Changsha, China). Real-time quantitative PCR (qPCR) was performed in triplicate via the Mastercycler® ep realplex system and TB Green™ Premix Ex Taq™ II (ACCURATE, Changsha, China). The specificity of the PCR products was analyzed by examining their melting curves. Relative gene expression levels were quantified via the $2^{\wedge(-\Delta\Delta Ct)}$ method.

2.14. Histopathological Analysis

The histological samples were subjected to hematoxylin and eosin (HE) staining for analysis. A portion of the samples was fixed in 4% paraformaldehyde and subsequently underwent a series of processes, including dehydration, clearing, and embedding in paraffin, to obtain 5-μm sections. Furthermore, sections of the colon embedded in paraffin were stained using the hematoxylin and eosin (HE) staining methods.

2.15. Statistical Analysis

SPSS was used for data analysis. Statistical analysis was performed using Student's unpaired *t*-test and one-way analysis of variance (ANOVA) to assess significant differences in the results. A significance level of 0.05 was used, and P values below this threshold were considered statistically insignificant. Data were expressed as mean ± deviation, and three independent replications were performed for each experiment.

3. Results

3.1. Construction and Validation of ΔESA_00986 and Complementary Strains

To investigate the molecular mechanism of *ESA_00986*, we generated an *ESA_00986* mutant strain via the homologous recombination technique. Figure 1A depicts the amplification sites targeted by the two primers (*ESA_00986* 1F/R, *ESA_00986* 2F/R) in the WT, Δ*ESA_00986*, and cpESA_00986 strains. PCR verification was conducted using two primers with DNA from the WT, ΔESA_00986, and cpESA_00986 strains as templates. The results of the PCR analysis are presented in Figure 1B. When the ESA_00986 1F/R primers were used, both the WT and cpESA_00986 strains exhibited a visible DNA band of 1166 bp; the ΔESA_00986 strain exhibited no DNA band. However, when using the ESA_00986 2F/R primers, the WT strain showed a DNA band of 1620 bp, whereas the ΔESA_00986 and cpESA_00986 strains displayed a DNA band of 600 bp. The PCR results showed successful construction of both the ΔESA_00986 and cpESA_00986 strains, indicating their suitability for further experiments.

A

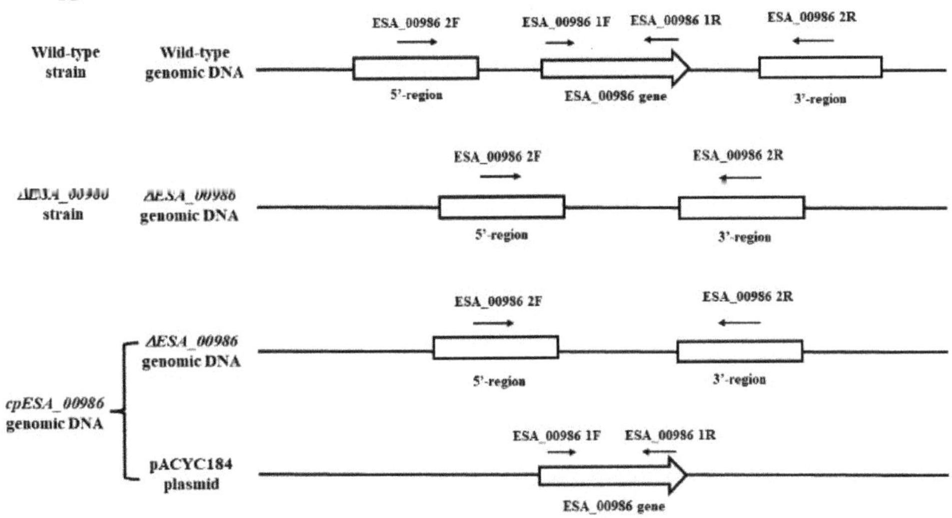

Figure 1. *Cont.*

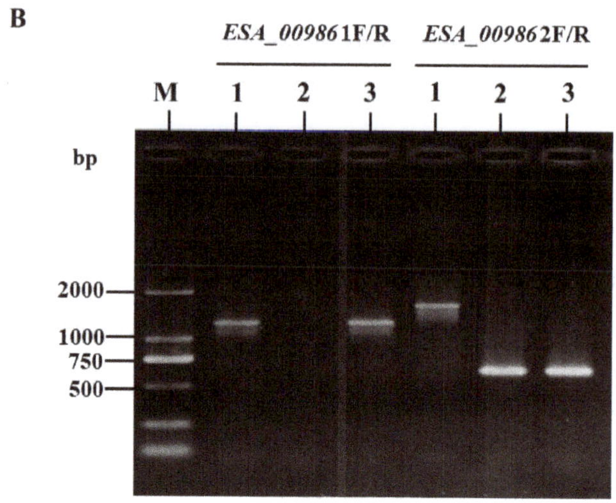

Figure 1. PCR validation of WT, ΔESA_00986, and cpESA_00986 strains (**A**) The positions of the two primer pairs of WT strain, ΔESA_00986 and cpESA_00986 strain. (**B**) PCR amplification results obtained via two primer pairs for the WT, ΔESA_00986, and cpESA_00986 strains. M, Marker; Lane 1, WT; Lane 2, ΔESA_00986; Lane 3, cpESA_00986.

3.2. *Effects of ESA_00986 Gene on Growth*

In order to assess the impact of *ESA_00986* gene deletion on the growth rate and the OD_{600} of the WT, the *ΔESA_00986*, and *cpESA_00986* strains were measured via a UV-Vis spectrophotometer over a 14-h period. The growth rate results are presented in Figure 2. Comparing the growth rate of the *ΔESA_00986* with that of the WT strain, no significant difference was observed. Likewise, there was no significant change in the OD_{600} of the *cpESA_00986* strain compared to the WT strain. These findings suggest that the deletion of the *ESA_00986* gene does not impact bacterial growth. They also indicate that *ESA_00986* is not essential for cell growth, thereby eliminating potential variations in growth patterns as a confounding factor in future experiments.

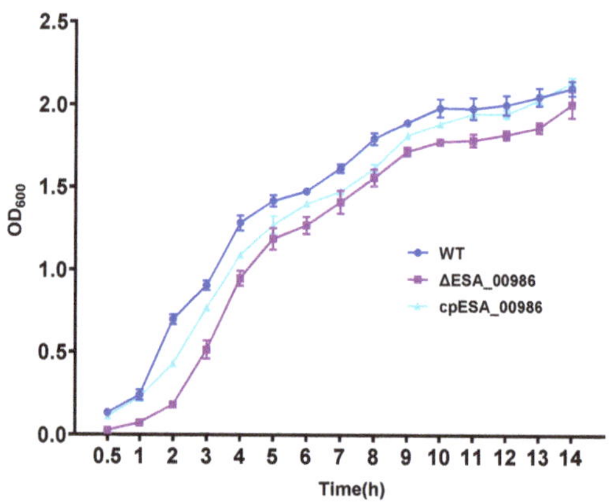

Figure 2. The growth curves of WT, *ΔESA_00986*, and *cpESA_00986* strains. Data are shown as mean ± S.D. Significant differences between *ΔESA_00986*/*cpESA_00986* groups and WT group were analyzed via one-way ANOVA. WT: wild-type strain; *ΔESA_00986*: mutant strain; *cpESA_00986*: complementary strain.

3.3. Effects of ESA_00986 Gene on Cell Adhesion/Invasion

ESA_00986 is predicted to be an Intimin/Invasin protein that is believed to play a role in bacterial adhesion to and invasion of host cells. The adhesion/invasion assay was performed to investigate the role of *ESA_00986* in the colonization of HCT-8 and HIEC-6 intestinal epithelial cell lines by *C. sakazakii*. Significant differences in colonization were observed between the WT, Δ*ESA_00986*, and *cpESA_00986* strains in both cell lines (Figure 3). The relative Δ*ESA_00986* strain adhesion/invasion rate of both HCT-8 and HIEC-6 cells was significantly lower compared to the WT strain. Furthermore, the *cpESA_00986* strain showed significantly higher adhesion/invasion compared to the Δ*ESA_00986* strain in both cell lines. Similarly, the relative invasion rate of the three strains showed the same trend. These findings indicate that the *ESA_00986* gene plays a crucial role in the colonization of *C. sakazakii* in vitro.

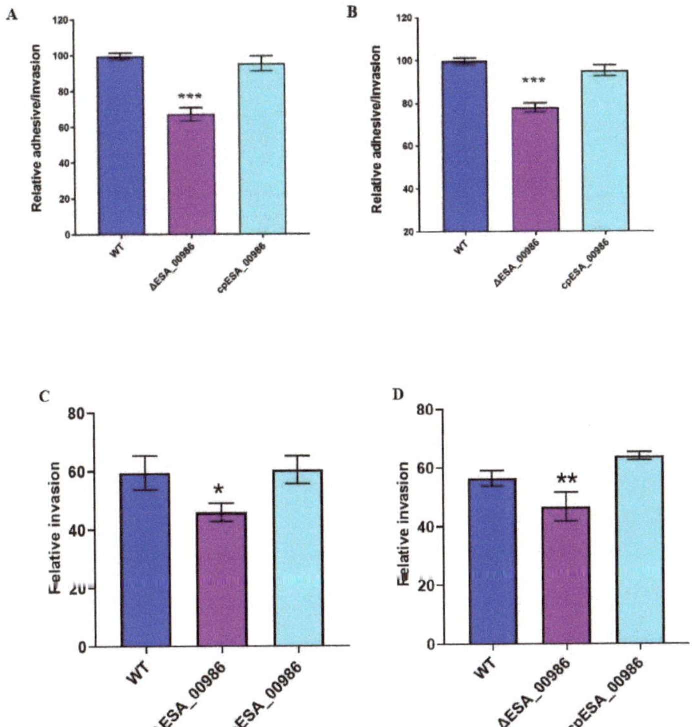

Figure 3. Effects of the *ESA_00986* gene on the adhesion and invasion. Data are shown as mean ± S.D. Significant differences between Δ*ESA_00986*/*cpESA_00986* groups and WT group was analyzed at * $p < 0.05$, ** $p < 0.01$, and *** $p < 0.001$ via one-way ANOVA. WT: wild-type strain; Δ*ESA_00986*: mutant strain; *cpESA_00986*: complementary strain. (**A**): the adhesion/invasion assay to the HCT-8 cells; (**B**): the adhesion/invasion assay to HIEC-6 cells; (**C**): the invasion assay to HCT-8 cells; (**D**): the invasion assay to HIEC-6 cells.

3.4. Effects of ESA_00986 Gene on Virulence of C. sakazakii In Vivo

As a predicted invasin, *ESA_00986* has the potential to influence the pathogenicity of *C. sakazakii*. Three separate groups of 48 h old rats were orally infected with WT, Δ*ESA_00986*, and *cpESA_00986* strains, and the survival of the animals was closely monitored for a period of 48 h following infection. Rats that were infected with the WT strain or *cpESA_00986* strains started experiencing mortality at 11 h, with 60% of the animals dying within 48 h in the group infected with the WT strain. In contrast, rats infected with the Δ*ESA_00986* strain exhibited delayed mortality, starting at 17 h; 80% of the infected rats survived for at least 48 h. The survival curve (Figure 4) exhibited a significant distinction in pathogenicity between the Δ*ESA_00986* and WT strains. Additionally, the colonization ability of *C. sakazakii* in rats was investigated in relation to the *ESA_00986* gene. The results revealed that the bacterial load of the Δ*ESA_00986* strain was lower than that of the WT strain in the blood, brain, liver, and spleen (Figure 5).

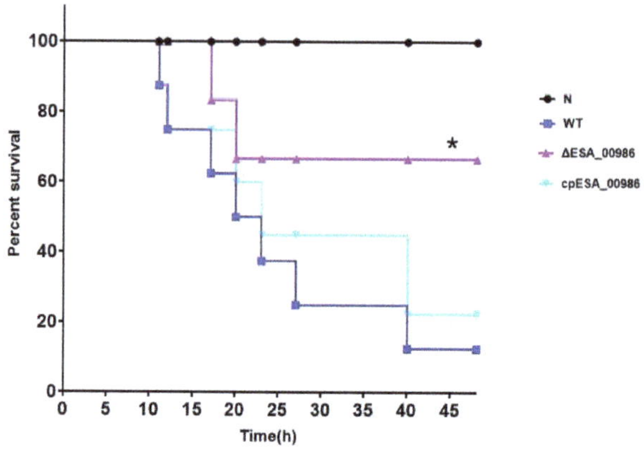

Figure 4. The survival curve of rats infected with *C. sakazakii* WT, Δ*ESA_00986* and *cpESA_00986* strains. Significant differences between Δ*ESA_00986*/*cpESA_00986* groups and WT group was analyzed at * $p < 0.05$ via one-way ANOVA. WT: wild-type strain; Δ*ESA_00986*: mutant strain; *cpESA_00986*: complementary strain.

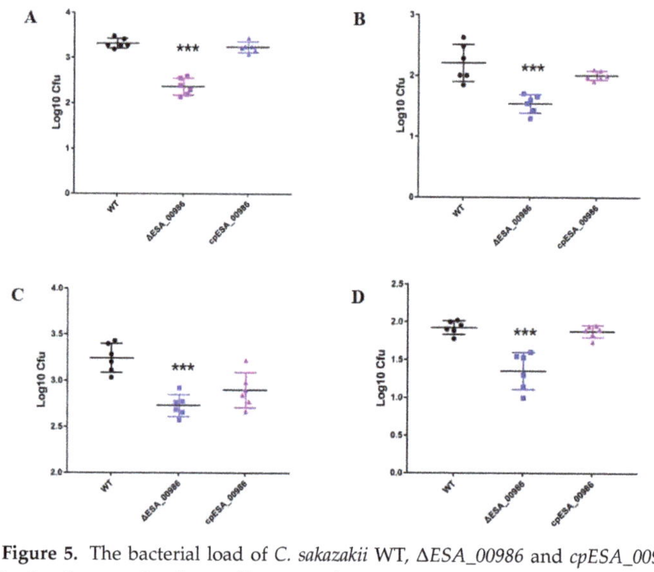

Figure 5. The bacterial load of *C. sakazakii* WT, Δ*ESA_00986* and *cpESA_00986* strains in blood, brain, liver, and spleen. Data are shown as mean ± S.D. Significant differences between Δ*ESA_00986*/*cpESA_00986* groups and WT group was analyzed at *** $p < 0.001$ via one-way ANOVA. WT: wild-type strain; Δ*ESA_00986*: mutant strain; *cpESA_00986*: complementary strain. (A): Brain; (B): Spleen; (C): Blood; (D): Liver.

3.5. Pathological Analysis

Histological analysis was conducted on brain and colon tissues collected from rats 48 h after bacterial infection. HE staining showed significant differences between the WT-infected group and normal group in brain tissue, including matrix dissolution and a loose, sponge-like structure (Figure 6A). In the Δ*ESA_00986*-infected group, the brain matrix exhibited slight looseness and showed signs of inflammatory cell infiltration. The *cpESA_00986*-infected group had the same results as the WT group. After 48 h of infection, the intestinal tissue of rats infected with the WT strain was examined and signs of villus dilation, necrosis, perforation, and destruction were found in the intestine (Figure 6B). In contrast, the severity of intestinal tissue damage was notably reduced in rats infected with the Δ*ESA_00986* strain. These findings suggest that knockout of the *ESA_00986* gene reduces the virulence of *C. sakazakii*.

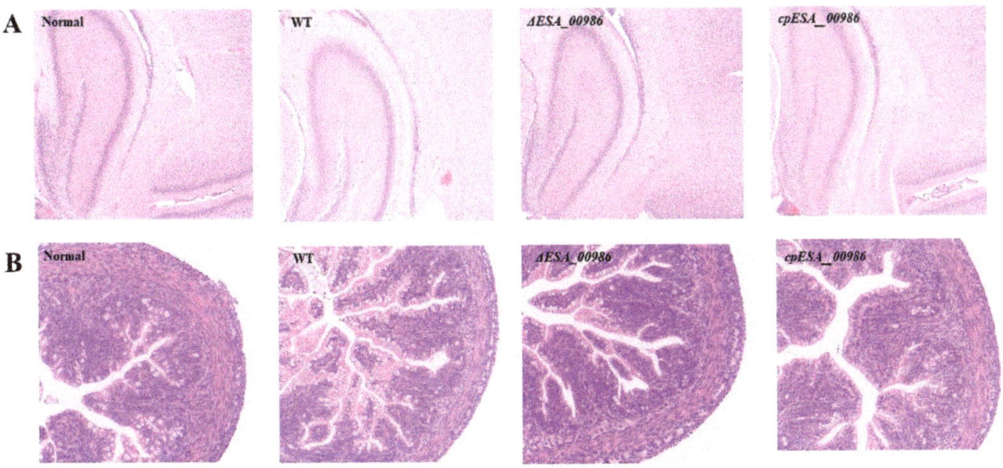

Figure 6. The brain tissue section (**A**) and the colon section (**B**) of *C. sakazakii* WT, Δ*ESA_00986*, and *cpESA_00986* strains. WT: wild-type strain-infected group; Δ*ESA_00986*: mutant strain-infected group; *cpESA_00986*: complementary strain-infected group; NC: normal group.

3.6. Effect of ESA_00986 on the Serum Inflammatory Cytokines

Inflammatory factor is an important cytokine involved in immune regulation during the inflammatory response and inflammatory diseases. When the body is injured and undergoes a stress response, pro-inflammatory cytokines (such as TNF-α, IFN-γ, IL-6, IL-1β, etc.) are produced for immune regulation. At the end of the 48-h toxicity experiment, the blood of suckling rats was collected, and the serum of rat was obtained via centrifugation. The inflammatory factor in the serum was measured by detection kit. As depicted in Figure 7, the serum contents of pro-inflammatory cytokines IL-6, TNF-α, IFN-γ and IL-1ß were significantly reduced in the mutant group compared with the wild-type group. There was no significant difference between the complement group and the wild group. Thus, the decrease in proinflammatory cytokine levels implies a reduction in inflammation caused by bacterial infection.

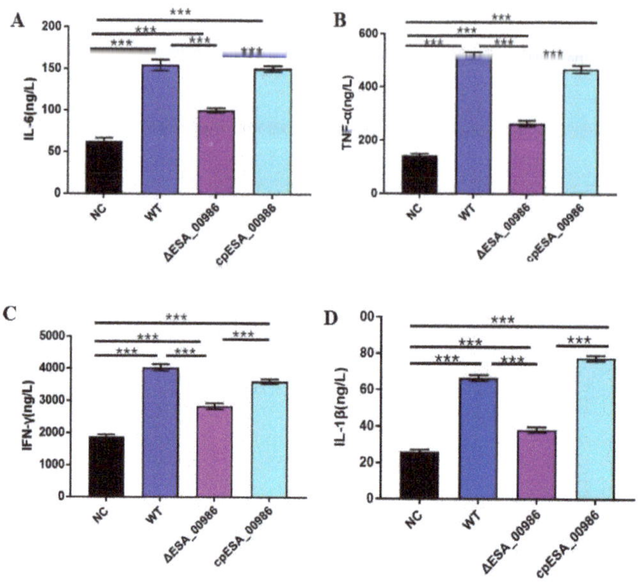

Figure 7. Inflammatory factor content in serum ((**A**): IL-6, (**B**): TNF-α, (**C**): IFN-γ, (**D**): IL-1ß)). Data are shown as mean ± S.D. Significant differences between Δ*ESA_00986*/*cpESA_00986* groups and WT group were analyzed at *** $p < 0.001$ via one-way ANOVA. WT: wild-type strain-infected group; Δ*ESA_00986*: mutant strain-infected group; *cpESA_00986:* complementary strain-infected group; NC: normal group.

3.7. Effect of ESA_00986 on the Expression of Genes Involved in Inflammation and Intestinal Integrity

The impact of *ESA_00986* on gene expression related to inflammation and intestinal integrity in the colon was assessed at the mRNA level, as depicted in Figure 8. After 48 h of infection, the Δ*ESA_00986* group exhibited a significant downregulation in the expression of inflammatory factors, including IL-6, IL-1β, IFN-γ, and TNF-α compared with the WT group. Conversely, the expression of genes associated with gut barrier function, such as zonula occludens-2 (ZO-2), claudins-1 (CLA-1), claudins-2 (CLA-2), and occludins-1 (OC-1), were upregulated in the colon of the Δ*ESA_00986* group. However, no significant difference was observed in the expression of zonula occludens-1 (ZO-1).

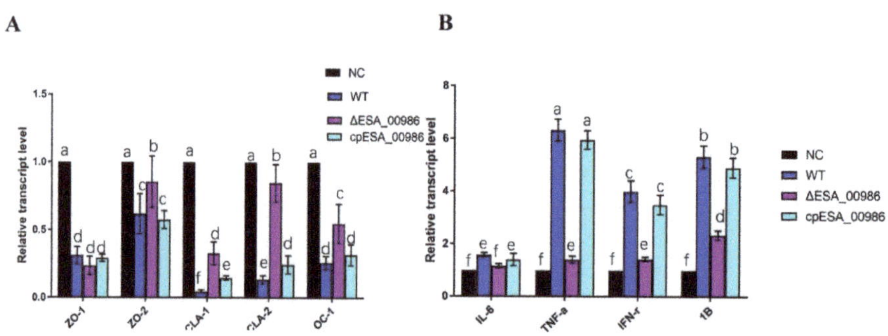

Figure 8. Gene expressions of tight-junction proteins (**A**) and inflammatory factors (**B**). Significant differences are indicated by different letters (a–f) ($p < 0.05$). WT: wild-type strain-infected group; Δ*ESA_00986*: mutant strain-infected group; *cpESA_00986:* complementary strains infected group; NC: normal group.

3.8. *Effects of ESA_00986 Gene on the Motility of C. sakazakii*

Motility, which is related to adhesion and invasion, was measured via swimming rings observed on the semi-solid medium. As is shown in Figure 9A, the size of the swim rings formed by WT, Δ*ESA_00986*, and *cpESA_00986* on 0.3% agar medium has changed. Compared with the WT strain, the *ESA_00986* mutant displayed a loss of swimming mobility, whereas *cpESA_00986* showed a partial restoration in swimming mobility. As depicted in Figure 9B, the average migration radius of the WT strain was approximately 2.76 ± 0.25 cm, that of Δ*ESA_00986* strain was approximately 1.73 ± 0.25 cm, and that of *cpESA_00986* was approximately 2.76 ± 0.30 cm. These findings provide evidence that the *ESA_00986* gene is crucial for the swimming mobility of *C. sakazakii*.

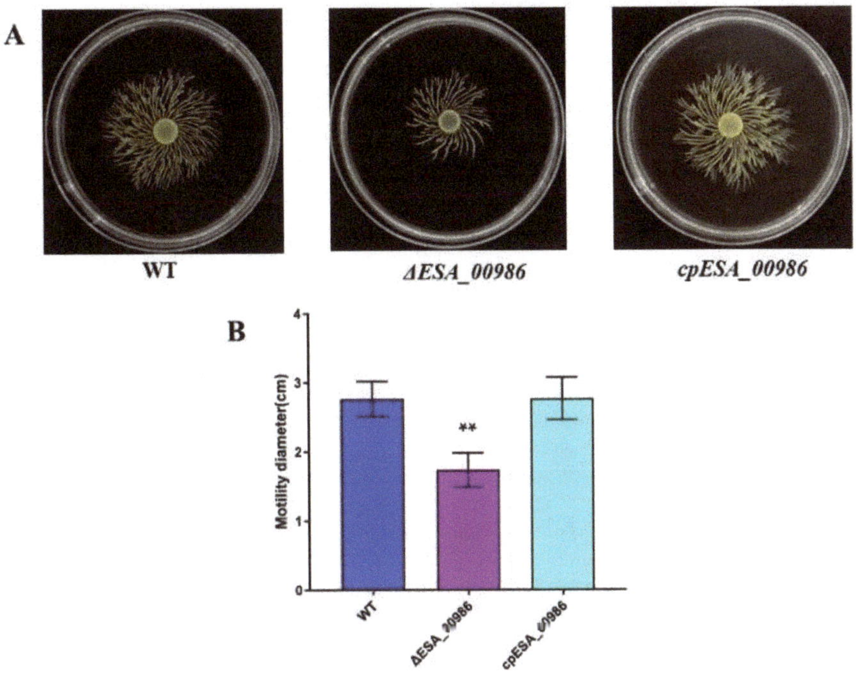

Figure 9. Mobility of the WT, Δ*ESA_00986* and *cpESA_00986* strains on semisolid medium. Data are shown as mean ± S.D. Significant differences between Δ*ESA_00986*/*cpESA_00986* groups and WT group were analyzed at ** $p < 0.01$ via one-way ANOVA. (**A**) The white translucent rings observed represent the migration rings of the bacteria. (**B**) The respective migration radii of the WT, Δ*ESA_00986*, and *cpESA_00986* strains.

3.9. *Effects of ESA_00986 Gene on Physiological Properties of Bacterial Surface*

In order to assess the membrane permeability of the Δ*ESA_00986* strain, bacteria were exposed to a hydrophobic fluorescent probe, 1-phenylnaphthylamine (NPN). An increase in NPN absorption was observed in the Δ*ESA_00986* strain (Figure 10A). To compare the surface hydrophobicity of the WT strain and the Δ*ESA_00986* strain, bacteria suspended in xylene mixture were trained for 3 h at room temperature; bacterial surface hydrophobicity was also reduced in the Δ*ESA_00986* strain (Figure 10B). In summary, the absence of *ESA_00986* led to significant changes in the permeability and hydrophobicity of the bacterial surface.

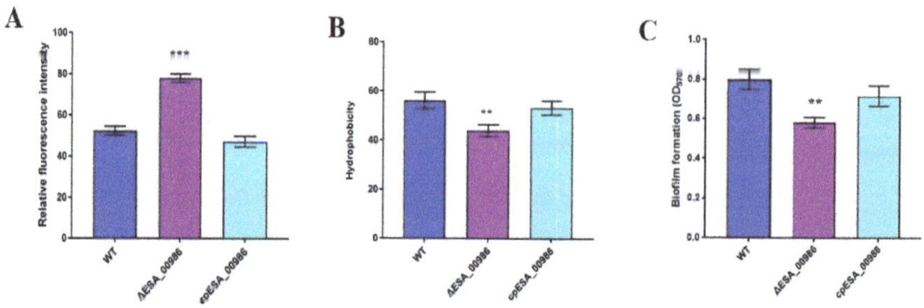

Figure 10. The membrane permeability (**A**), surface hydrophobicity (**B**) and biofilm formation (**C**) of *C. sakazakii* WT, Δ*ESA_00986* and *cpESA_00986* strains, respectively. Data are shown as mean ± S.D. Significant differences between Δ*ESA_00986*/*cpESA_00986* groups and WT group were analyzed at ** $p < 0.01$, *** $p < 0.001$ via one-way ANOVA. WT: wild-type strain; Δ*ESA_00986*: mutant strain; *cpESA_00986*: co mplementary strain.

3.10. Effects of ESA_00986 Gene on Biofilm Synthesis

In this study, we investigated the impact of the *ESA_00986* gene on biofilm formation in *C. sakazakii* ATCC BAA-894. Quantitative analysis was performed using crystal violet staining to assess the extent of biofilm formation. The WT strain and Δ*ESA_00986* mutant strain showed significant differences in biofilm formation within 72 h. Compared with the WT strain, biofilm formation was significantly reduced in the Δ*ESA_00986* mutant (Figure 10C). These findings indicate that the absence of the *ESA_00986* gene significantly impacts the formation of biofilms in *C. sakazakii* within 72 h. The *ESA_00986* gene has played a positive role in biofilm formation.

4. Discussion

Previously, we compared the adhesion and infection ability of 20 strains of *C. sakazakii* on HCT-8 cells, and found two strains with the strongest and weakest infection ability. Later, we conducted a comparative proteomic study on them and found an unknown and interesting protein, ESA_00986. The expression of this protein was much higher in strains with strong adhesion ability than in strains with low adhesion ability. This unknown protein (ESA_00986) from *C. sakazakii* BAA-894 showed homology with *Yersinia* and *Salmonella* intimin/invasin proteins. Therefore, we speculate that *ESA_00986* gene products may play a role in the pathogenesis of *C. sakazakii.*

In order to determine whether the *ESA_00986* gene is required for adherence to and invasion of host cells and for systemic infection, we examined the influence of the *ESA_00986* gene on the virulence of *C. sakazakii* in vitro using cell adhesion and invasion assays, and in vivo using a neonatal rat infection model. It was found that the mutant has defects in its ability to adhere to and infect intestinal epithelial cells, and the load of mutant strain in blood and tissues was significantly reduced. The ability of bacterial pathogens to adhere to and invade host cells, including macrophages and epithelial cells, plays a crucial role in bacterial pathogenesis. This ability is essential for systemic infection by and transmission of the pathogen. Before bacterial pathogens can successfully invade a host, they must first adhere to the surfaces of epithelial cells [36]. Invasins are proteins that allow bacteria to penetrate cells; they usually adhere to a cell before invading eukaryotic cells and also can help pathogens colonize, maintain, and spread pathogens within the host organism [37]. Deletion of the invasion gene *ychO* in avian pathogenic *E. coli* decreased its ability to adhere to and infect chicken fibroblasts [38]. In the present study, the results suggest that *ESA_00986* plays an important role in adhesion and infection by *C. sakazakii* and knocking out *ESA_00986* can reduce the toxicity of *C. sakazakii* in vivo, indicating that *ESA_00986* encodes an important virulence factor.

In this study, the animal experimental results showed that compared with the wild-type group, the levels of inflammatory factors (TNF-α, IFN-γ, IL-6, IL-1β) were reduced in serum and intestine of the mutant group. Inflammatory factors are an important class of cytokines that are involved in immune regulation during the occurrence of inflammatory response and inflammatory diseases. They are rapidly produced in large quantities under inflammatory conditions and are thought to be involved in a variety of inflammatory processes and induce a variety of inflammatory diseases [39,40]. IL-6, IL-1β, TNF-α, and IFN-γ are crucial regulators of NF-κB-mediated inflammation and play a critical role in the pathogenesis of colitis [41–44].

In comparison with the WT group, the expression of genes encoding ZO-2, occludin and claudin-1 in the intestinal tissue of the mutant group was upregulated, while their expression in the complementary strain group was consistent with the WT group. As demonstrated via HE staining of intestinal tissues, compared with the WT group, the mutant group also showed mild symptoms such as shedding of the villi of colon tissue, dilation of intestinal tissue, and thinning or destruction of the intestinal wall. The integrity of the intestinal epithelial barrier and the stability of the intestinal environment are widely recognized as crucial factors in maintaining intestinal health and promoting host resistance against pathogen invasion [45]. Tight junctions are integral components of the intestinal epithelial barrier and play an important role in preventing bacterial infection [46–48]. Tight-junction proteins ZO-1 and ZO-2 are primary proteins involved in the formation and maintenance of tight junctions. As markers of intestine mechanical barrier, they are closely associated with the integrity of epithelial cells. Occludin is also a crucial protein involved in maintaining the stability and function of tight junctions, contributing to the integrity of the barrier. Claudin-1 is a transmembrane protein that serves as a key component of tight junctions and plays a vital role in their structure and function. Tight junctions can prevent *Campylobacter jejuni* from invading laterally from epithelial cells. They play a major role in bacterial invasion. Pro-inflammatory cytokines, TNF-α, can induce tight junctions to break down and promote the invasion of *C. jejuni* [49,50]. Enteropathogenic *E. coli* can promote infection by breaking the tight connections of the small intestine's epithelial cells [51]. A previous study showed that increasing expression of claudins and occludin can enhance intestinal barrier function of mouse and reduce *Salmonella* infection [52]. In agreement with these studies, the protein ESA_00986 can damage the intestinal barrier via inhibition the related genes expression (ZO-2, claudins and occludin) and enhance the intestinal infection ability of *C. sakazakii.*

In this study, it was found that compared with the wild-type, the biofilm forming ability of Δ*ESA_00986* was reduced. Biofilms are commonly described as aggregates of microbial cells that adhere to both biological and abiotic surfaces, forming complex structures, consisting of a variety of major biomacromolecules that act as defensive barriers and important adhesion bases [53]. Numerous studies have demonstrated the important role of biofilms in the adhesion and invasion of human epithelial cells by pathogenic bacteria. Biofilms provides a protective environment that enhances bacterial attachment, colonization, and subsequent invasion of host cells, contributing to the pathogenicity of these bacteria [54]. The ability of bacteria to attach to surfaces and form biofilms contributes to successive infections [55]. Naziri et al. reported the role of *uropathogenic Escherichia coli* biofilms in the pathogenesis of initial attachment and invasion of the human urinary tract [56]. Hojjatolah Zamani et al. reported biofilm formation in *uropathogenic Escherichia coli* were association with adhesion factor genes, such as *papAH, bmaE* and *sfaS* [57]. *K. pneumoniae* biofilms is associated with colonization of the gastrointestinal, respiratory, and urinary tracts, as well as the development of invasive infections, particularly in immunocompromised patients [58]. *Cronobacter* has the ability to adhere to diverse surfaces and form biofilms, which enables it to withstand different stress conditions, enhance adhesion, and promote pathogenesis [59,60]. It is reported that *ompF* and *bcsR* of *C. sakazakii* can regulate adhesion/invasion by influencing biofilm synthesis [14,34]. As several members of the intimin/invasion protein family, compared with the wild-type, the biofilm forming ability of Δ*ESA_00986* was also reduced. Consistent with these studies, the protein ESA_00986 may play a role in adhesion/invasion of *C. sakazakii* via regulating biofilm biosynthesis.

5. Conclusions

In this study, we demonstrated for the first time that deletion of *ESA_00986*, a gene encoding a virulence factor with Ig-like domains. The deletion of *ESA_00986* can lead to a notable reduction in invasion of epithelial cells and dissemination into the rat tissue. *ESA_00986* may played a positive role in adhesion/invasion through upregulation the intestinal inflammation via the NF-κB signaling pathway, damaging the intestinal barrier via inhibition the related genes expression and enhancing the biofilm synthesis. The ESA_00986 protein can be one of the best targets for future drug and vaccine development due to the interaction with the host immune system. This study provides valuable insights in the molecular mechanism and physiological function of the *ESA_00986* gene in *C. sakazakii*.

Author Contributions: Conceptualization, P.L. and X.D.; Data curation, Y.F.; Methodology, Y.F. and P.L.; Validation, Y.F., D.Z., C.Z., J.J. and X.J.; Writing—original draft, Y.F.; Writing—review & editing Y.F., P.L. and X.D. All authors have read and agreed to the published version of the manuscript.

Funding: This work was supported by National Natural Science Foundation of China (31972167) and Key R&D program of Hebei Province (20372801D).

Data Availability Statement: The data presented in this study are available on request from the corresponding author.

Conflicts of Interest: The authors declare no conflict of interest.

References

1. Feeney, A.; Kropp, K.A.; O'connor, R.; Sleator, R.D. *Cronobacter sakazakii*: Stress survival and virulence potential in an opportunistic foodborne pathogen. *Gut Microbes* **2014**, *5*, 711–718. [CrossRef] [PubMed]
2. Iversen, C.; Lane, M.; Forsythe, S. The growth profile, thermotolerance and biofilm formation of Enterobacter sakazakii grown in infant formula milk. *Lett. Appl. Microbiol.* **2004**, *38*, 378–382. [CrossRef] [PubMed]
3. Kucerova, E.; Clifton, S.W.; Xia, X.-Q.; Long, F.; Porwollik, S.; Fulton, L.; Fronick, C.; Minx, P.; Kyung, K.; Warren, W.; et al. Genome sequence of *Cronobacter sakazakii* BAA-894 and comparative genomic hybridization analysis with other Cronobacter species. *PLoS ONE* **2010**, *5*, e9556. [CrossRef]
4. Jing, C.E.; Du, X.J.; Li, P.; Wang, S. Transcriptome analysis of *Cronobacter sakazakii* ATCC BAA-894 after interaction with human intestinal epithelial cell line HCT-8. *Appl. Microbiol. Biotechnol.* **2016**, *100*, 311–322. [CrossRef] [PubMed]
5. Fakruddin; Rahaman, M.; Ahmed, M.M.; Hoque, M. Stress tolerant virulent strains of *Cronobacter sakazakii* from food. *Biol. Res.* **2014**, *47*, 63. [CrossRef] [PubMed]
6. Yan, Q.; Condell, O.; Power, K.; Butler, F.; Tall, B.; Fanning, S. *Cronobacter* species (formerly known as *Enterobacter sakazakii*) in powdered infant formula: A review of our current understanding of the biology of this bacterium. *J. Appl. Microbiol.* **2012**, *113*, 1–15. [CrossRef] [PubMed]
7. Carvalho, G.G.; Calarga, A.P.; Teodoro, J.R.; Queiroz, M.M.; Astudillo-Trujillo, C.A.; Levy, C.E.; Brocchi, M.; Kabuki, D.Y. Isolation, comparison of identification methods and antibiotic resistance of *Cronobacter* spp. in infant foods. *Food Res. Int.* **2020**, *137*, 109643. [CrossRef]
8. Yan, Q.; Wang, J.; Gangiredla, J.; Cao, Y.; Martins, M.; Gopinath, G.R.; Stephan, R.; Lampel, K.; Tall, B.D.; Fanning, S. Comparative Genotypic and Phenotypic Analysis of *Cronobacter* species Cultured from Four Powdered Infant Formula Production Facilities: Indication of Pathoadaptation along the Food Chain. *Appl. Environ. Microbiol.* **2015**, *81*, 4388–4402. [CrossRef]
9. Jaradat, Z.W.; Al Mousa, W.; Elbetieha, A.; Al Nabulsi, A.; Tall, B.D. *Cronobacter* spp.—Opportunistic food-borne pathogens. A review of their virulence and environmental-adaptive traits. *J. Med. Microbiol.* **2014**, *63 Pt 8*, 1023–1037. [CrossRef]
10. Giri, C.P.; Shima, K.; Tall, B.D.; Curtis, S.; Sathyamoorthy, V.; Hanisch, B.; Kim, K.S.; Kopecko, D.J. *Cronobacter* spp. (previously *Enterobacter sakazakii*) invade and translocate across both cultured human intestinal epithelial cells and human brain microvascular endothelial cells. *Microb. Pathog.* **2012**, *52*, 140–147. [CrossRef]
11. Franco, A.A.; Hu, L.; Grim, C.J.; Gopinath, G.; Sathyamoorthy, V.; Jarvis, K.G.; Lee, C.; Sadowski, J.; Kim, J.; Kothary, M.H.; et al. Characterization of putative virulence genes on the related RepFIB plasmids harbored by *Cronobacter* spp. *Appl. Environ. Microbiol.* **2011**, *77*, 3255–3267. [CrossRef]
12. Hartmann, I.; Carranza, P.; Lehner, A.; Stephan, R.; Eberl, L.; Riedel, K. Genes involved in *Cronobacter sakazakii* biofilm formation. *Appl. Environ. Microbiol.* **2010**, *76*, 2251–2261. [CrossRef] [PubMed]

13. Townsend, S.; Barron, J.C.; Loc-Carrillo, C.; Forsythe, S. The presence of endotoxin in powdered infant formula milk and the influence of endotoxin and *Enterobacter sakazakii* on bacterial translocation in the infant rat. *Food Microbiol.* **2007**, *24*, 67–74. [CrossRef] [PubMed]
14. Gao, J.; Han, Z.; Li, P.; Zhang, H.; Du, X.; Wang, S. Outer Membrane Protein F Is Involved in Biofilm Formation, Virulence and Antibiotic Resistance in *Cronobacter sakazakii*. *Microorganisms* **2021**, *9*, 2338. [CrossRef] [PubMed]
15. Nair, M.K.M.; Venkitanarayanan, K.; Silbart, L.K.; Kim, K.S.; Awadallah, M.A.; Ahmed, H.A.; Merwad, A.M.; Elez, R.M.A.; Saleh, K.M.; Li, Z.; et al. Outer membrane protein A (OmpA) of *Cronobacter sakazakii* binds fibronectin and contributes to invasion of human brain microvascular endothelial cells. *Foodborne Pathog. Dis.* **2009**, *6*, 495–501. [CrossRef]
16. Singh, N.; Goel, G.; Raghav, M. Insights into virulence factors determining the pathogenicity of *Cronobacter sakazakii*. *Virulence* **2015**, *6*, 433–440. [CrossRef]
17. Kim, K.; Kim, K.P.; Choi, J.; Lim, J.A.; Lee, J.; Hwang, S.; Ryu, S. Outer membrane proteins A (OmpA) and X (OmpX) are essential for basolateral invasion of *Cronobacter sakazakii*. *Appl. Environ. Microbiol.* **2010**, *76*, 5188–5198. [CrossRef]
18. Chandrapala, D.; Kim, K.; Choi, Y.; Senevirathne, A.; Kang, D.-H.; Ryu, S.; Kim, K.-P. Putative Inv is essential for basolateral invasion of Caco-2 cells and acts synergistically with OmpA to affect in vitro and in vivo virulence of *Cronobacter sakazakii* ATCC 29544. *Infect. Immun.* **2014**, *82*, 1755–1765. [CrossRef]
19. Li, P.; Zong, W.; Zhang, Z.; Lv, W.; Ji, X.; Zhu, D.; Du, X.; Wang, S. Effects and molecular mechanism of flagellar gene flgK on the motility, adhesion/invasion, and desiccation resistance of *Cronobacter sakazakii*. *Food Res. Int.* **2023**, *164*, 112418. [CrossRef]
20. Choi, Y.; Kim, K.P.; Kim, K.; Choi, J.; Shin, H.; Kang, D.H.; Ryu, S. Possible roles of LysR-type transcriptional regulator (LTTR) homolog as a global regulator in *Cronobacter sakazakii* ATCC 29544. *Int. J. Med. Microbiol.* **2012**, *302*, 270–275. [CrossRef]
21. Reis, R.S.; Horn, F. Enteropathogenic *Escherichia coli*, Samonella, Shigella and Yersinia: Cellular aspects of host-bacteria interactions in enteric diseases. *Gut Pathog.* **2010**, *2*, 8. [CrossRef] [PubMed]
22. Luo, Y.; Frey, E.A.; Pfuetzner, R.A.; Creagh, A.L.; Knoechel, D.G.; Haynes, C.A.; Finlay, B.B.; Strynadka, N.C.J. Crystal structure of enteropathogenic *Escherichia coli* intimin-receptor complex. *Nature* **2000**, *405*, 1073–1077. [CrossRef] [PubMed]
23. Hamburger, Z.A.; Brown, M.S.; Isberg, R.R.; Bjorkman, P.J. Crystal structure of invasin: A bacterial integrin-binding protein. *Science* **1999**, *286*, 291–295. [CrossRef] [PubMed]
24. Jerse, A.E.; Yu, J.; Tall, B.; Kaper, J.B. A genetic locus of enteropathogenic *Escherichia coli* necessary for the production of attaching and effacing lesions on tissue culture cells. *Proc. Natl. Acad. Sci. USA* **1990**, *87*, 7839–7843. [CrossRef]
25. Isberg, R.R.; Leong, J.M. Multiple beta 1 chain integrins are receptors for invasin, a protein that promotes bacterial penetration into mammalian cells. *Cell* **1990**, *60*, 861–871. [CrossRef]
26. Sadana, P.; Mönnich, M.; Unverzagt, C.; Scrima, A. Structure of the *Y. pseudotuberculosis* adhesin InvasinE. *Protein Sci.* **2017**, *26*, 1182–1195. [CrossRef]
27. Bodelón, G.; Palomino, C.; Fernández, L. Immunoglobulin domains in *Escherichia coli* and other enterobacteria: From pathogenesis to applications in antibody technologies. *FEMS Microbiol. Rev.* **2013**, *37*, 204–250. [CrossRef]
28. Seo, K.S.; Kim, J.W.; Park, J.Y.; Viall, A.K.; Rohde, H.N.; Schnider, D.R.; Lim, S.Y.; Hong, J.B.; Hinnebusch, B.J.; O'Loughlin, J.L.; et al. Role of a new intimin/invasin-like protein in Yersinia pestis virulence. *Infect. Immun.* **2012**, *80*, 3559–3569. [CrossRef]
29. Clark, M.A.; Hirst, B.H.; Jepson, M.A. Inoculum composition and Salmonella pathogenicity island 1 regulate M-cell invasion and epithelial destruction by Salmonella typhimurium. *Infect. Immun.* **1998**, *66*, 724–731. [CrossRef]
30. Ji, X.; Lu, P.; Xue, J.; Zhao, N.; Zhang, Y.; Dong, L.; Zhang, X.; Li, P.; Hu, Y.; Wang, J.; et al. The lipoprotein NlpD in *Cronobacter sakazakii* responds to acid stress and regulates macrophage resistance and virulence by maintaining membrane integrity. *Virulence* **2021**, *12*, 415–429. [CrossRef]
31. Liu, J.; Zhang, D.; Lian, S.; Gu, X.; Hou, Q.; Xia, P.; Zhu, G. Mechanism of nitrite transporter NirC in motility, biofilm formation, and adhesion of avian pathogenic *Escherichia coli*. *Arch. Microbiol.* **2021**, *203*, 4221–4231. [CrossRef] [PubMed]
32. Rashid, M.H.; Kornberg, A. Inorganic polyphosphate is needed for swimming, swarming, and twitching motilities of Pseudomonas aeruginosa. *Proc. Natl. Acad. Sci. USA* **2000**, *97*, 4885–4890. [CrossRef] [PubMed]
33. Andriantsoanirina, V.; Teolis, A.-C.; Xin, L.X.; Butel, M.J.; Aires, J. Bifidobacterium longum and Bifidobacterium breve isolates from preterm and full term neonates: Comparison of cell surface properties. *Anaerobe* **2014**, *28*, 212–215. [CrossRef]
34. Gao, J.X.; Li, P.; Du, X.J.; Han, Z.H.; Xue, R.; Liang, B.; Wang, S. A Negative Regulator of Cellulose Biosynthesis, bcsR, Affects Biofilm Formation, and Adhesion/Invasion Ability of *Cronobacter sakazakii*. *Front. Microbiol.* **2017**, *8*, 1839. [CrossRef] [PubMed]
35. Ganan, M.; Campos, G.; Muñoz, R.; Carrascosa, A.; de Pascual-Teresa, S.; Martinez-Rodriguez, A. Effect of growth phase on the adherence to and invasion of Caco-2 epithelial cells by Campylobacter. *Int. J. Food Microbiol.* **2010**, *140*, 14–18. [CrossRef]
36. Pizarro-Cerdá, J.; Cossart, P. Bacterial Adhesion and Entry into Host Cells. *Cell* **2006**, *124*, 715–727. [CrossRef]
37. Ribet, D.; Cossart, P. How bacterial pathogens colonize their hosts and invade deeper tissues. *Microbes Infect.* **2015**, *17*, 173–183. [CrossRef] [PubMed]
38. Pilatti, L.; de Paiva, J.B.; Rojas, T.C.G.; Leite, J.L.; Conceição, R.A.; Nakazato, G.; da Silveira, W.D. The virulence factor ychO has a pleiotropic action in an Avian Pathogenic *Escherichia coli* (APEC) strain. *BMC Microbiol.* **2016**, *16*, 35. [CrossRef]
39. Siljan, W.W.; Holter, J.C.; Nymo, S.H.; Husebye, E.; Ueland, T.; Aukrust, P.; Mollnes, T.E.; Heggelund, L. Cytokine responses, microbial aetiology and short-term outcome in community-acquired pneumonia. *Eur. J. Clin. Investg.* **2018**, *48*, e12865. [CrossRef]
40. Esmon, C.T. The impact of the inflammatory response on coagulation. *Thromb. Res.* **2004**, *114*, 321–327. [CrossRef]

41. Papasian, C.J.; Silverstein, R.; Gao, J.J.; Bamberger, D.M.; Morrison, D.C. Anomalous role of tumor necrosis factor alpha in experimental enterococcal infection. *Infect. Immun.* **2002**, *70*, 6628–6637. [CrossRef] [PubMed]
42. Leendertse, M.; Heikens, E.; Wijnands, L.M.; van Luit-Asbroek, M.; Teske, G.J.D.; Roelofs, J.J.; Bonten, M.J.M.; van der Poll, T.; Willems, R.J.L. Enterococcal surface protein transiently aggravates Enterococcus faecium-induced urinary tract infection in mice. *J. Infect. Dis.* **2009**, *200*, 1162–1165. [CrossRef] [PubMed]
43. Zou, J.; Shankar, N. Surface protein Esp enhances pro-inflammatory cytokine expression through NF-κB activation during enterococcal infection. *Innate Immun.* **2016**, *22*, 31–39. [CrossRef] [PubMed]
44. Watkins, R.L.; Pallister, K.B.; Voyich, J.M. The SaeR/S gene regulatory system induces a pro-inflammatory cytokine response during Staphylococcus aureus infection. *PLoS ONE* **2011**, *6*, e19939. [CrossRef] [PubMed]
45. Okumura, R.; Takeda, K. Roles of intestinal epithelial cells in the maintenance of gut homeostasis. *Exp. Mol. Med.* **2017**, *49*, e338. [CrossRef] [PubMed]
46. Lu, R.Y.; Yang, W.X.; Hu, Y.J. The role of epithelial tight junctions involved in pathogen infections. *Mol. Biol. Rep.* **2014**, *41*, 6591–6610. [CrossRef]
47. Awad, W.A.; Hess, C.; Hess, M. Enteric Pathogens and Their Toxin-Induced Disruption of the Intestinal Barrier through Alteration of Tight Junctions in Chickens. *Toxins* **2017**, *9*, 60. [CrossRef]
48. Zheng, M.; Sun, S.; Zhou, J.; Liu, M. Virulence factors impair epithelial junctions during bacterial infection. *J. Clin. Lab. Anal.* **2021**, *35*, e23627. [CrossRef]
49. Hatayama, S.; Shimohata, T.; Amano, S.; Kido, J.; Nguyen, A.Q.; Sato, Y.; Kanda, Y.; Tentaku, A.; Fukushima, S.; Nakahashi, M.; et al. Cellular Tight Junctions Prevent Effective Campylobacter jejuni Invasion and Inflammatory Barrier Disruption Promoting Bacterial Invasion from Lateral Membrane in Polarized Intestinal Epithelial Cells. *Front. Cell Infect. Microbiol.* **2018**, *8*, 15. [CrossRef]
50. Chen, M.L.; Ge, Z.; Fox, J.G.; Schauer, D.B. Disruption of tight junctions and induction of proinflammatory cytokine responses in colonic epithelial cells by Campylobacter jejuni. *Infect. Immun.* **2006**, *74*, 6581–6589. [CrossRef]
51. Thanabalasuriar, A.; Koutsouris, A.; Weflen, A.; Mimee, M.; Hecht, G.; Gruenheid, S. The bacterial virulence factor NleA is required for the disruption of intestinal tight junctions by enteropathogenic *Escherichia coli*. *Cell Microbiol.* **2010**, *12*, 31–41. [CrossRef]
52. Zhang, Y.-G.; Wu, S.; Xia, Y.; Sun, J. Salmonella infection upregulates the leaky protein claudin-2 in intestinal epithelial cells. *PLoS ONE* **2013**, *8*, e58606. [CrossRef] [PubMed]
53. Jamal, M.; Ahmad, W.; Andleeb, S.; Jalil, F.; Imran, M.; Nawaz, M.A.; Hussain, T.; Ali, M.; Rafiq, M.; Kamil, M.A. Bacterial biofilm and associated infections. *J. Chin. Med. Assoc.* **2018**, *81*, 7–11. [CrossRef] [PubMed]
54. Guerra, M.E.S.; Destro, G.; Vieira, B.; Lima, A.S.; Ferraz, L.F.C.; Hakansson, A.P.; Darrieux, M.; Converso, T.R. Klebsiella pneumoniae Biofilms and Their Role in Disease Pathogenesis. *Front. Cell Infect. Microbiol.* **2022**, *12*, 877995. [CrossRef] [PubMed]
55. Del Pozo, J.L. Biofilm-related disease. *Expert Rev. Anti Infect. Ther.* **2018**, *16*, 51–65. [CrossRef] [PubMed]
56. Naziri, Z.; Kilegolan, J.A.; Moezzi, M.S.; Derakhshandeh, A. Biofilm formation by uropathogenic *Escherichia coli*: A complicating factor for treatment and recurrence of urinary tract infections. *J. Hosp. Infect.* **2021**, *117*, 9–16. [CrossRef]
57. Zamani, H.; Salehzadeh, A. Biofilm formation in uropathogenic *Escherichia coli*: Association with adhesion factor genes. *Turk. J. Med. Sci.* **2018**, *48*, 162–167. [CrossRef]
58. Piperaki, E.T.; Syrogiannopoulos, G.A.; Tzouvelekis, L.S.; Daikos, G.L. Klebsiella pneumoniae: Virulence, Biofilm and Antimicrobial Resistance. *Pediatr. Infect. Dis. J.* **2017**, *36*, 1002–1005. [CrossRef]
59. Lehner, A.; Riedel, K.; Eberl, L.; Breeuwer, P.; Diep, B.; Stephan, R. Biofilm formation, extracellular polysaccharide production, and cell-to-cell signaling in various *Enterobacter sakazakii* strains: Aspects promoting environmental persistence. *J. Food Prot.* **2005**, *68*, 2287–2294. [CrossRef]
60. Kim, H.; Ryu, J.-H.; Beuchat, L.R. Attachment of and biofilm formation by *Enterobacter sakazakii* on stainless steel and enteral feeding tubes. *Appl. Environ. Microbiol.* **2006**, *72*, 5846–5856. [CrossRef]

foods

MDPI

Article

Evaluation of Hygiene Practice for Reducing *Campylobacter* Contamination on Cutting Boards and Risks Associated with Chicken Handling in Kitchen Environment

Honggang Lai [1,2], Yuanyue Tang [2,3], Fangzhe Ren [2,3], Xin-an Jiao [2,3] and Jinlin Huang [2,3,*]

[1] School of Tourism and Cuisine, Yangzhou University, Yangzhou 225001, China; lhghstc@163.com
[2] Jiangsu Key Lab of Zoonosis, Yangzhou University, Yangzhou 225009, China; tangyy@yzu.edu.cn (Y.T.); fzren@yzu.edu.cn (F.R.); jiao@yzu.edu.cn (X.-a.J.)
[3] Joint International Research Laboratory of Agriculture and Agri-Product Safety, Ministry of Education of China, Yangzhou University, Yangzhou 225009, China
* Correspondence: jinlin@yzu.edu.cn; Tel.: +86-514-8799-0579

Abstract: Cutting boards can serve as potential carriers for the cross-contamination of pathogens from chicken to other surfaces. This study aimed to assess chefs' handling practices of cutting boards across five provinces in China and identify the key factors contributing to unsafe cutting board usage, including cleaning methods and handling practices. Handling practices associated with cutting boards were examined through a web-based survey (N = 154), while kitchen environment tests were conducted to investigate the splashing or survival of *Campylobacter*, inoculated in chicken or on cutting boards, to mimic the practices of chefs. Among chefs in the five provinces of China, wood and plastic cutting boards were the most commonly used for preparing chicken meat. Approximately 33.7% of chefs washed boards with running tap water, 31.17% of chefs washed boards with detergent, and 24.03% of chefs cleaned boards by scraping them with a knife after preparing other meats or chicken. The study tested 23 cutting boards from commercial kitchens for *Campylobacter* presence before and after chicken preparation and cleaning. Among these, 17 were cleaned with a knife, 5 with running tap water, and only 1 with disinfectant. Results showed that cleaning with a knife significantly reduced *Campylobacter* presence on cutting boards ($p < 0.05$), while the three main cleaning methods were inadequate in eliminating contamination to a safe level. In kitchen environment tests, contaminated chicken was chopped on cutting boards, with a maximum distance of 60 cm for low contamination, and 120 cm for medium and high contamination levels. This suggested a contamination risk exposure area ranging from 60 cm to 120 cm. *Campylobacter* survival on surfaces of wood, plastic, and stainless steel was also tested, with plastic surfaces showing the longest survival time (4.5 h at 15 °C and 3.5 h at 25 °C) In comparison, survival time on stainless steel or wood surfaces was only 3 h, implying a cross-contamination risk exposure period of 3 to 4.5 h after chicken preparation. In conclusion, based on the current study data, the practices employed by chefs play an important role in *Campylobacter* transfer in the kitchen environment. The presence of *Campylobacter* on cutting boards even after wiping or droplet splashing highlights its potential as a source of cross-contamination in the kitchen environment. So, chefs in China should reinforce their hygiene culture and adopt effective cutting board cleaning practices to prevent pathogen contamination.

Keywords: cutting boards; hygiene practice; *Campylobacter*; risk exposure

Citation: Lai, H.; Tang, Y.; Ren, F.; Jiao, X.-a.; Huang, J. Evaluation of Hygiene Practice for Reducing *Campylobacter* Contamination on Cutting Boards and Risks Associated with Chicken Handling in Kitchen Environment. *Foods* **2023**, *12*, 3245. https://doi.org/10.3390/foods12173245

Academic Editors: Gary Dykes, Ewen C. D. Todd and Efstathios Giaouris

Received: 16 July 2023
Revised: 14 August 2023
Accepted: 26 August 2023
Published: 29 August 2023

1. Introduction

Campylobacter spp., which are commonly present in chicken meat, represent a major cause of *Campylobacter* foodborne diseases [1–3]. The primary factor contributing to these diseases is cross-contamination between contaminated chicken meat and cooking utensils [4,5]. *Campylobacter* spp. are microaerobic bacteria that pose challenges to cultivation under normal atmospheric conditions [6]. Cross-contamination and inadequate kitchen hygiene practices, such as improper handwashing or inadequate cleaning of surfaces and utensils, play a significant role in *Campylobacter* transmission [7]. Therefore, preventing cross-contamination is crucial. After handling poultry meat, the effectiveness of cleaning may not meet consumer expectations, particularly since microorganisms do not leave visible traces of dirt that can be easily detected [8,9]. Despite efforts to control *Campylobacter* the global incidence of *Campylobacter*-related illness has not shown a significant decrease over time [10].

Campylobacter can easily spread from contaminated chicken meat to various kitchen equipment, including cutting boards, clothing, and knives [11]. Different materials, such as wood, plastic, or stainless steel, are commonly used for cutting boards in domestic and food service kitchens. However, wooden cutting boards have been found to harbor microbial contaminants and pose challenges in terms of effective cleaning [12]. Some studies demonstrated that cutting boards played a role in cross-contamination of foodborne pathogens from poultry meat to cucumber [8,13]. The porous nature of wooden surfaces allows bacteria to penetrate easily, increasing the risk of cross-contamination incidents [12]. Furthermore, in China, the national food safety standard GB 14,934 requires that foodborne pathogens should not be detectable on food contact surfaces such as cutleries, boards, and knives [14].

In China, it has been reported that 77.41% of chicken carcass samples and 37.37% of kitchen surfaces show *Campylobacter* spp. contamination in commercial kitchens [11]. This study aimed to address the knowledge gap concerning the variability in chefs' behavioral preferences across different provinces, as well as the spread and survival potential of *Campylobacter* and cross-contamination practices during the preparation of raw chicken on cutting boards in the kitchen environment, and study the effectiveness of different hygiene procedures for reducing *Campylobacter* contamination on boards and risks while chicken handling in the kitchen environment. This work was conducted as part of a comprehensive research project and ran parallel to ongoing baseline microbiological surveys on chicken handling in various provinces of China.

2. Materials and Methods

2.1. Web-Based Survey among Chefs in Eastern China

A web-based survey was conducted to investigate hygiene and handling practices for chicken preparation in commercial kitchens across different regions of China (Jiangsu Province. Guangdong Province, Fujian Province, Guizhou Province, and Hunan Province) with the aim of capturing geographical and cultural variations. The study protocols were reviewed and approved by the Yangzhou University Human Research Ethics Committee (Permit No. YZUHL20210114). The study included chefs over the age of 18 who were responsible for chicken handling in commercial kitchens. Participants were recruited through the Internet for the study in five provinces of China. The aims and objectives of the study were explained to each participant, and the confidentiality of their information was verbally confirmed. The recruited participants were given the choice to complete the questionnaire using Wenjuan Star, an online platform, with options for immediate or later completion. Data collection took place between July and August 2021.

The questionnaire consisted of 12 multiple-choice questions specifically designed to gather information about self-reported practices related to the last instance when chefs prepared raw poultry in a commercial kitchen. The use of a multiple-choice approach aimed to minimize subjectivity and ensure clarity in measuring and addressing various aspects of the topic. A total of 154 completed questionnaires were collected from chefs located in five provinces of China, as presented in Table 1. The majority of respondents in the kitchens were male. The age group of 18–35 was well-represented in all five provinces. Education levels varied among the respondents, and notably, the proportion of individuals with tertiary qualifications was significantly higher in Guizhou and Hunan provinces compared to the other provinces.

Table 1. Socio-demographic characteristics of the survey participants %.

	Total Sample (n = 154)	Jiangsu Province (n = 36)	Guangdong Province (n = 48)	Fujian Province (n = 33)	Guizhou Province (n = 17)	Hunan Province (n = 20)
1. Gender						
Male	70.78	69.44	79.17	72.73	64.71	55.00
Female	29.22	30.56	20.83	27.27	35.29	45.00
2. Age						
18–25	89.61	75.00	97.92	93.94	82.35	95.00
26–35	5.84	16.67	0.00	6.06	0.00	5.00
36–45	1.95	2.78	0.00	0.00	11.76	0.00
46–60	2.60	5.56	2.08	0.00	5.88	0.00
3. Education						
Junior high school education	3.25	0.00	8.33	0.00	5.88	0.00
Senior high school education	3.90	8.33	2.08	0.00	11.76	0.00
Vocational education	48.05	27.78	79.17	57.58	17.65	20.00
College education	39.61	50.00	8.33	36.36	64.71	80.00
Graduate education	5.19	13.89	2.08	6.06	0.00	0.00

2.2. Sample Collection and Campylobacter Examination Method

In Jiangsu province, China, a total of 23 commercial kitchens were visited for sampling. During the sampling process, the specialized chefs were instructed to handle chicken carcasses following their regular daily preparation routines in the designated area for raw meat preparation. Prior to chicken preparation, two sterilized cotton balls soaked in physiological saline solution were used to wipe 250 cm^2 of the exterior surface and 250 cm^2 of the interior surface of the chicken carcasses. Samples from cutting boards were collected at three specific stages: before handling the chicken, after handling the chicken, and during the cleaning procedure. For cutting boards, two sterilized cotton balls soaked in physiological saline solution were used to wipe a surface area of approximately 100 cm^2, as described in a previous study [11,15].

The enumeration and isolation of *Campylobacter* spp. were conducted using the plating method. The original solution was diluted 10-fold in a PBS solution, and 100 µL of each dilution was spread onto *Campylobacter*-selective agar base (modified CCDA; Oxoid, UK) plates supplemented with selective antibiotics. The plates were then incubated under microaerobic conditions (5% O_2, 10% CO_2, and 85% N_2) at 42 °C for 36–48 h. Each dilution was plated in duplicate to ensure accuracy. All colonies exhibiting the characteristic morphology of *Campylobacter* were counted within the countable dilution range of 15–300 colony-forming units (CFU) per plate. For further identification, up to five *Campylobacter* colonies were selected from the plates. For species identification, multiplex PCR was conducted using specific primers as recorded in a previous study [11]. The PCR amplification targeted the 16S rRNA gene for all *Campylobacter* species, the mapA gene for *Campylobacter jejuni*, and the ceuE gene for *Campylobacter coli*.

2.3. Simulating the Handling of Contaminated Chicken in the Kitchen Environment

To reduce bacterial loads on the chicken samples, the samples underwent irradiation at the Yangzhou Gamma-Ray Center using a dosage of 25 kGy. After irradiation, the samples were carefully placed in an ice box and transported to the laboratory for future utilization. In the laboratory, each bag containing the whole chicken was immersed in a 500 mL bag containing the *Campylobacter* ATCC12662 cultures (10^6, 10^7, 10^8 log_{10}CFU/mL). The bags were then shaken for 30 min at 120 rpm to ensure even distribution of the bacteria on the surface of the whole chicken samples. After the incubation period, the whole chicken samples were removed from the bags and transferred to a clean bag for a duration of 2 min. Subsequently, the samples were immediately placed into a sterile bag, sealed, and stored at a temperature of 4 °C for later use. The chickens were then transported to a model kitchen at Yangzhou University, where they were chopped into smaller pieces by a group of volunteers consisting of culinary students attending Yangzhou University. The study protocols were reviewed and approved by the Yangzhou University Human Research Ethics Committee (Permit No. YZUHL20210146). Following the handling of the chicken by volunteers, the sample sites indicated in Figure 1 were wiped using two sterilized cotton balls soaked in physiological saline solution. Cotton balls were utilized to wipe a surface area of approximately 100 cm^2 in order to collect any potential bacterial contaminants.

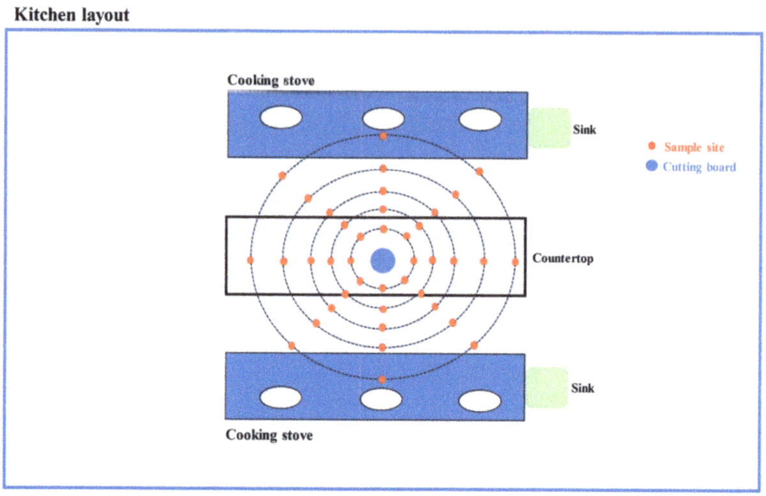

Figure 1. Schematic diagram of sampling points at monitoring locations during simulated fresh chicken processing in the model kitchen. Blue circles represent cutting boards, red circles represent monitoring points. The distance between each point is 30 cm, forming a cross shape with equidistantly spaced monitoring points.

2.4. Simulating the Handling of Contaminated Chicken in a Kitchen Environment

Chicken juice, also referred to as meat exudate, was prepared following the methodology described in a previous study [16]. To initiate the preparation process, frozen whole chickens were procured from various supermarkets located in Yangzhou City, China. The frozen whole chickens were allowed to thaw overnight at room temperature. Subsequently, the exudate, or the juice released from the meat, was collected. To remove any debris, the collected exudate was subjected to centrifugation. To ensure sterility, a sterile polyether sulfone syringe filter with a pore size of 0.2 µm (Millipore) was used for filtration. The resulting chicken juice was divided into smaller aliquots and stored at a temperature of −20 °C until it was ready for use in subsequent experiments or analyses.

First, the cultured standard strain ATCC12662 broth was diluted separately with PBS (phosphate-buffered saline) and chicken juice. The dilution was then adjusted to achieve an optical density (OD) value of 0.05, corresponding to a concentration of approximately 10^7 CFU/mL. Next, 500 µL of the bacterial suspension was inoculated onto the surface of different materials, such as stainless steel, plastic, or wood. Each material had dimensions of 5 cm × 5 cm. After inoculation, the material samples were on a shaking table at temperatures of 25 °C and 15 °C, respectively. A time interval of 0.5 h was allowed between measurements. At each designated time interval, three pieces of the inoculated materials were removed and the number of *Campylobacter* on their surfaces was measured. This was achieved by diluting the bacterial suspension obtained from each piece and performing a plate count method.

2.5. Calculations and Statistics

The information obtained from the completed questionnaires was recorded and managed using Microsoft Excel 2010 for efficient data handling and organization. Subsequently, descriptive data analysis was performed using Stata 11.0 software. This involved calculating frequencies and percentages of responses within each category. The results of this analysis were presented in tabular form, providing a clear and concise summary of the data, which allows for easy interpretation and understanding of the distribution of responses across different categories. The chi-square test, a statistical test commonly employed to assess the association between categorical variables, was utilized in the analysis.

The data analysis was performed using IBM SPSS v.21 software. Loads between samples were compared using the independent sample *t*-test. A level of significance of 0.05 was applied for all statistical comparisons.

3. Results

3.1. Chefs' Self-Reported Handling Practices While Preparing Chicken

In the web survey, a total of 154 chefs who used cutting boards for chicken preparation were consulted (Table 2). The survey aimed to gather information about the types of cutting boards used, the methods of washing hands and knives, and overall hygiene practices. Among the surveyed chefs, similar proportions reported using plastic boards (50%) and wood boards (44.16%). A small percentage of respondents (1.95%) used stainless steel boards, while 3.90% reported using other types of boards. Differences in cutting board preferences were observed across different provinces in China. In Jiangsu Province, wood boards were the most common choice, with 55.56% of respondents using them. In Guangdong Province, plastic boards were the most prevalent, with 62.50% of respondents using them. In Fujian Province, 6.06% of chefs used stainless steel boards, while in Jiangsu province, the usage was 2.78%. Regarding other materials used for cutting boards, 4.17% of respondents in Guangdong Province, 10% in Hunan Province, and 11.76% in Fujian Province reported using them. These findings highlighted the variations in cutting board preferences among chefs in different provinces, with wood and plastic being the most commonly used materials. The data are summarized in Table 2, which provides a comprehensive overview of the distribution of cutting board types across the surveyed provinces.

Table 2. Answer frequencies of hygiene practices of chefs reported while preparing chicken.

Items	Total Sample (*n* = 154)	Jiangsu Province (*n* = 36)	Guangdong Province (*n* = 48)	Fujian Province (*n* = 33)	Guizhou Province (*n* = 17)	Hunan Province (*n* = 20)
Type of catering service						
Hotel	58.44	50.00	72.92	72.73	17.65	50.00
Restaurant	35.71	44.44	20.83	24.24	70.59	45.00
Fast food restaurant	5.84	5.56	6.25	3.03	11.76	5.00
Type of meat						
Fresh chicken	33.12	19.44	41.67	39.39	29.41	30.00
Frozen chicken	16.23	22.22	12.50	18.18	11.76	15.00
Chicken slaughter in Wet market	38.96	41.67	33.33	30.30	58.82	45.00
Chicken breast or leg	11.69	16.67	12.50	12.12	0.00	10.00
Whether the whole chicken had been eviscerated or not						
Yes	87.01	94.44	83.33	81.82	88.24	90.00
No	12.99	5.56	16.67	18.18	11.76	10.00
Before cutting, whether or not the chicken had been washed						
Yes	91.56	88.89	95.83	87.88	88.24	95.00
No	8.44	11.11	4.17	12.12	11.76	5.00

Table 2. *Cont.*

Items	Total Sample (n = 154)	Jiangsu Province (n = 36)	Guangdong Province (n = 48)	Fujian Province (n = 33)	Guizhou Province (n = 17)	Hunan Province (n = 20)
Type of cutting board for chicken preparation						
Plastic	50.00	41.67	62.50	48.48	41.18	45.00
Wood	44.16	55.56	33.33	45.45	47.06	45.00
Stainless steel	1.95	2.78	0.00	6.06	0.00	0.00
Other	3.90	0.00	4.17	0.00	11.76	10.00
Before chicken preparation, whether or not boards used for cutting other meat						
Yes	40.91	38.89	39.58	36.36	41.18	55.00
No	59.09	61.11	60.42	63.64	58.82	45.00
After other meat preparation, method of cleaning boards						
Washing with running tap water	33.12	33.33	45.83	24.24	11.76	35.00
Scrapping with knife	31.17	19.44	25.00	42.42	41.18	40.00
Washing with detergent	35.71	47.22	29.17	33.33	47.06	25.00
While preparing chicken, whether or not cloths used to clean cutting board						
Yes	48.05	61.11	41.67	45.45	41.18	50.00
No	51.95	38.89	58.33	54.55	58.82	50.00
Cleaning method for cloths						
Washing with running tap water	17.53	19.44	18.75	21.21	5.88	15.00
Washing with detergent	51.95	50.00	56.25	42.42	64.71	50.00
Boiling	29.22	30.56	20.83	36.36	29.41	35.00
No washing	1.30	0.00	4.17	0.00	0.00	0.00
After chicken preparation, method of cleaning boards						
Washing with running tap water	33.77	38.89	41.67	39.39	5.88	20.00
Scrapping with knife	24.03	19.44	16.67	18.18	41.18	45.00
Washing with detergent	41.56	41.67	39.58	42.42	52.94	35.00
No washing	0.65	0.00	2.08	0.00	0.00	0.00
After touching the chicken, (usual) method of cleaning hands						
No washing	3.90	8.33	6.25	0.00	0.00	0.00
Washing with running tap water	41.56	27.78	39.58	51.52	41.18	55.00
Washing with disinfectant	54.55	63.89	54.17	48.48	58.82	45.00
After chicken preparation, (usual) method of cleaning the knife						
Washing with running tap water	40.91	50.00	41.67	45.45	11.76	40.00
Washing with detergent	51.95	38.89	54.17	48.48	82.35	50.00
Clean with cloth	6.49	8.33	4.17	6.06	5.88	10.00
No washing	0.65	2.78	0.00	0.00	0.00	0.00

It was found that the cutting boards were often used for cutting other types of meat in advance of being used to prepare chicken meat. Among the surveyed chefs, the highest percentage of prior use of boards for cutting other types of meat was reported in Hunan Province (55%), followed by Fujian Province (36.36%). When it came to cleaning the boards after cutting other meats, different methods were reported. A total of 33.12% of chefs washed the boards with running tap water, 31.17% scrapped the boards with a knife, and 35.71% washed the boards with detergent. In Guangdong province, the preferred method was washing with running tap water (45.83%), followed by scrapping with a knife in Fujian (42.42%), Guizhou (41.18%), and Hunan province (40.00%). In Jiangsu (47.22%) and Guizhou province (47.06%), a higher percentage of chefs reported the use of detergent for cleaning the boards. After the chicken preparation process, 48% of chefs reported using cloths to clean chicken juice off the kitchen surface. Regarding the cleaning of cutting boards, after chicken preparation, 41.67%, 39.39%, and 33.77% of chefs in Guangdong, Fujian, and Jiangsu province, respectively, cleaned the boards with running tap water. In

Hunan and Guizhou Provinces, 45% and 41.18% of chefs, respectively, kept the boards clean by scrapping them with a knife. It was worth noting that the majority of chefs (54.55%) expressed a preference for washing cutting boards with disinfectant. Additionally, half of the chefs reported choosing to wash their knives with disinfectant, which was considered a safe handling practice. These findings shed light on the cleaning practices and preferences of chefs regarding cutting boards and knives after handling different types of meats. Proper cleaning and disinfection methods were crucial to ensuring food safety and preventing cross-contamination in the kitchen.

3.2. *Campylobacter Contamination of Cutting Boards in Commercial Kitchens during Chicken Preparation*

Table 3 presents the different methods used to clean cutting boards during chicken preparation. In 17 kitchens, chefs cleaned their boards with knives. Prior to chicken preparation, boards from 4 out of the 17 kitchens tested positive for *Campylobacter*. However, after chicken preparation, positive tests for *Campylobacter* on the boards significantly increased to 94.12%. Despite following cleaning procedures, boards from 9 out of 17 kitchens still tested positive for *Campylobacter*. Fortunately, the average *Campylobacter* loads on the boards noticeably decreased from 2.97 ± 0.94 $Log_{10}CFU/100\ cm^2$ to 2.10 ± 0.56 $Log_{10}CFU/100\ cm^2$ ($p < 0.05$). In five kitchens, no *Campylobacter* was detected on any of their boards prior to cutting chicken. However, after chicken preparation, all of the boards tested positive for *Campylobacter*, with an average load of 3.44 ± 0.85 $Log_{10}CFU/100\ cm^2$. When the boards were cleaned with running tap water, three out of the five boards were still contaminated with *Campylobacter*, with an average load of 2.85 ± 0.47 $Log_{10}CFU/100\ cm^2$. Only one kitchen used disinfectant to clean its boards. Before cutting, *Campylobacter* was not detected on the boards of that kitchen. Unfortunately, after chicken preparation, the boards were contaminated with *Campylobacter* at a load of 3.76 ± 0.00 $Log_{10}CFU/100\ cm^2$. Even after cleaning with disinfectant, the boards still tested positive for *Campylobacter*, with an average load of 1.6 ± 0.00 $Log_{10}CFU/100\ cm^2$.

Table 3. Different cleaning methods for reducing *Campylobacter* loads on cutting boards in 23 commercial kitchens in eastern China.

Cleaning Method	Stage	Total	Positive	Detection Rate %	Average Loads ($Log_{10}CFU/100\ cm^2$)
cleaning with a knife (n = 17)	chicken	222	172	77.48	2.98 ± 0.93
	before cutting	17	4	23.53	2.45 ± 0.68 [a]
	after cutting	17	16	94.12	2.97 ± 0.94 [b]
	after cleaning process	17	9	52.94	2.10 ± 0.56 [a]
cleaning with running tap water (n = 5)	chicken	67	58	86.57	3.25 ± 0.68
	before cutting	5	0	0	-
	after cutting	5	5	100	3.44 ± 0.85 [a]
	after cleaning process	5	3	60	2.85 ± 0.47 [a]
cleaning with disinfectant (n = 1)	chicken	12	12	100	4.03 ± 0.31
	before cutting	1	0	0	-
	after cutting	1	1	100	3.76 ± 0.00 [a]
	After cleaning process	1	1	100	1.6 ± 0.00 [a]

Different letters indicate significant differences among the groups ($p < 0.05$).

3.3. *Risk Area of Campylobacter Spread during Simulated Chicken Preparation*

Figure 2 displayed the variation in *C. jejuni* levels in chicken carcasses that were initially contaminated with similar concentrations of ATCC12662 (10^7 $CFU/100\ cm^2$ high; 10^6 $CFU/100\ cm^2$ medium; or 10^5 $CFU/100\ cm^2$ low). This difference in contamination levels persisted across all three groups, indicating that the initial contamination level had a consistent impact on the subsequent *C. jejuni* levels.

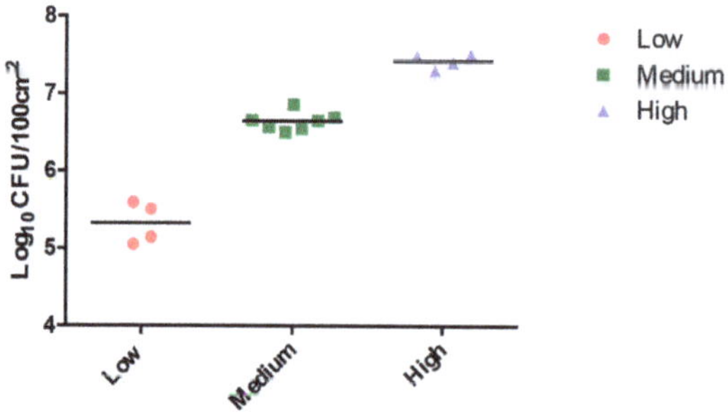

Figure 2. Concentration of *C. jejuni* in chicken carcasses ($\log_{10}$ CFU/100 cm^2) initially contaminated with the same concentration of ACTCC 12,662 at low, medium, or high levels ($\log_{10}$CFU/mL), measured in 15 meat samples analyzed after treatment.

As Figure 3a,b show, a significant negative correlation between the distance and *Campylobacter* contamination on sample sites in the kitchen was found. The results indicated that as the distance increased, the contamination of *Campylobacter* on the sample sites decreased. Notably, in a model kitchen environment, where the maximum detection distance for the low pollution group was 60 cm and for the medium and high pollution groups was 120 cm, there was a similar average number of positive sample sites. However, when the distance reached 60 cm, the reduction in *Campylobacter* loads was not statistically significant ($p > 0.05$).

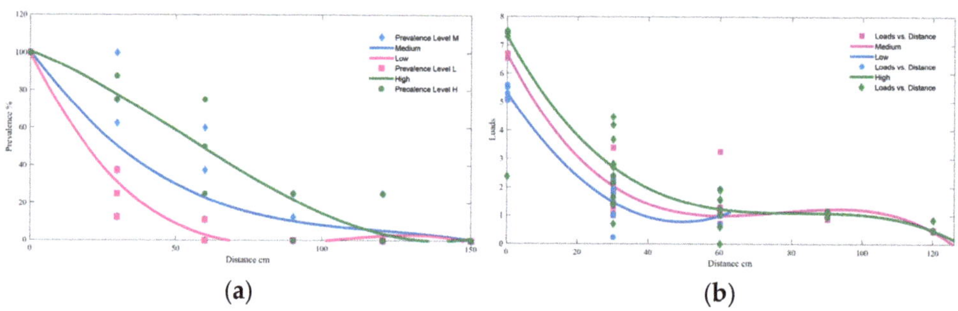

Figure 3. Detection rate (**a**) and Loads (**b**) of *C. jejuni* ($\log_{10}$CFU/100 cm^2) in cotton samples compared to the level of *C. jejuni* ($\log_{10}$CFU/100 cm^2) on the chicken carcasses for 40 sample sites in a model kitchen after cutting chicken.

3.4. Risk Time of *Campylobacter* Survival on Different Surfaces

To simulate the survival of *Campylobacter* on boards made of different materials during chicken preparation in a kitchen environment, *Campylobacter* was added to a suspension containing chicken juice and PBS.

The presence of chicken juice significantly increased the survival time of *Campylobacter* on various abiotic surfaces at both 15 °C and 25 °C. As Figure 4 demonstrates, there was a clear trend of a longer survival time for *Campylobacter* in the presence of chicken juice compared to PBS. In all cases involving chicken juice, the longest survival time was observed on plastic surfaces, reaching up to 4.5 h at 15 °C (Figure 4b) and 3.5 h at 25 °C (Figure 4d). On stainless steel and wood surfaces, the survival time was shorter, with *Campylobacter* surviving for only 3 h. Notably, *Campylobacter* exhibited a faster decline in survival on wood surfaces compared to plastic and stainless steel surfaces at 15 °C.

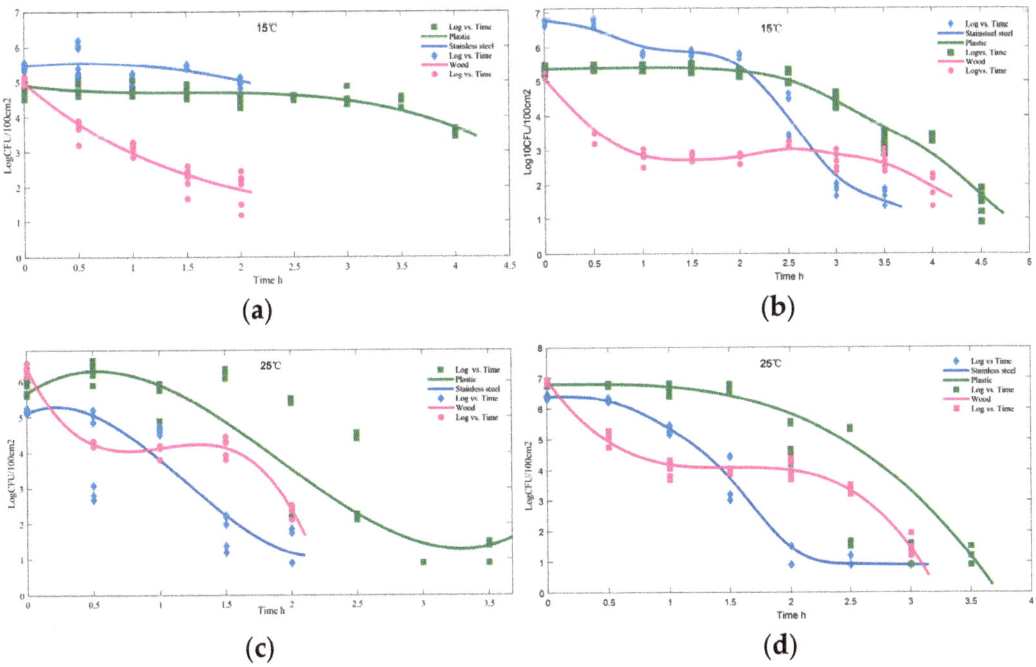

Figure 4. Survival of *Campylobacter* on boards made of different materials. The cultured standard strain ATCC12662 broth was diluted with PBS (**a**,**c**) and chicken juice (**b**,**d**), and then 500 μL of bacterial suspension was inoculated on the surface of different materials Finally, the materials were placed in an incubator at 15 °C (**a**,**b**) or 25 °C (**c**,**d**). Mean values are shown.

4. Discussion

In the present study, we conducted a survey among chefs from commercial kitchens across five provinces in China to gather information on their routine practices regarding the usage and hygiene of cutting boards. We then evaluated different cleaning procedures to assess their effectiveness in eliminating *Campylobacter* contamination on the boards. Additionally, in laboratory experiments, we simulated chicken preparation on cutting boards and tested the survival of *Campylobacter*. In this section, we will discuss the findings in relation to the risks associated with chicken preparation based on our experiments.

In the UK, it was found that 75% of men and 17% of women do not consistently wash their hands after raw food preparation; similarly, in the United States, data show that 20% of people do not wash their hands with soap [17]. In Egypt and Iraq, alarming statistics revealed that 90% of consumers do not wash their hands promptly after handling chicken. This lack of hand hygiene after poultry handling posed a significant risk of the spread of bacteria and foodborne illnesses [18]. In our study, we found that 41.56% of chefs reported washing their hands with running tap water, while 54.55% of chefs stated that they wash their hands with disinfectant. These findings highlighted different handwashing practices among chefs in the study population. Eriksson's hypothesis suggested that hands play a crucial role in the transmission of *Campylobacter*. According to this hypothesis, thorough handwashing using antibacterial substances is considered critically important in preventing the spread of *Campylobacter* [5]. However, it is unfortunate that not everyone in commercial kitchens in China recognizes the importance of proper hand hygiene. Adequate hand hygiene practices, including thorough handwashing with antibacterial substances, are crucial in food preparation settings to prevent the spread of bacteria and foodborne illnesses. It is encouraging to note that a significant majority of consumers in New Zealand, specifically 97%, expressed their intention to use separate knives and cutting boards when preparing chicken compared to other food materials [19]. It is concerning to note that in two Middle Eastern countries, a significant portion of respondents, specifically 28.8% and 6.5%, respectively, expressed their unwillingness to use separate cutting boards when handling chicken [20]. Our study revealed concerning findings regarding the practices of chefs in five provinces in China in relation to cutting boards used for chicken preparation. Specifically, 40.91% of chefs reported using the same cutting boards for cutting other types of meat before preparing chicken. Additionally, only 35.71% of chefs chose to wash the cutting boards with detergent (Table 2). In a study conducted in Norway, it was found that 8% of respondents reported using a cloth as their preferred utensil for cleaning [21]. A significant percentage of chefs in five provinces in China (ranging from 41.18% to 61.11%) expressed a preference for using cloths to clean cutting boards during chicken preparation. The presence of *Campylobacter* has been detected in cloths following the preparation of raw poultry [22,23]. Kitchen cloths can potentially serve as vehicles for cross-contamination of pathogens, transferring them from food spills to other foods or food contact surfaces. Indeed, the education of chefs and kitchen staff is crucial in ensuring food safety in commercial kitchens. The study conducted by Mihalache [24] highlighted the importance of making chefs aware of the critical moments when they need to clean their hands, utensils, and surfaces.

Cutting boards made from a variety of materials, including wood and plastic, are commonly used in wet markets worldwide. However, wood cutting boards have gained attention as potential hazardous surfaces only since 1990 [12]. Research has shown that wood cutting boards can harbor microbial contaminants and present challenges in terms of effective cleaning. In the current study, plastic and wood cutting boards were prevalent in commercial kitchens. It is important to note that wooden surfaces are porous and can facilitate bacterial penetration, thereby increasing the risk of cross-contamination incidents [25]. The study also revealed that cutting boards retained significant levels of *Campylobacter* even after the initial wiping. In fact, the amount of *Campylobacter* detected in the second wiping surpassed the limit of detection (100 CFU/mL) for all samples contaminated with ST-918 and for 18 out of 20 samples contaminated with ST-257 [5]. This finding suggests that traditional cleaning methods may not effectively eliminate *Campylobacter* from cutting boards. Furthermore, the use of disinfectants for cleaning cutting boards has been debated. Some studies have indicated that the use of disinfectants can potentially contribute to the emergence of disinfectant-resistant bacteria and may not completely eradicate antibiotic-resistant bacteria [26]. In our study, despite chefs choosing to clean the boards with disinfectants in one commercial kitchen, *Campylobacter* was still detected. This finding can be attributed to the formation of biofilms by foodborne pathogens on food contact surfaces [27]. In a study conducted in France, the adhesion ability of selected *Campylobacter* isolates to inert surfaces was investigated to explore its association with their transferability. It was found that all the characterized isolates from chicken skin samples demonstrated adhesion to inert surfaces, with more than 90% (25/27) of the isolates exhibiting a moderate to high adhesion ability [28]. So, we should reconsider and revise the strategy for controlling bacterial contamination on cutting boards in kitchen settings. One potential approach, the use of composite materials to cover cutting boards and other polymeric surfaces in meat processing environments, could inhibit the growth of foodborne pathogens, and should be recommended [29].

A study emphasized that contaminated water from the process of washing raw poultry had the potential to travel a significant distance, reaching up to 28 inches (71 cm) on both sides of the sink and 20 inches (51 cm) in front of the sink. Disturbingly, a portion of the chefs in our study (12.99%) reported not eviscerating the whole chicken, and a majority (91.56%) preferred washing the chicken before cutting, both of which could be considered unsafe practices leading to droplet splashing containing *Campylobacter*. During the simulation of chicken preparation in the kitchen, it was observed that *Campylobacter* could be splashed as far as 60–120 cm away from the cutting boards, indicating a high-risk zone extending up to

120 cm where other materials could be exposed to a high risk of contamination (Figure 3). In a previous study, it was reported that sinks and their immediate surroundings (within 0–15 cm) exhibited the highest frequency of contamination following the preparation of chicken thighs inoculated with E. coli DH5-α [30]. Similarly, our study demonstrates that cutting boards and their nearest vicinity (0–30 cm) have the highest frequency of *Campylobacter* contamination after chicken chopping. It is widely acknowledged that cross-contamination is the primary route of *Campylobacter* transmission in the kitchen [8,11,31–34]. However, the findings suggest that droplet splashing can also be an important route for the spread of *Campylobacter* in the Chinese kitchen environment, requiring increased attention. Therefore, it is crucial to design and equip kitchen layouts that effectively prevent the spread of bacteria, thereby interrupting potential transmission pathways.

Campylobacter does not survive for extended periods on food contact surfaces such as equipment, countertops, cutting boards, or kitchen utensils [35]. However, in the presence of wet and cold refrigeration conditions, *Campylobacter* could survive on dry surfaces for several days. In our study, *Campylobacter* was found to survive on different materials used to make cutting boards for a risk exposure time of 3 to 4.5 h when exposed to chicken juice, which can enhance the probability of transmission from boards to other materials (Figure 4). The presence of other co-contaminants has been suggested to enhance the survival of *Campylobacter* in adverse environmental conditions [36]. Some common bacterial species associated with poultry-contaminated boards, such as *Aeromonas* spp., *Brochothrix campestris, Enterobacter cloacae, Pseudomonas putida, Serratia marcescens, Staphylococcus aureus,* and *Streptococcus* spp., have been identified as emerging pathogens or food spoilage organisms [12]. *S. aureus* has been shown to enhance the survival of *Campylobacter* strains under adverse conditions, including low temperatures and aerobic environments [37]. It is important to consider the potential presence of a microbial community on cutting boards, which may contribute to the longer survival of bacteria in the kitchen environment compared to in laboratory tests. The findings of our study suggest that the three cleaning methods evaluated are not effective in completely eliminating *Campylobacter* contamination on cutting boards. Therefore, the effectiveness of cleaning measures and the issue of bacterial disinfectant-resistance must be carefully considered. Additionally, attention shall be given to the hygiene of knives, hands, and cloths to avoid secondary contamination of cutting boards and food materials.

5. Conclusions

The objective of this study was to assess the usage and hygiene practices associated with cutting boards among chefs in five provinces of China. The study aimed to evaluate the impact of different cleaning methods and identify the areas and time intervals after chicken preparation that pose a risk for foodborne illnesses. The findings indicated that wood and plastic cutting boards are widely used in commercial kitchens across the five provinces in China studied, and commonly employed when cutting various materials, including chicken. The study also examined the different cleaning methods employed by chefs, revealing that 41.56% of chefs chose to wash their cutting boards with detergent. Among the provinces, the occurrence of this cleaning practice was higher in Jiangsu Province (41.67%), Guangdong Province (39.58%), Fujian Province (42.42%), and Hunan Province (52.94%). However, even with this cleaning method, there still may be a risk of *Campylobacter* spreading in their kitchens. The paper highlighted the potential sources of cross-contamination in the kitchen environment, such as droplet splashing of *Campylobacter* or the survival of the pathogen on cutting boards even after cleaning. These factors contributed to the risk of *Campylobacter* transmission. In conclusion, it is recommended that chefs in China should reconsider their strategies for effectively cleaning cutting boards to prevent pathogen contamination. Specifically, attention should be given to the risk of *Campylobacter* splashing and surviving on cutting boards. By addressing these concerns, the probability of *Campylobacter* spreading to humans can be significantly reduced, thereby mitigating the risk of foodborne illnesses in commercial kitchens.

Author Contributions: Conceptualization, H.L., X.-a.J. and J.H.; methodology, H.L. and F.R.; software, H.L. and J.H.; interpretation of data, H.L., Y.T. and J.H.; formal analysis, H.L. and Y.T.; original draft preparation, H.L. and Y.T.; review and editing, H.L., X.-a.J. and J.H.; supervision, X.-a.J. and J.H.; funding acquisition, J.H. All authors have read and agreed to the published version of the manuscript.

Funding: The present study was supported by the National Natural Science Foundation of China (NSFC 32172939), the Taishan Industry Leading Talents Project in Shandong Province (tscy20190113), and the Natural Science Foundation of the Higher Education Institutions of Jiangsu Province, China (20KJB180017), Yangzhou University Interdisciplinary Research Foundation for Veterinary Medicine Discipline of Targeted Support (yzuxk202003).

Data Availability Statement: The data presented in this study are available on request from the corresponding author.

Acknowledgments: We thank the staff of commercial kitchens in five provinces of China for completing our questionnaire and assisting us in the process of sampling.

Conflicts of Interest: The authors declare no conflict of interest.

References

1. EFSA. The European Union summary report on trends and sources of zoonoses, zoonotic agents and food-borne outbreaks in 2017. *EFSA J.* **2018**, *16*, e05500.
2. EFSA. The European Union One Health 2020 Zoonoses Report. *EFSA J.* **2021**, *19*, 6971.
3. Pogreba-Brown, K.; Barrett, E. Campylobacter and ethnicity-A Case-Case analysis to determine differences in disease presentation and risk factors. *Foodborne Pathog. Dis.* **2018**, *15*, 277–284. [CrossRef]
4. Luber, P.; Brynestad, S.; Topsch, D.; Scherer, K.; Bartelt, E. Quantification of *Campylobacter* species cross-contamination during handling of contaminated fresh chicken parts in kitchens. *Appl. Environ. Microb.* **2006**, *72*, 66–70. [CrossRef]
5. Eriksson, D.; Råhlén, E.; Bergenkvist, E.; Skarin, M.; Fernström, L.L.; Rydén, J.; Hansson, I. Survival of *Campylobacter jejuni* in frozen chicken meat and risks associated with handling contaminated chicken in the kitchen. *Food Control* **2023**, *145*, 109471. [CrossRef]
6. Silva, J.; Leite, D.; Fernandes, M.; Mena, C.; Gibbs, P.A.; Teixeira, P. *Campylobacter* spp. as a Foodborne Pathogen: A Review. *Front. Microbiol.* **2011**, *2*, 200. [CrossRef]
7. Facciolàr, A.; Riso, R.; Avventuroso, E.; Visalli, G.; Laganà, P. *Campylobacter*: From microbiology to prevention. *J. Prev. Med. Hyg.* **2017**, *58*, E79–E92.
8. Cardoso, M.J.; Ferreir, V.; Truninger, M.; Mena, C.; Gibbs, P.A.; Teixeira, P. Cross-contamination events of *Campylobacter* spp. in domestic kitchens associated with consumer handling practices of raw poultry. *Int. J. Food Microbiol.* **2021**, *338*, 108984. [CrossRef]
9. Møretrø, T.; Martens, L.; Teixeira, P.; Ferreira, V.B.; Maia, R.; Maugesten, T.; Langsrud, S. Is visual motivation for cleaning surfaces in the kitchen consistent with a hygienically clean environment? *Food Control* **2020**, *111*, 107077. [CrossRef]
10. Takeoka, K.; Abe, H.; Koyama, K.; Koseki, S. Experimentally observed *Campylobacter jejuni* survival kinetics in chicken meat products during model gastric digestion tended to be lower than model predictions. *Food Microbiol.* **2022**, *102*, 103932. [CrossRef]
11. Lai, H.; Tang, Y.; Ren, F.; Li, Z.; Li, F.; Cui, C.; Jiao, X.; Huang, J. An investigation into the critical factors influencing the spread of *Campylobacter* during chicken handling in commercial kitchens in China. *Microorganisms* **2021**, *9*, 1164. [CrossRef]
12. Ngan, W.Y.; Rao, S.; Chan, L.C.; Sekoai, P.T.; Pu, Y.; Yao, Y.; Fung, A.H.Y.; Habimana, O. Impacts of wet market modernization levels and hygiene practices on the microbiome and microbial safety of wooden cutting boards in Hong Kong. *Microorganisms* **2020**, *8*, 1941. [CrossRef] [PubMed]
13. Bai, Y.; Lin, X.; Zhu, J.; Cui, S.; Guo, L.; Yan, S.; Wang, W.; Xu, J.; Li, F. Quantification of cross-contamination of *Campylobacter jejuni* during food preparation in a model kitchen in China. *J. Food Prot.* **2021**, *84*, 850–856. [CrossRef]
14. *GB 14934-2016*; National Food Safety Standards for Disinfection of Catering Utensils. National Health and Family Planning Commission: Beijing, China, 2017.
15. Erkmen, O. Isolation and counting of *Campylobacter jejuni*. In *Microbiological Analysis of Foods and Food Processing Environments*; Erkmen, O., Ed.; Elsevier Inc.: London, UK, 2022; pp. 181–192.
16. Birk, T.; Ingmer, H.; Andersen, M.T.; Jørgensen, K.; Brøndsted, L. Chicken juice, a food-based model system suitable to study survival of *Campylobacter jejuni*. *Lett. Appl. Microbiol.* **2004**, *38*, 66–71. [CrossRef]
17. Redmond, E.C.; Griffith, C.J. Consumer food handling in the home: A review of food safety studies. *J. Food Prot.* **2003**, *66*, 130–161. [CrossRef]
18. Evans, E.W.; Redmond, E.C. Behavioral Observation and Microbiological Analysis of Older Adult Consumers' Cross-Contamination Practices in a Model Domestic Kitchen. *J. Food Prot.* **2018**, *81*, 569–581. [CrossRef]
19. Allan, P.D.; Palmer, C.; Chan, F.; Lyons, R.; Nicholson, O.; Rose, M.; Hales, S.; Baker, M.G. Food safety labelling of chicken to prevent campylobacteriosis: Consumer expectations and current practices. *BMC Public Health* **2018**, *18*, 414. [CrossRef]

20. Habib, I.; Harb, A.; Hansson, I.; Vågsholm, I.; Osama, W.; Adnan, S.; Anwar, M.; Agamy, N.; Boqvist, S. Challenges and opportunities towards the development of risk assessment at the consumer phase in developing countries-The case of *Campylobacter* cross-contamination during handling of raw chicken in two middle eastern countries. *Pathogens* **2020**, *9*, 62. [CrossRef]
21. Møretrø, T.; Moen, B.; Almli, V.L.; Teixeira, P.; Ferreira, V.B.; Asli, A.W.; Nilsen, C.; Langsrud, S. Dishwashing sponges and brushes: Consumer practices and bacterial growth and survival. *Int. J. Food Microbiol.* **2021**, *337*, 108928. [CrossRef]
22. Mattick, K. The survival of foodborne pathogens during domestic washing-up and subsequent transfer onto washing-up sponges, kitchen surfaces and food. *Int. J. Food Microbiol.* **2003**, *85*, 213–226. [CrossRef]
23. Lai, H.; Tang, Y.; Wang, Z.; Ren, F.; Kong, L.; Jiao, X.; Huang, L. Handling practice as a critical point influencing the transmission route of *campylobacter* throughout a commercial restaurant kitchen in China. *Food Control* **2022**, *139*, 109056. [CrossRef]
24. Mihalache, O.A.; Moretro, T.; Borda, D.; Dumitrascu, L.; Neagu, C.; Nguyen-The, C.; Maitre, I.; Didier, P.; Teixeira, P.; Lopes Junqueira, L.O.; et al. Kitchen layouts and consumers' food hygiene practices: Ergonomics versus safety. *Food Control* **2022**, *131*, 108433. [CrossRef]
25. Adetunji, V.O.; Isola, T.O. Crystal violet binding assay for assessment of biofilm formation by *listeria monocytogenes* and *listeria* spp. on wood, steel and glass surfaces. *Glob. Vet.* **2011**, *6*, 6–10.
26. Bock, L.J.; Wand, M.E.; Sutton, J.M. Varying activity of chlorhexidine-based disinfectants against *Klebsiella pneumoniae* clinical isolates and adapted strains. *J. Hosp. Infect.* **2016**, *93*, 42–48. [CrossRef]
27. Zhao, X.; Zhao, F.; Wang, J.; Zhong, N. Biofilm formation and control strategies of foodborne pathogens: Food safety perspectives. *RSC Adv.* **2017**, *7*, 36670–36683. [CrossRef]
28. Guyard-Nicodème, M.; Tresse, O.; Houard, E.; Jugiau, F.; Courtillon, C.; EL Manaa, K.; Laisney, M.J.; Chemaly, M. Characterization of *Campylobacter spp.* transferred from naturally contaminated chicken legs to cooked chicken slices via a cutting board. *Int. J. Food Microbiol.* **2013**, *164*, 7–14. [CrossRef]
29. Serov, D.A.; Baimler, I.V.; Burmistriv, D.E.; Baryshev, A.S.; Yanykin, D.V.; Astashev, M.E.; Simakin, A.V.; Gudkov, S.V. The development of new nanocomposite polytetrafluoroethylene/Fe(2)O(3) NPs to prevent bacterial contamination in meat industry. *Polymers* **2022**, *14*, 4880. [CrossRef]
30. Kirchner, M.K. *The Effect of Consumer Behaviors on Cross-Contamination While Preparing Meals in a Consumer Kitchen*; North Carolina State University: Raleigh, NC, USA, 2020.
31. DE Boer, E.; Hahne, M. Cross-contamination with *Campylobacter jejuni* and *Salmonella* spp. from Raw Chicken Products During Food Preparation. *J. Food Prot.* **1990**, *53*, 1067–1068. [CrossRef]
32. Gorman, R.; Bloomfield, S.; Adley, C.C. A study of cross-contamination of food-borne pathogens in the domestic kitchen in the Republic of Ireland. *Int. J. Food Microbiol.* **2002**, *76*, 143–150. [CrossRef]
33. Mylius, S.D.; Nauta, M.J.; Havelaar, A.H. Cross-contamination during food preparation: A mechanistic model applied to chicken-borne *Campylobacter*. *Risk Anal.* **2007**, *27*, 803–813. [CrossRef]
34. Van Asselt, E.D.; de Jong, A.E.; de Jong, R.; Nauta, M.J. Cross-contamination in the kitchen: Estimation of transfer rates for cutting boards, hands and knives. *J. Appl. Microbiol.* **2008**, *105*, 1392–1401. [CrossRef] [PubMed]
35. Hakeem, M.J.; Lu, X. Survival and control of *Campylobacter* in poultry production environment. *Front. Cell Infect. Microbiol.* **2020**, *10*, 615049. [CrossRef] [PubMed]
36. Teh, K.H.; Flint, S.; French, N. Biofilm formation by *Campylobacter jejuni* in controlled mixed-microbial populations. *Int. J. Food Microbiol.* **2010**, *143*, 118–124. [CrossRef]
37. Karki, A.B.; Ballard, K.; Harper, C.; Sheaff, R.J.; Fakhr, M.K. *Staphylococcus aureus* enhances biofilm formation, aerotolerance, and survival of *Campylobacter* strains isolated from retail meats. *Sci. Rep.* **2021**, *11*, 13837. [CrossRef]

foods

MDPI

Article

Characterization of Molecular Chaperone GroEL as a Potential Virulence Factor in *Cronobacter sakazakii*

Dongdong Zhu [1], Yufei Fan [1], Xiaoyi Wang [1], Ping Li [1], Yaping Huang [1], Jingbo Jiao [1], Chumin Zhao [1], Yue Li [1], Shuo Wang [1,2] and Xinjun Du [1,*]

[1] State Key Laboratory of Food Nutrition and Safety, College of Food Science and Engineering, Tianjin University of Science and Technology, Tianjin 300457, China; duiduizdd@mail.tust.edu.cn (D.Z.); 17908002@mail.tust.edu.cn (Y.F.); rebeccamargaret@msn.cn (X.W.); zoelxx@tust.edu.cn (P.L.); huangyp555@163.com (Y.H.); 18753391899@163.com (J.J.); zcm3278@163.com (C.Z.); i_liyue2020@163.com (Y.L.); s.wang@tust.edu.cn (S.W.)

[2] Tianjin Key Laboratory of Food Science and Health, School of Medicine, Nankai University, Tianjin 300071, China

* Correspondence: xjdu@tust.edu.cn; Tel./Fax: +86-22-60912484

Abstract: The molecular chaperone GroEL of *C. sakazakii*, a highly conserved protein encoded by the gene *grol*, has the basic function of responding to heat shock, thus enhancing the bacterium's adaptation to dry and high-temperature environments, which poses a threat to food safety and human health. Our previous study demonstrated that GroEL was found in the bacterial membrane fraction and caused a strong immune response in *C. sakazakii*. In this study, we tried to elucidate the subcellular location and virulent effects of GroEL. In live *C. sakazakii* cells, GroEL existed in both the soluble and insoluble fractions. To study the secretory mechanism of GroEL protein, a non-reduced Western immunoblot was used to analyze the form of the protein, and the result showed that the exported GroEL protein was mainly in monomeric form. The exported GroEL could also be located on bacterial surface. To further research the virulent effect of *C. sakazakii* GroEL, an indirect immunofluorescence assay was used to detect the adhesion of recombinant GroEL protein to HCT-8 cells. The results indicated that the recombinant GroEL protein could adhere to HCT-8 cells in a short period of time. The recombinant GroEL protein could activate the NF-κB signaling pathway to release more pro-inflammatory cytokines (TNF-α, IL-6 and IL-8), downregulating the expression of tight-junction proteins (claudin-1, occluding, ZO-1 and ZO-2), which collectively resulted in dose-dependent virulent effects on host cells. Inhibition of the *grol* gene expression resulted in a significant decrease in bacterial adhesion to and invasion of HCT-8 cells. Moreover, the deficient GroEL also caused slow growth, decreased biofilm formation, defective motility and abnormal filamentation of the bacteria. In brief, *C. sakazakii* GroEL was an important virulence factor. This protein was not only crucial for the physiological activity of *C. sakazakii* but could also be secreted to enhance the bacterium's adhesion and invasion capabilities.

Keywords: *Cronobacter sakazakii*; GroEL; export; virulent effects

Citation: Zhu, D.; Fan, Y.; Wang, X.; Li, P.; Huang, Y.; Jiao, J.; Zhao, C.; Li, Y.; Wang, S.; Du, X. Characterization of Molecular Chaperone GroEL as a Potential Virulence Factor in *Cronobacter sakazakii*. *Foods* **2023**, *12*, 3404. https://doi.org/10.3390/foods12183404

Academic Editor: Giorgia Perpetuini

Received: 15 August 2023
Revised: 4 September 2023
Accepted: 6 September 2023
Published: 12 September 2023

1. Introduction

Cronobacter sakazakii, an opportunistic Gram-negative pathogen mainly found to be associated with milk and cheese protein powders, was reported to cause both intestinal and systemic human diseases, such as meningitis, bacteremia and necrotizing enterocolitis through complex bacterium–host interactions [1]. Studies have shown that its main infection targets are neonates, infants and elderly individuals [2]. In 2010, two mexican infants were infected with *C. sakazakii* and developed bloody diarrhea [3]. In 2016, a premature female infant was diagnosed with sepsis caused by *C. sakazakii* ultimately leading to cerebral liquefaction and necrosis [4]. *C. sakazakii* is not thermotolerant (survives between 6 and 45 °C) but can proliferate in dry stress, even at a low water activity level of 0.3 [5]. It was reported that 18.75% of quick-frozen food collected from 39 cities in china were contaminated with *C. sakazakii* in 576 samples, wherein the contamination of frozen flour products was up to 44.34% [6]. In addition, opened powdered formula and breast pump parts were highly susceptible to contamination with *C. sakazakii* [7].

To date, many studies have been performed to investigate the pathogenic mechanism of this bacterium, and the usual virulence factors in Gram-negative bacteria, including lipopolysaccharide (LPS), enterotoxin, flagella and some outer membrane proteins, were reported to be involved in the pathogenicity of *C. sakazakii* [8–11]. Other virulence-associated factors of *C. sakazakii*, such as the bcsABC operon controlling cellulose synthesis, methyl-accepting chemotaxis protein Mcp and UDP-N-acetylglucosamine acyltransferase's ligand Labp, are also reported to contribute to the pathogenicity of the bacterium [12–14]. However, although these virulence-associated factors are shown to be associated with *C. sakazakii* infection, the key virulence determinants of *C. sakazakii* and the details of this pathogen's infection of host cells are largely unknown. Recently, many microbiological secretome studies have shown the release of various cytoplasmic proteins into the medium supernatant, especially evolutionarily conserved proteins, including glycolytic enzymes, translation factors and chaperones [15]. The secretory protein was crucial for bacteria to obtain nourishment, adapt to the environment and infect the host. Whether these exported proteins of cytoplasm are caused by accidental bacterial lysis or secreted through a special pathway remains hotly debated [16–18], while studies have shown that these proteins can adhere to the surface of microorganisms and have immunogenicity [19]. The virulence-associated roles of these proteins require more detailed studies.

In our previous work, the molecular chaperone GroEL was shown to be expressed at a high level in *C. sakazakii* and to cause a strong immune response [20]. The basic function of GroEL is to participate in the regulation of the heat shock response to maintain cellular homeostasis and form a nanocage structure with the co-chaperonin protein GroES to rescue the proteins from inappropriate folding and aggregation [21]. GroEL mutation was reported to cause the accumulation of a large number of newly translated peptides [22]. In addition, GroEL is also a familiar member in the secretome of eukaryotes and prokaryotes. Studies have shown that the exported GroEL of different prokaryotes can adhere to different human cells [23,24], and GroEL can produce immune protection in immune response and reduce the degree of lesions and the mortality of related diseases to a certain extent [25]. Although GroEL can act as an adhesin to aggravate specific tissue diseases caused by several corresponding prokaryotes, whether it is an important virulence factor in related diseases caused by *C. sakazakii* is still unknown.

Therefore, the purpose of this study was to identify whether *C. sakazakii* can export GroEL and whether GroEL serves as a virulence factor of this bacterium. This study indicated that *C. sakazakii* can export monomeric GroEL. The GroEL protein was able to adhere to the intestinal epithelium cells and caused adverse effects. Inhibition of the *grol* gene expression not only caused physical defects in *C. sakazakii* but also reduce the bacterial adhesion to and invasion of hosts. In brief, we proved that *C. sakazakii* GroEL is an important virulence factor, and this study will provide new insight into the pathogenicity of *C. sakazakii*.

2. Materials and Methods

2.1. Bacterial Strains and Growth Conditions

In this study, *Escherichia coli* BL21 (DE3) and *C. sakazakii* ATCC29544 strains stored at the Tianjin University of Science and Technology were used. Both bacterial strains were grown in Luria–Bertani (LB) medium at 37 °C under constant shaking unless otherwise stated.

2.2. Bacterial Fractionation

The bacterial fractionation was performed using Goulhen's method with some modifications [26]. Briefly, *C. sakazakii* was cultured to an OD_{600} of 1.0 (1×10^{10} CFU/mL), followed by centrifugation to separate the medium supernatant from the bacterial cells.

The crude outer membrane vesicle (OMV) was harvested using the saturated ammonium sulfate precipitation method from the medium supernatant and then ultracentrifuged at 250,000× g for 1 h at 4 °C using a supercentrifuge (Beckman, Brea, CA, USA) to obtain the purified OMV. The ultracentrifuged supernatant was rich in other secreted proteins. The periplasmic fractions were isolated through the osmotic shock method. The residual pellet was broken through mild sonication and then centrifuged to separate the crude membrane envelope and the cytoplasmic fractions. The crude membrane envelope was re-suspended in 2% Triton X-100 containing 10 mM $MgCl_2$ after washing three times with TBS buffer. The suspension was ultracentrifuged at 200,000× g for 2 h to separate the cytoplasmic-membrane-rich cell envelope (supernatant) and the outer-membrane-rich cell envelope (pellet).

2.3. *Identification of GroEL and Its Forms in Different Bacterial Fractions*

The different bacterial fractions were separated through SDS-PAGE (12% acrylamide), and then, the protein samples were transferred to a polyvinylidene fluoride membrane (0.22 μm, 200 mA, 90 min). The membrane was blocked with 10% skim milk for 3 h in a 50 mM TBS buffer, followed by incubation with the primary antibody (anti-GroEL monoclonal antibody, Abcam, Cambridge, UK) for 2 h at room temperature. After washing three times with TBS buffer containing 0.1% Tween 20, the membrane was subsequently incubated with the horseradish-peroxidase-linked secondary antibody. HRP substrate (Millipore, Boston, MA, USA) was added, and the result was detected using an enhanced chemiluminescence detection system. As for the non-reduced Western immunoblot, we adopted Qamra's method but with some modifications [27]. Briefly, the protein samples were separated using SDS-PAGE (7% acrylamide) and then transferred to a nitrocellulose membrane (0.45 μm). Electrophoresis was carried out at 18 V for 18 h. The following procedures of the non-reduced Western immunoblot were the same as those described above.

2.4. *Determination of the Bacterial Surface GroEL*

Detection of the GroEL localizing on the *C. sakazakii* surface was carried out using an indirect immunofluorescence assay (IIF). In brief, the tetracycline gene of the pACYC-184 plasmid was completely replaced with the *grol* gene containing the His tag using a ClonExpress® II One Step Cloning Kit (Vazyme, Nanjing, China), and then, the recombinant plasmid was transformed into *C. sakazakii*. The recombinant bacteria were cultured to OD_{600} 1.0, and subsequently, the bacteria were washed three times with a PBS buffer. The bacteria were treated with an anti-His-Tag monoclonal antibody (ABclonal, Wuhan, China) (60 min incubation, 1:100 in PBS) and subsequently with the secondary antibody goat anti-mouse IgG (H+L)-FITC (Sungene Biotech, Shanghai, China) (60 min incubation, 1:2000 in PBS). The fluorescence signals were detected using a confocal microscope (Lecia Tcs Sp8, Weztlar, Germany). *C. sakazakii* containing an empty pACYC-184 plasmid was used as control. The primers for amplifying the *grol* gene (F1, R1) and the pACYC-184 backbone vector (F2, R2) are shown in Table S1.

2.5. *Adhesive Ability of GroEL to HCT-8 Cells*

E. coli BL21 (DE3) containing recombinant pET-26b-*grol* was used to express recombinant GroEL (rGroEL), followed by the purification of rGroEL as described previously [28]. IIF was used to determine the adhesion of GroEL to HCT-8 cells. Briefly, the human enterocyte-like epithelial HCT-8 cells (ATCC) were maintained in RPMI 1640 medium (Gibco, Stockrick, CA, USA) containing 10% fetal bovine serum (Gibco, Stockrick, CA, USA) and then transferred to a coverglass-bottom dish for further culture until monolayer cells covered the dish bottom. After discarding the culture medium, the cells were washed three times with phosphate-buffered saline (PBS) (in each step of the subsequent treatment, the cells were washed in the same way), followed by fixing the cells with ice-cold 4% formaldehyde for 15 min and incubation with purified rGroEL at 37 °C for 30 min in serum-free RPMI 1640 medium. Subsequently, the cells were incubated with an anti-His-Tag monoclonal antibody (60 min incubation, 1:100 in PBS) and then with the secondary antibody goat anti-mouse IgG (H+L)-FITC (60 min incubation, 1:2000 in PBS). The cell nuclei were stained using 4, 6-diamidino-2-phenylindole (DAPI). The control group was prepared by using the eluent from the His-Tag Purification Resin, which was loaded with total proteins of *E. coli* BL21 (DE3) containing empty plasmid pET-26b (isoconcentration imidazole eluent) to replace rGroEL, and the other steps were the same as described above. Fluorescence was detected with a confocal microscope.

2.6. Determination of GroEL Protein Virulent Effects

The virulent effects of GroEL were evaluated by determining the viability of cells as described previously with some modifications [26]. Briefly, HCT-8 cells were plated on 96-well plates to yield monolayer cells, followed by adding purified rGroEL and culturing the cells for another 48 h at 37 °C in serum-free RPMI 1640 medium. The growth of HCT-8 cells was studied using 0.1% crystal violet (CV) staining and the absorbance was measured at OD_{595}. The viability of HCT-8 cells was analyzed by using a 3-(4, 5-dimethylthiazol-2-yl)-2, 5-diphenyl tetrazolium bromide assay (MTT assay), and the absorbance was detected at OD_{495}. The eluent of *E. coli* BL21 (DE3) containing empty plasmid pET-26b was used instead of rGroEL as a control. Then, 5 μg/mL rGroEL was treated using 2 μg/mL anti-GroEL monoclonal antibody at 4 °C for 2 h before stimulating the host cells (cure group).

2.7. Detection of Cytokines

The pro-inflammatory cytokines were analyzed using an enzyme linked immunosorbent assay. Briefly, HCT-8 cells were plated on 6-well plates to yield monolayer cells, followed by adding purified rGroEL (3 μg/mL, final concentration), culturing the cells for 24 h, and identifying the cytokines in the supernatant. To verify whether the specific release of cytokines was due to rGroEL bioactivity, rGroEL pretreated with 1 μg/mL anti-GroEL monoclonal antibody for 2 h at 4 °C was used as a negative control to stimulate the host cells (anti-GroEL group). The eluent of *E. coli* BL21 (DE3) containing empty plasmid pET-26b was also used as a negative control (ck group). Tumor necrosis factor-α (TNF-α) was measured using ELISA kits (Jiancheng, Nanjing, China). Interleukin-6 (IL-6) and interleukin-8 (IL-8) assays were undertaken by the Beijing Sinouk Institute of Biological Technology.

2.8. Signaling Pathway Analysis

Cell culture followed the steps in Section 2.7. The culture medium was discarded and the remnants were washed three times with a PBS buffer, following which a weak RIPA buffer (Sigma) was used to digest them. This was followed by centrifugation at 4 °C (12,000× *g*, 15 min) to remove cell debris, and then, phosphatase Inhibitor (Merck, Darmstadt, Germany) and PMSF (Sigma, St Louis, MO, USA) were added to avoid protein degradation. The detection of signaling pathway proteins was carried out using a Western immunoblot. The monoclonal antibodies including anti-NF-Kb, anti-IκBα, anti-p-IκBα (Abcam, Cambridge, UK) and anti-beta actin (preserved in our laboratory) were used in this process, respectively. The specific method followed the steps in Section 2.3. As for the difference in the tight-junction protein between the GroEL-treated group and the control group, the total RNA of each group was extracted, followed by the construction of cDNA using the PrimeScript™ II Reverse Transcriptase kit (Takara, Kyoto, Japan). Subsequently, real-time quantitative PCR (RT-qPCR) was used to detect the transcription levels of the relative genes. The transcription level of the relative genes was analyzed using the $2^{-\Delta\Delta Ct}$ method. $2^{-\Delta\Delta Ct}$ method (please delete it).

2.9. C. sakazakii grol Gene Silencing

The detailed protocol for inhibiting *grol* gene expression was performed as previously reported with some modifications [29]. Briefly, the protospacer adjacent motif (PAM) was designed as 'GGACACGCCGTCTTTGGTGA' to target the non-template DNA strand of the *C. sakazakii grol* gene. The PAM motif was linked to the pTargetF plasmid (spectinomycin resistant) with the primers F3 and R3 by reverse PCR (Table S1). The plasmid of the pdCas9-bacteria (chloramphenicol resistant) donated inactive cas9 protein. In the case of adding spectinomycin and chloramphenicol at the same time, the bacteria containing the two aforementioned plasmids were cultured in LB medium with (CRISPRi-treated group) or without (untreated group, to remove the interference of spectinomycin and chloramphenicol) 1 μM anhydrotetracycline.

2.10. Bacterial Viability Assay

The MTT assay method was used to measure the bacterial viability [30]. Briefly, the bacteria were cultured to an OD_{600} of 0.6, followed by collection of the bacteria, three washes with PBS buffer, and dilution of the bacteria with fresh LB medium to an OD_{600} of 0.1. Subsequently, 200 μL bacterial suspension and 20 μL MTT (the final concentration was 0.5%) were transferred precisely in a 37 °C preheated 1.5 mL tube, followed by manually mixing the tube for a few seconds to initiate the reduction reaction and then incubating the mixture at 37 °C for 20 min to produce formazan crystals. Finally, the crystals were dissolved in dimethyl sulfoxide (DMSO), and the absorbance was measured at OD_{550}.

2.11. Bacterial Adhesion and Invasion

Briefly, HCT-8 cells were cultured as described for the experiment to evaluate GroEL virulent effects. Logarithmic-phase bacteria were washed three times with PBS buffer to remove excess antibiotics. Next, equal amounts of different bacteria (untreated group, CRISPRi-treated group) were added into 96-well plates and incubated for 30 min. Host cells were washed three times with PBS buffer and treated with Triton X-100, followed by counting of the adherent bacteria. As for the bacterial invasion, the bacteria and HCT-8 cells were co-incubated for 90 min, followed by using 100 μg/mL gentamicin to destroy the bacteria adhering to the surface of host cells and then three washes with PBS buffer. Finally, the HCT-8 cells were treated with Triton X-100 to release the invaded bacteria and then the number of bacteria was determined.

2.12. Biofilm Formation Experiment

The biofilm formation experiment was analyzed through fluorescence staining. Logarithmic-phase bacteria (100 μL) were transferred to a coverglass-bottom dish and cultured statically for 3 days at 37 °C to establish a biofilm, followed by fixing the biofilm with 4% glutaraldehyde overnight at 4 °C. After removing the supernatant and washing three times with PBS, the biofilm was stained using SYBR Green I at room temperature for 30 min in the dark, followed by removing the fluorescence dye and washing three times with PBS. Finally, the morphology of the biofilm was observed under a confocal laser scanning microscope (Leica TCS SP8, Weztlar, Germany).

2.13. Bacterial Motility and Morphology

Bacterial motility was observed on an LB medium (0.3% agar powder). Briefly, different groups of logarithmic-phase bacteria (untreated group, CRISPRi-treated group) were added to semi-solid LB medium in an equal amount. After culturing at 25 °C or 37 °C for 12 h, the bacterial motility was observed. The morphological characteristics of the bacteria were investigated using a scanning electron microscope (SEM). Before measurement, the samples were coated with a gold layer, followed by observation of the samples in an SEM (SU1510, Hitachi, Tokyo, Japan) operating at 4700× magnification.

2.14. Statistical Analysis

The SPSS 18.0 software was used for statistical analysis of the data. The significant differences of the results were assessed using the unpaired *t*-test or Duncan's test. A threshold below *p*-values of 0.05 was considered statistically significant (N * $p < 0.05$, ** $p < 0.01$, *** $p < 0.001$, **** $p < 0.0001$). At least three independent replicates were conducted for each experiment, and the results were expressed as mean ± deviation.

3. Results

3.1. Subcellular Localization and Quaternary Structure of GroEL

GroEL is a mainly cytosol-localized protein of *C. sakazakii* with an important role in the regulation of biological process of the bacteria and may contribute to the interaction of *C. sakazakii* with the host when the protein occurs extracellularly. In order to further understand the virulent roles of GroEL, the subcellular localization of this protein was investigated in detail. The localizations of GroEL in different fractions of *C. sakazakii* were confirmed based on the Western immunoblot using a GroEL-specific monoclonal antibody. As shown in Figure 1A,B, GroEL was found in all bacterial fractions, and most of it was retained in the cytoplasm. Some GroEL was also found in the periplasm and the ultracentrifuged supernatant (the fraction that removed the OMV). In the insoluble fractions (cytoplasmic membrane, outer membrane and OMV), the highest amount of the protein was detected in the OMV fraction. In order to identify whether GroEL exists on the bacterial surface, the overexpression of recombinant GroEL was detected on whole *C. sakazakii* cells through IIF. Irradiated bacteria that could recombinantly express GroEL emitted green fluorescence (Figure 1C), and there was no fluorescence signal in the control group, indicating that the GroEL was located on the surface of the bacterium. In order to know the forms of exported GroEL, the fractions of the periplasm and OMV were separated and tested using non-reduced SDS-PAGE and detected using a Western immunoblot. As shown in Figure 2A, the abundant monomer (60 kDa) and dimer (120 kDa) of GroEL were detected in the periplasmic fraction. Meanwhile, in the OMV fraction, most of this protein was mainly in the form of a monomer, and only trace amounts of dimeric GroEL (120 kDa) were detected, indicating that *C. sakazakii* exported GroEL protein mainly in the monomeric form.

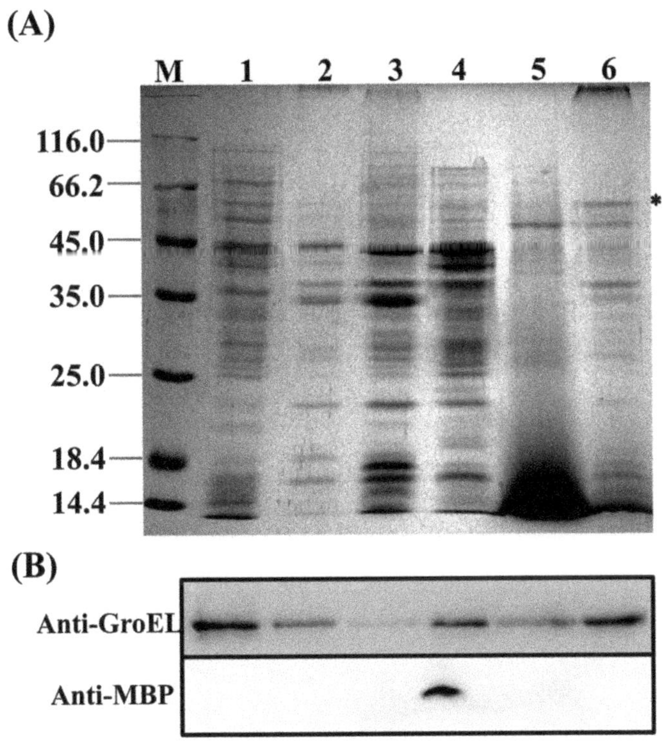

Figure 1. *Cont.*

(C)

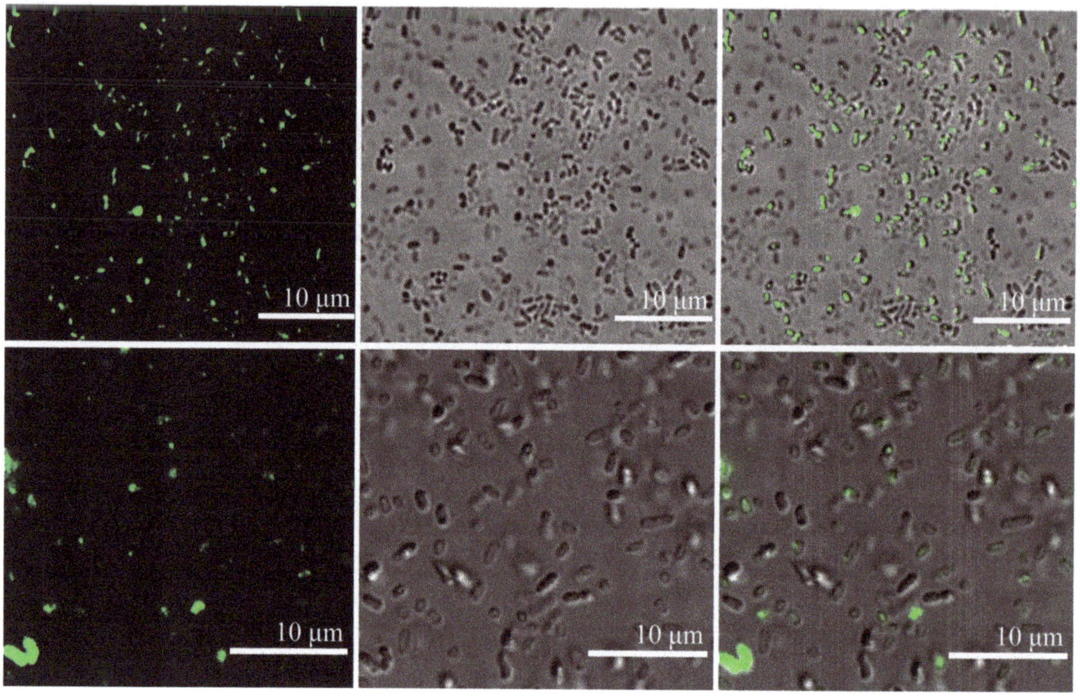

Figure 1. Localization of the GroEL protein in *C. sakazakii*. (**A**) SDS-PAGE of different fractions of *C. sakazakii* and the band of GroEL (60 kDa) is labeled with an asterisk; each lane contains 10 μg proteins except for ultracentrifuged supernatant fraction (30 μg). M, protein marker; Lane 1, cytoplasm; Lane 2, cytoplasmic membrane; Lane 3, outer membrane; Lane 4, periplasm; Lane 5, ultracentrifuged supernatant; Lane 6, OMV. (**B**) Identification of GroEL (top) and the periplasmic maltose binding protein MBP (dwon) in various bacterial fractions using Western immunoblot, the band of GroEL (60 kDa) is labeled with an asterisk (please delete it). MBP (43.4 kDa) was used to evaluate bacterial lysis, and anti-MBP monoclonal antibody (Abclonal, Wuhan, China) was used to identify this protein. Lane 1, cytoplasm (10 μg); Lane 2, cytoplasmic membrane (10 μg); Lane 3, outer membrane (10 μg); Lane 4, periplasm (10 μg); Lane 5, ultracentrifuged supernatant (30 μg); Lane 6, OMV (10 μg). (**C**) Indirect immunofluorescence detection of GroEL protein associated with bacterial surface. The fluorescence signal was taken from the visual field at 400× (top) and 630× magnification (bottom), respectively. * $p < 0.05$.

3.2. Virulent Effects of rGroEL Protein on HCT-8 Cells

In order to verify the exact virulent roles of GroEL, the recombinant expressed protein (rGroEL) was prepared, and the virulence was investigated. After expressing recombinant GroEL protein in *E. coli* BL21 (DE3) and purifying it through affinity chromatography, the Western immunoblot was used to identify purified rGroEL. As shown in Figure 2B, the molecular weight of the rGroEL is approximately 60 kDa. The purifed rGroEL only contained monomeric and dimeric forms, which were identified by Gel permeation chromatography. After co-incubation with host cells for 30 min, the distribution of rGroEL in HCT-8 cells was analyzed using IIF and the results are shown in Figure 3A, which was taken using a confocal microscope with 400× magnification. The whole cell surface was seen to be full of green fluorescence, indicating that *C. sakazakii* GroEL could rapidly adhere to HCT-8 cells in 30 min. The adverse effects of *C. sakazakii* GroEL on HCT-8 cells was evaluated by measuring the viability of residual HCT-8 cells after co incubation with rGroEL in a serum-free RPMI 1640 medium. With an increase in rGroEL concentration, both the viability and the number of host cells in the culture decreased. When the rGroEL concentration reached 10 μg/mL, the viability and the number of host cells reduced to 3.46% ($p < 0.001$; Figure 3B) and 4.71% ($p < 0.0001$; Figure 3C) relative to the control group, respectively. As for the cure group, the viability and the number of host cells recovered 33.82% ($p < 0.001$) and 21.03% ($p < 0.001$) compared with the group that was treated with 5 μg/mL rGroEL, respectively. These data indicate that *C. sakazakii* GroEL had a harmful effect on host cells and could induce dose-dependent apoptosis or cell necrosis, especially at a high dose.

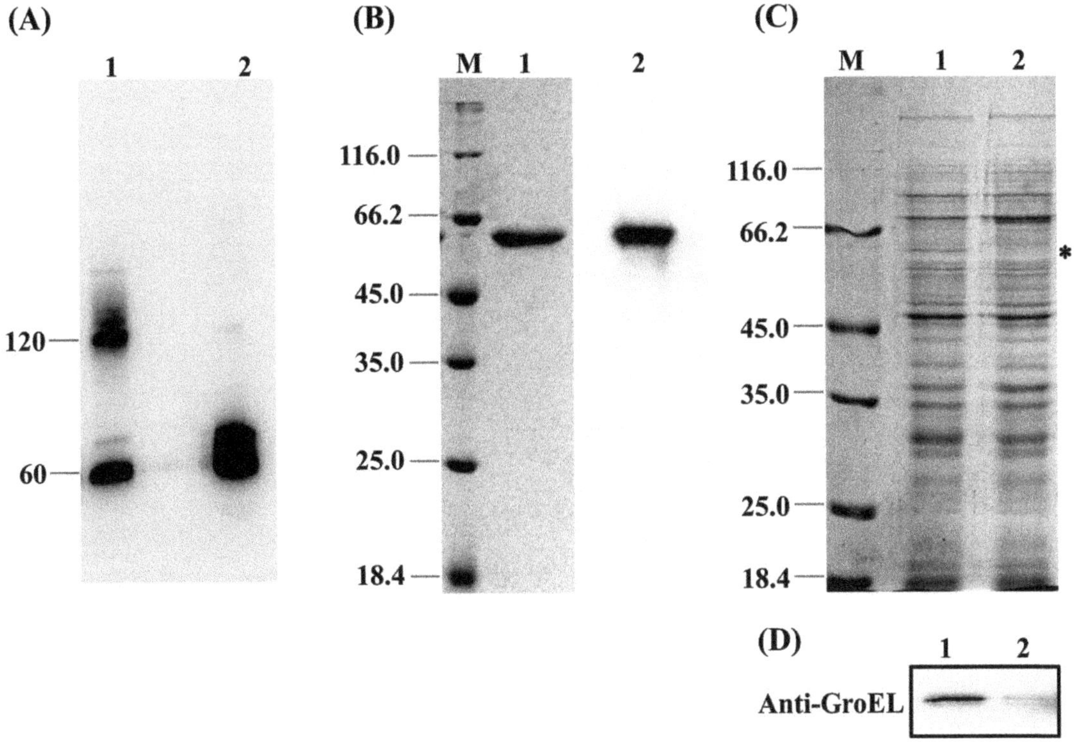

Figure 2. (**A**) The quaternary structure of GroEL in periplasmic fraction and OMV. Lane 1, periplasm (40 μg); Lane 2, OMV (60 μg). (**B**) Purification of recombinant GroEL protein through affinity chromatography. M, protein marker; Lane 1, SDS-PAGE profile depicting purification of rGroEL that has removed the signal peptide of plasmid vector; Lane 2, identification of rGroEL using Western immunoblot. (**C**) SDS-PAGE profile analyzing the expression level of GroEL protein of bacteria, and the band of GroEL (60 kDa) is labeled with an asterisk. M, Marker; Lane 1, untreated group (10 μg); Lane 2, CRISPRi-treated group (10 μg). (**D**) Quantification of relative expression of *grol* gene using Western immunoblot. Lane 1, untreated group (10 μg); Lane 2, CRISPRi-treated group (10 μg). * $p < 0.05$.

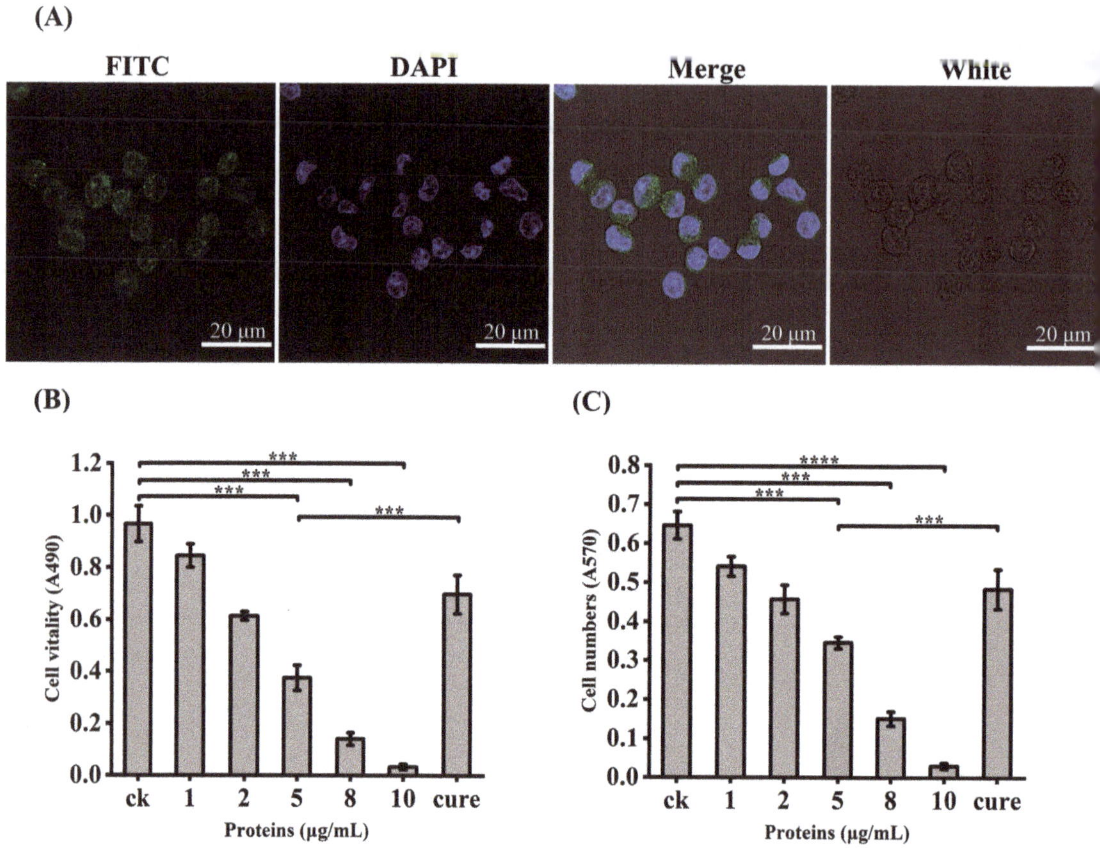

Figure 3. GroEL virulent effect determination. (**A**) IIF for the localization of recombinant GroEL in HCT-8 cells, 400× magnification; FITC, identification of recombinant GroEL on the surface of HCT-8 cells using FITC fluorescence staining; DAPI, fluorescent staining of cell nuclei by DAPI; White, observation of HCT-8 cells at light microscope, Merge, the overlaying of the above three images. (**B**) changes in the vitality of HCT-8 cells after GroEL treatment; (**C**) changes in the number of HCT-8 cells after GroEL treatment. Three independent replicates were conducted for each experiment. *** $p < 0.001$, **** $p < 0.0001$.

3.3. Nuclear Factor Kappa-B (NF-κB) Activization and Downregulation of Tight-Junction Proteins in HCT-8 Cells under rGroEL Stimulation

To explore whether *C. sakazakii* GroEL has ability to activate intracellular signal transduction in HCT-8 cells, the expression of related proteins of NF-κB pathways was detected using a Western immunoblot. As shown in Figure 4A, there was no significant difference in the total expression of NF-κB inhibitor alpha (IκBα) under the addition of rGroEL protein, while rGroEL could promote the expression of phospho-IκBα (p-IκBα), which was an activated hallmark of NF-κB. In addition, rGroEL stimulated a concomitant increase in NF-κB, indicating that *C. sakazakii* GroEL could activate NF-κB pathways to regulate the release of pro-inflammatory cytokines. In the meantime, the relative gene transcription of tight-junction proteins were significantly reduced in rGroEL-treated cells. The quantities of gene transcription were detected through RT-qPCR, and the relative primers are shown in Table S1. As shown in Figure 4B, the gene transcriptional levels including those of claudin-1(CLDN-1), occluding (OCLN), ZO-1 and ZO-2 decreased by 46.59% ($p < 0.01$),

23.61% ($p < 0.05$), 33.28% ($p < 0.05$), 35.62% ($p < 0.01$), respectively, which indicated that *C. sakazakii* GroEL was able to assist the bacteria to invade host cells easily.

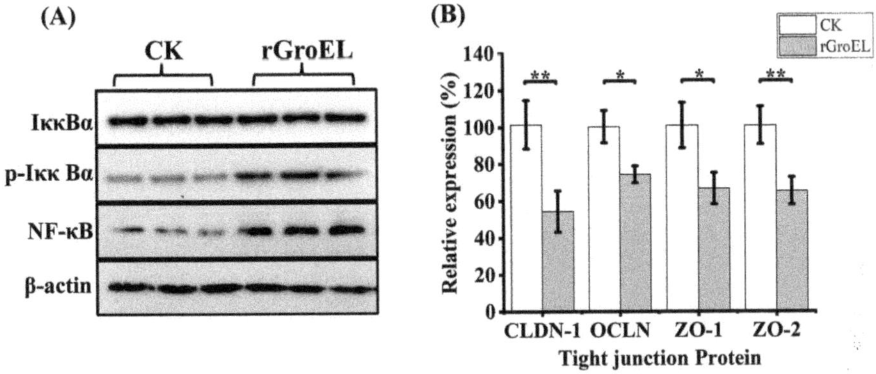

Figure 4. GroEL contributes to NF-κB activation and intestinal cell permeability. (**A**) Immunoblot of GroEL-induced relative signaling proteins of NF-κB pathways levels in HCT-8 cells. β-Actin was used as control. (**B**) RT-qPCR detected the transcription level of related tight-junction protein genes of HCT-8 cells. The gene glyceraldehyde-3-phosphate dehydrogenase (GAPDH) was used as control. And each experiment was repeated at least three times. * $p < 0.05$, ** $p < 0.01$.

3.4. Cytokine Release from HCT-8 Cells under rGroEL Stimulation

The mechanism by which rGroEL induces cell apoptosis or cell necrosis was further investigated by using an immunoenzymatic method to measure the pro-inflammatory cytokines, including TNF-α, IL-6 and IL-8, in the cell supernatant. As shown in Table 1, 3 μg/mL rGroEL could stimulate HCT-8 cells to produce 439.35 ± 14.71 pg/mL TNF-α, which was 33.68% higher than that of the control group ($p < 0.0001$). Meanwhile, 13.29 ± 0.57 pg/mL IL-6 and 11.23 ± 0.64 pg/mL IL-8 were released, which were 25.26% ($p < 0.0001$) and 110.69% ($p < 0.001$) higher than those of the control group, respectively. However, the production of the three cytokines could not be significantly ($p > 0.05$) improved when the HCT-8 cells were stimulated with rGroEL treated with a specific anti-GroEL monoclonal antibody.

Table 1. Determination of pro-inflammatory factors released by HCT-8 cells under the stimulation of GroEL.

Pro-Inflammatory Cytokines (pg/mL)	Control	rGroEL	rGroEL + Anti-GroEL
TNF-α	328.66 ± 21.36	439.35 ± 14.71	344.97 ± 28.47
IL-6	10.61 ± 0.38	13.29 ± 0.57	10.89 ± 1.32
IL-8	5.33 ± 0.55	11.23 ± 0.64	6.23 ± 1.43

3.5. CRISPRi-Mediated grol Gene Silencing and Bacterial Viability

In order to further examine the virulent roles of GroEL, the *grol* gene was silenced in *C. sakazakii* using the CRISPRi method. Expression of the GroEL protein was detected using SDS-PAGE. As shown in Figure 2C,D, the CRISPRi-treated group versus the untreated group, 80% downregulation in the GroEL protein expression was observed based on the analysis of the Western blot gray intensity. In order to add the same amount of viable bacteria in the following experiments, we measured the bacterial viability using the MTT assay after CRISPRi treatment. As shown in Figure S2B, the bacterial viability of the CRISPRi-treated group was 327.35 ± 19.34 MRU/mL OD_{600}, and there was no significant difference in viability among all the groups in the logarithmic growth phase, suggesting that *grol* gene silencing did not dramatically decrease the bacterial viability and that adding an equal amount of bacteria in the subsequent experiments could ensure the same viability for different bacterial groups.

3.6. Effects of GroEL on C. sakazakii Adhesion and Invasion

The foregoing adhesion-based analysis suggested that GroEL could interact with HCT-8 cells. Whether GroEL is an important participant in the interaction with HCT-8 cells in active *C. sakazakii* cells was determined by comparing the adhesion or invasion of the *grol* gene knockout strain with the control groups. As shown in Figure 5A, only 2.3×10^3 CFU/well of CRISPRi-treated bacteria could adhere to host cells, accounting for 31.84% ($p < 0.001$) of the untreated group. In terms of bacterial invasion (Figure 5B), 4.18×10^6 CFU/well of bacteria invaded the HCT-8 cells, covering 57.1% ($p < 0.001$) of the untreated group.

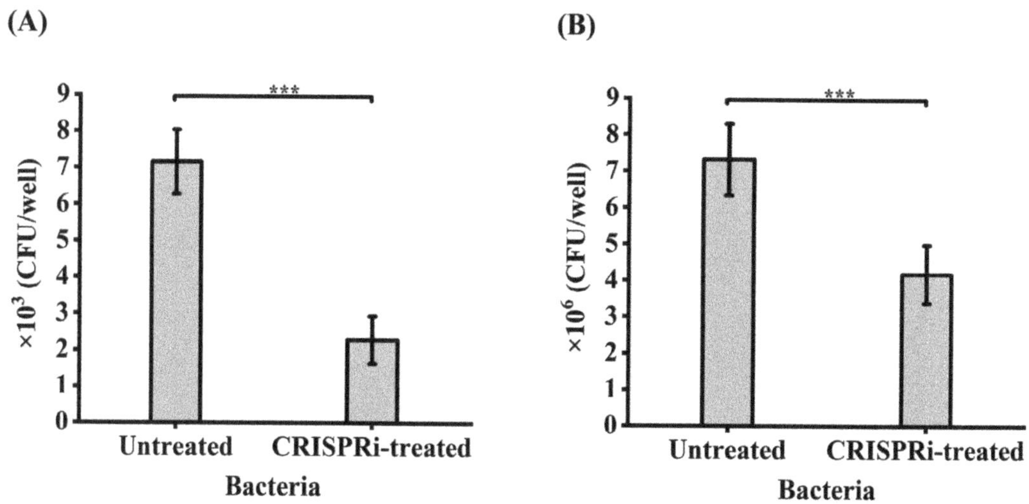

Figure 5. Adhesion (**A**) and invasion (**B**) of bacteria to HCT-8 cells. Each experiment was repeated three times. *** $p < 0.001$.

3.7. Effects of GroEL on Biofilm Formation, Motility and Morphology of C. sakazakii

The biofilm formation of *C. sakazakii* was evaluated through SYBR Green I staining. The results are shown in Figure 6A, the biofilm showed green fluorescence under 3D confocal scanning microscopy. The biofilm in the CRISPR-treated group showed a scattered form and faint green fluorescence, in contrast to a bright and compact biofilm in the untreated group. Through grayscale analysis, the fluorescence of the CRISPRi-treated group was approximately 57.14% that of the untreated group ($p < 0.001$), indicating the defective biofilm forming capacity of the CRISPRi-treated group. The effect of *grol* gene silencing on bacterial motility was investigated by adding an equal amount of different bacteria to semi-solid LB medium and culturing for 12 h. As shown in Figure 6B, the CRISPRi-treated group exhibited defective motility at 37 °C. As for bacterial morphology, the SEM images of bacteria were taken from the visual field at 4700× magnification (Figure 6C). Compared with the control group, the CRISPRi-treated group showed filamentation, with the number of abnormal bacteria accounting for about 3–5% and the length of the filamentous cell reaching about 10–40 μm.

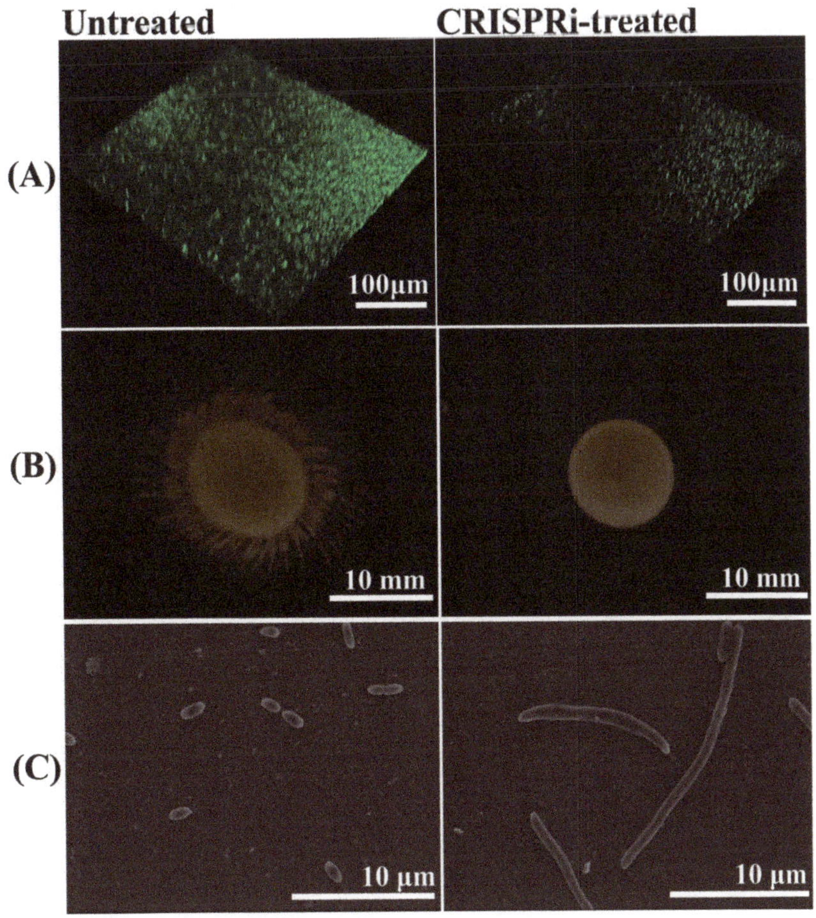

Figure 6. (**A**) Quantification of biofilm via SYBR Green I stain; (**B**) bacterial motility assay at 37 °C; (**C**) bacterial morphological observation through SEM.

4. Discussion

C. sakazakii is a redefined pathogen, and its virulence determinants are not well understood, although several common virulence factors studied in Gram-negative bacteria have been shown to affect the pathogenicity of *C. sakazakii* [8–14]. In this study, we elucidated that the GroEL protein was a potential virulence factor of *C. sakazakii* from three aspects including the secretion of GroEL, its virulence mechanism and its effects on bacterial character.

Most research has focused on the chaperone activity of GroEL in recent years, while in this study, the protein was found have the capability to overcome terrible obstacles composed by the *C. sakazakii* envelope and may act as a potent virulence factor in disease initiation. It is necessary that most GroEL is retained in the cytoplasm because the bacteria require the chaperone activity of GroEL to maintain protein homeostasis in the cytoplasm. Other fractions associated with GroEL indicate the complicated process of secretion of this protein. However, GroEL was mainly exported as a monomer, although trace amounts of dimeric GroEL (120 kDa) were detected in the OMV fraction (Figures 2 and S1). Zhao has reported that GroEL secretion requires an intact dimeric protein complex in *Bacillus subtilis* [31]; however, we consider that the dimeric GroEL should stem from slow reconstitution of the monomer because the GroEL sequence contains the information for its folding, assembly, and function [32]. Furthermore, *C. sakazakii's* exported monomeric GroEL appears to be more energy efficient and more flexible to recognize host receptors than the bulky dimer. The GroEL in the ultracentrifuged supernatant fraction had the same molecular weights as that in the OMV fraction, including 50, 60 and 70 kDa, while the 50 kDa GroEL was absent in the outer membrane fraction. We speculate that the 70 kDa band may contain an unknown 10 kDa protein to assist GroEL transmembrane transport because GroEL does not carry the classical signal peptide and anchor sequence. The 50 kDa band may be a degradation product. It has been reported that a GroEL nanocage could bind to the cytoplasmic membrane protein SecA, which is the core protein of the Sec system, and participate in transmembrane transport of secretory proteins by promoting secA release from the membrane in *E. coli* [33]. We consider that the misfolded GroEL might be unfolded by the GroEL–GroES nanocage and captured by SecA and subsequently secreted by the Sec system. The exported GroEL was able to exist on the bacterial surface and thus strengthen the colonization intensity of *C. sakazakii*.

To date, the toxicity of GroEL in prokaryotic organisms has been studied preliminarily only in several bacteria that cause specific tissue diseases, such as *Porphyromonas gingivalis*, in which GroEL could aggravate the alveolar inflammation and bone loss of rats, leading to the occurrence of periodontal disease [34], and *Mycobacterium tuberculosis*, in which the *grol* gene mutant could not produce granulomatous inflammation in animal experiments [35]. To our knowledge, no reports are available regarding the GroEL virulent effects of *C. sakazakii*. In this study, the *C. sakazakii* GroEL was proved to have dose-dependent virulent effects on HCT-8 cells. Furthermore, *C. sakazakii* GroEL could activate the NF-κB signaling pathway to produce more pro-inflammatory cytokines, including TNF-α, IL-6 and IL-8, which could promote a serious inflammatory reaction and hamper intestinal mucosal barrier function. Previous studies have shown that the levels of IL-6 and IL-8 were significantly increased in intestinal epithelial cells after barrier disruption [36,37]. The increased release of these cytokines, especially pleiotropic cytokine TNF-α with the potential to induce apoptosis, was closely related to the expression and the cellular redistribution of epithelial junctional proteins [38], facilitating myosin light-chain kinase (MLCK)–mediated opening of the epithelial barrier [39,40]. Here, the stimulation of *C. sakazakii* GroEL also caused the low expression of some tight-junction proteins including claudin-1, occluding, ZO-1 and ZO-2 in HCT-8 cells. The cell barrier became permeable with the reduction in the tight-junction proteins, especially at inflammatory sites [41]. These reports indicate that *C. sakazakii* GroEL may induce the dysfunction of intestinal epithelial cells by stimulating host cells to produce excessive pro-inflammatory cytokines, leading to the occurrence of bacterial translocation and inflammatory bowel diseases [40].

The GroEL–GroES system could capture more than 250 substrate proteins, and most of these proteins are involved in bacterial survival or pathogenicity [42]. When the expression of GroEL was inhibited to about 80%, *C. sakazakii* vitality was not affected, and only a slower growth of bacteria was observed (Figure S2A), indicating a small amount of GroEL formed nanocages to sustain the survival of the bacteria. Biofilms have been shown to be implicated in the infection and environmental persistence of *C. sakazakii* [1], and defective biofilm formation was also found in the CRISPRi-treated group in this study. The protein kinase PrkC regulates the phosphorylation of GroEL, which contributes to the formation of GroEL-GroES nanocage, and then the GroEL nanocage mediates the folding of relevant phosphokinases, thereby promoting bacteria to form more abundant biofilms [43,44]. This beneficial cycle is damaged by *grol* gene silencing, which impairs the formation of biofilms in *C. sakazakii*. The CRISPRi-treated group also showed defective motility despite no involvement of GroEL in the folding of flagellin [42]. However, there was an abundant inclusion body of DnaK/DnaJ protein on the low expression of GroEL in *E. coli*. The DnaK/DnaJ/GrpE system held an extensive substrate network, and studies have shown that the DnaK mutant of *Clostridium difficile* lacks flagella and motility [45,46]. Therefore, the decreased motility in the CRISPRi-treated *C. sakazakii* may be caused by the impaired DnaK/DnaJ/GrpE system. The *grol* gene silencing also aggravated *C. sakazakii* filamentation, which may be attributed to the cell-division proteins FtsE [47]. Additionally, the bulkily filamentous cells will further damage the flexibility of the bacteria and thus impair the motility of *C. sakazakii*. However, the production of LPS and lipid A were not affected (Figure S3A,B) despite the involvement of some substrate proteins of GroEL in carbohydrate/lipid transport and metabolism [39]. In short, the abovementioned physiological defects caused by deficient GroEL will weaken the pathogenicity of *C. sakazakii*.

5. Conclusions

In conclusion, this study proved that the GroEL protein is a potential virulence factor. *C. sakazakii* could mainly export monomeric GroEL protein through its OMVs. The exported GroEL could exist on the surface of the bacteria and was able to quickly adhere to human enterocyte-like epithelial cells, indicating that GroEL contributed to bacterial adhesion and colonization. GroEL could individually activate the NF-κB signaling pathway to induce the inflammatory response and downregulated the expression of some tight-junction proteins, ultimately causing necrosis of the host cell, which would be beneficial for the bacteria to cross the intestinal barrier. In addition, other properties associated with the pathogenicity of the bacteria, including biofilm formation, motility and morphological character, were controlled by the expression of GroEL. These abovementioned facts indicate that GroEL is important for bacterial pathogenicity.

Supplementary Materials: The following supporting information can be downloaded at: https://www.mdpi.com/article/10.3390/foods12183404/s1. Figure S1: The quaternary structure of GroEL in various fractions. Based on the content of GroEL in the cytoplasm, the proportion of GroEL in other fractions is calculated through gray-scale scanning, so as to calculate the loading quantity of other fractions of protein samples. Lane 1, cytoplasm (10 μg); Lne 2, cytoplasmic membrane (31 μg); Lane 3, outer membrane (125 μg); Lane 4, periplasm (19 μg); Lane 5, ultracentrifuged supernatant (64 μg); Lane 6, OMV (13 μg). Figure S2: (A) Bacterial growth curve (37 °C); (B) MTT experiment assays of the viability vitality of logarithmic-phase bacteria which is in logarithmic phase. ns: no significant difference. Figure S3: Determination of the production of LPS, and two groups of LPS samples with the equal amounts were loaded. (A) Quantification of LPS production; (B) Silver staining assays of LPS components. Lane1, Untreated group; Lane 2, CRISPRI-treated group. ns: no significant difference. Table S1: Primers used in this study.

Author Contributions: Conceptualization, S.W. and X.D.; Validation, D.Z., Y.F., X.W., Y.H., J.J., C.Z. and Y.L.; Data curation, D.Z.; Writing—original draft, D.Z.; Writing—review & editing, D.Z., P.L. and X.D. All authors have read and agreed to the published version of the manuscript.

Funding: This work was supported by the National Natural Science Foundation of China (31972167), the Key R&D program of Hebei Province (20372801D) and the National Key R&D Program of China (2018YFC1603800).

Data Availability Statement: The data used to support the findings of this study can be made available by the corresponding author upon request.

Conflicts of Interest: The authors declare no conflict of interest.

References

1. Henry, M.; Fouladkhah, A. Outbreak history, biofilm formation, and preventive measures for control of *Cronobacter sakazakii* in infant formula and infant care settings. *Microorganisms* **2019**, *7*, 77. [CrossRef] [PubMed]
2. Lai, K.K. *Enterobacter sakazakii* infections among neonates, infants, children, and adults. *Medicine* **2001**, *80*, 113–122. [CrossRef] [PubMed]
3. Jackson, E.E.; Flores, J.; Fernández-Escartín, E.P.; Forsythe, S.J. Reevaluation of a suspected *Cronobacter sakazakii* outbreak in Mexico. *J. Food Prot.* **2015**, *78*, 1191–1196. [CrossRef]
4. Bowen, A.; Wiesenfeld, H.C.; Kloesz, J.L.; Pasculle, A.W.; Nowalk, A.J.; Brink, L.; Elliot, E.; Martin, H.; Tarr, C.L. Notes from the field: *Cronobacter sakazakii* infection associated with feeding extrinsically contaminated expressed human milk to a premature infant—Pennsylvania, 2016. *MMWR. Morb. Mortal. Wkly. Rep* **2017**, *66*, 761–762. [CrossRef]
5. Gurtler, J.B.; Kornacki, J.L.; Beuchat, L.R. *Enterobacter sakazakii*: A coliform of increased concern to infant health. *Int. J. Food. Microbiol.* **2005**, *104*, 1–34. [CrossRef]

6. Li, Q.; Li, C.; Chen, L.; Cai, Z.; Wu, S.; Gu, Q.; Zhang, Y.; Wei, X.; Zhang, J.; Yang, X.; et al. *Cronobacter* spp. isolated from quick-frozen foods in china: Incidence, genetic characteristics, and antibiotic resistance. *Foods* **2023**, *12*, 3019. [CrossRef]
7. Haston, J.C.; Miko, S.; Cope, J.R.; McKeel, H.; Walters, C.; Joseph, L.A.; Griswold, T.; Katz, L.S.; Andújar, A.A.; Tourdot, L.; et al. *Cronobacter sakazakii* infections in two infants linked to powdered infant formula and breast pump equipment—United States, 2021 and 2022. *MMWR. Morb. Mortal. Wkly. Rep.* **2023**, *72*, 223–226. [CrossRef]
8. Chandrapala, D.; Kim, K.; Choi, Y.; Senevirathne, A.; Kang, D.H.; Ryu, S.; Kim, K.-P. Putative Inv is essential for basolateral invasion of Caco-2 cells and acts synergistically with OmpA to affect in vitro and in vivo virulence of *Cronobacter sakazakii* ATCC 29544. *Infect. Immun.* **2014**, *82*, 1755–1765. [CrossRef]
9. Pagotto, F.J.; Nazarowec-White, M.; Bidawid, S.; Farber, J.M. *Enterobacter sakazakii*: Infectivity and enterotoxin production in vitro and in vivo. *J. Food. Prot.* **2003**, *66*, 370–375. [CrossRef]
10. Cruz-Córdova, A.; Rocha-Ramírez, L.M.; Ochoa, S.A.; González-Pedrajo, B.; Espinosa, N. and Eslava, C. Flagella from five *Cronobacter* species induce pro-inflammatory cytokines in macrophage derivatives from human monocytes. *PLoS ONE* **2012**, *7*, e52091. [CrossRef] [PubMed]
11. Kim, K.; Kim, K.P.; Choi, J. Outer membrane proteins A (OmpA) and X (OmpX) are essential for basolateral invasion of *Cronobacter sakazakii*. *Appl. Environ. Microbiol.* **2010**, *76*, 5188–5198. [CrossRef] [PubMed]
12. Hu, L.; Grim, C.J.; Franco, A.A.; Jarvis, K.G.; Sathyamoorthy, V.; Kothary, M.H.; McCardell, B.A.; Tall, B.D. Analysis of the cellulose synthase operon genes, bcsA, bcsB, and bcsC in *Cronobacter* species: Prevalence among species and their roles in biofilm formation and cell-cell aggregation. *Food. Microbiol.* **2015**, *52*, 97–105. [CrossRef] [PubMed]
13. Choi, Y.; Kim, S.; Hwang, H.; Kim, K.P.; Kang, D.H.; Ryu, S. Plasmid-encoded MCP is involved in virulence, motility, and biofilm formation of *Cronobacter sakazakii* ATCC 29544. *Infect. Immun.* **2015**, *83*, 197–204. [CrossRef]
14. Kim, S.; Yoon, H.; Ryu, S. New virulence factor CSK29544_02616 as LpxA binding partner in *Cronobacter sakazakii*. *Sci. Rep.* **2018**, *8*, 835. [CrossRef] [PubMed]
15. Li, M.; Rosenshine, I.; Tung, S.L.; Wang, X.H.; Friedberg, D.; Hew, C.L. Comparative proteomic analysis of extracellular proteins of enterohemorrhagic and enteropathogenic *Escherichia coli* strains and their ihf and ler mutants. *Appl. Environ. Microbiol.* **2004**, *70*, 5274–5282. [CrossRef]
16. Aguilera, L.; Ferreira, E.; Giménez, R.; Fernández, F.J.; Taulés, M.; Aguilar, J.; Vega, M.C.; Badia, J.; Baldomà, L. Secretion of the housekeeping protein glyceraldehyde-3-phosphate dehydrogenase by the LEE-encoded type III secretion system in enteropathogenic *Escherichia coli*. *Int. J. Biochem. Cell Biol.* **2012**, *44*, 955–962. [CrossRef] [PubMed]
17. Terrasse, R.; Amoroso, A.; Vernet, T.; Di Guilmi, A.M. *Streptococcus pneumoniae* GAPDH is released by cell lysis and interacts with peptidoglycan. *PLoS ONE* **2015**, *10*, e0125377. [CrossRef]
18. Ebner, P.; Luqman, A.; Reichert, S.; Hauf, K.; Popella, P.; Forchhammer, K.; Otto, M.; Götz, F. Non-classical protein excretion is boosted by PSMα-induced cell leakage. *Cell Rep.* **2017**, *20*, 1278–1286. [CrossRef] [PubMed]
19. Daubenspeck, J.M.; Liu, R.; Dybvig, K. Rhamnose links moonlighting proteins to membrane phospholipid in mycoplasmas. *PLoS ONE* **2016**, *11*, e0162505. [CrossRef]
20. Wang, J.; Du, X.J.; Lu, X.N.; Wang, S. Immunoproteomic identification of immunogenic proteins in *Cronobacter sakazakii* strain BAA-894. *Appl. Microbiol. Biotechnol.* **2013**, *97*, 2077–2091. [CrossRef]
21. Yan, X.; Shi, Q.; Bracher, A.; Miličić, G.; Singh, A.K.; Hartl, F.U. GroEL ring separation and exchange in the chaperonin reaction. *Cell* **2018**, *172*, 605–617. [CrossRef]
22. Chapman, E.; Farr, G.W.; Usaite, R.; Furtak, K.; Fenton, W.A.; Chaudhuri, T.K.; Hondorp, E.R.; Matthews, R.G.; Wolf, S.G.; Yates, J.R.; et al. Global aggregation of newly translated proteins in an *Escherichia coli* strain deficient of the chaperonin GroEL. *Proc. Natl. Acad. Sci. USA* **2006**, *103*, 15800–15805. [CrossRef]
23. Pantzar, M.; Teneberg, S.; Lagergård, T. Binding of *Haemophilus ducreyi* to carbohydrate receptors is mediated by the 58.5-kDa GroEL heat shock protein. *Microbes. Infect.* **2006**, *8*, 2452–2458. [CrossRef]
24. Nalbant, A.; Kant, M. Bacterial heat shock protein GroEL (Hsp64) exerts immunoregulatory effects on T Cells by utilizing apoptosis. *PLoS ONE* **2016**, *11*, e0164380. [CrossRef]
25. Hagiwara, M.; Kurita-Ochiai, T.; Kobayashi, R.; Hashizume-Takizawa, T.; Yamazaki, K.; Yamamoto, M. Sublingual vaccine with GroEL attenuates atherosclerosis. *J. Dent. Res.* **2014**, *93*, 382–387. [CrossRef] [PubMed]
26. Goulhen, F.; Hafezi, A.; Uitto, V.J.; Hinode, D.; Nakamura, R.; Grenier, D.; Mayrand, D. Subcellular localization and cytotoxic activity of the GroEL-like protein isolated from *Actinobacillus actinomycetemcomitans*. *Infect. Immun.* **1998**, *66*, 5307–5313. [CrossRef] [PubMed]
27. Qamra, R.; Srinivas, V.; Mande, S.C. *Mycobacterium tuberculosis* GroEL homologues unusually exist as lower oligomers and retain the ability to suppress aggregation of substrate proteins. *J. Mol. Biol.* **2004**, *342*, 605–617. [CrossRef] [PubMed]
28. Song, J.R.; Fu, Y.W.; Li, P.; Du, T.; Du, X.J.; Wang, S. *Cronobacter sakazakii* protective effect of recombinant proteins of during pregnancy on the offspring. *Front. Cell Infect. Microbiol.* **2020**, *10*, 15. [CrossRef] [PubMed]
29. Qi, L.S.; Larson, M.H.; Gilbert, L.A.; Doudna, J.A.; Weissman, J.S.; Arkin, A.P.; Lim, W.A. Repurposing CRISPR as an RNA-guided platform for sequence-specific control of gene expression. *Cell* **2013**, *152*, 1173–1183. [CrossRef]
30. Wang, H.; Cheng, H.; Wang, F.; Wei, D.; Wang, X. An improved 3-(4,5-dimethylthiazol-2-yl)-2,5-diphenyl tetrazolium bromide (MTT) reduction assay for evaluating the viability of *Escherichia coli* cells. *J. Microbiol. Methods* **2010**, *82*, 330–333. [CrossRef]

31. Zhao, L.; Chen, J.; Sun, J.; Zhang, D. Multimer recognition and secretion by the non-classical secretion pathway in *Bacillus subtilis*. *Sci. Rep.* **2017**, *7*, 44023. [CrossRef]
32. Ybarra, J.; Horowitz, P.M. Refolding and reassembly of active chaperonin GroEL after denaturation. *J. Biol. Chem* **1995**, *270*, 22113–22115. [CrossRef]
33. Bochkareva, E.S.; Solovieva, M.E.; Girshovich, A.S. Targeting of GroEL to SecA on the cytoplasmic membrane of *Escherichia coli*. *Proc. Natl. Acad. Sci. USA* **1998**, *95*, 478–483. [CrossRef]
34. Lin, F.Y.; Hsiao, F.P.; Huang, C.Y.; Shih, C.M.; Tsao, N.W.; Yang, S.F.; Chang, N.C.; Hung, S.L.; Lin, Y.-W.; Tsai, C.S. *Porphyromonas gingivalis* GroEL induces osteoclastogenesis of periodontal ligament cells and enhances alveolar bone resorption in rats. *PLoS ONE* **2014**, *9*, e102450. [CrossRef]
35. Hu, Y.; Henderson, B.; Lund, P.A.; Tormay, P.; Ahmed, M.T.; Gurcha, S.S.; Besra, G.S.; Anthony, R.M. *Mycobacterium tuberculosis* mutant lacking the groEL homologue cpn60.1 is viable but fails to induce an inflammatory response in animal models of infection. *Infect. Immun.* **2008**, *76*, 1535–1546. [CrossRef]
36. Dann, S.M.; Spehlmann, M.E.; Hammond, D.C.; Iimura, M.; Hase, K.; Choi, L.J.; Hanson, E.; Eckmann, L. IL-6-dependent mucosal protection prevents establishment of a microbial niche for attaching/effacing lesion-forming enteric bacterial pathogens. *J. Immunol.* **2008**, *180*, 6816–6826. [CrossRef]
37. Akbari, P.; Braber, S.; Alizadeh, A.; Verheijden, K.A.; Schoterman, M.H.; Kraneveld, A.D.; Fink-Gremmels, J. Galacto-oligosaccharides Protect the intestinal barrier by maintaining the tight junction network and modulating the inflammatory responses after a challenge with the mycotoxin deoxynivalenol in human Caco-2 cell monolayers and B6C3F1 mice. *J. Nutr.* **2015**, *145*, 1604–1613. [CrossRef]
38. Grabinger, T.; Bode, K.J.; Demgenski, J.; Seitz, C.; Delgado, M.E.; Kostadinova, F.; Reinhold, C.; Etemadi, N.; Wilhelm, S.; Schweinlin, M.; et al. Inhibitor of apoptosis protein-1 regulates tumor necrosis factor-mediated destruction of intestinal epithelial cells. *Gastroenterology* **2017**, *152*, 867–879. [CrossRef]
39. Marchiando, A.M.; Shen, L.; Graham, W.V.; Edelblum, K.L.; Duckworth, C.A.; Guan, Y.; Montrose, M.H.; Turner Jerrold, R.; Watson, A.J.M. The epithelial barrier is maintained by in vivo tight junction expansion during pathologic intestinal epithelial shedding. *Gastroenterology* **2011**, *140*, 1208–1218. [CrossRef]
40. Drolia, R.; Tenguria, S.; Durkes, A.C.; Turner, J.R.; Bhunia, A.K. *Listeria* adhesion protein induces intestinal epithelial barrier dysfunction for bacterial translocation. *Cell Host. Microbe* **2018**, *23*, 470–484. [CrossRef]
41. Noth, R.; Lange-Grumfeld, J.; Stuber, E.; Kruse, M.L.; Ellrichmann, M.; Hasler, R.; Hampe, J.; Bewig, B.; Rosenstiel, P.; Schreiber, S.; et al. Increased intestinal permeability and tight junction disruption by altered expression and localization of occludin in a murine graft versus host disease model. *BMC. Gastroenterol.* **2011**, *11*, 109. [CrossRef]
42. Kerner, M.J.; Naylor, D.J.; Ishihama, Y.; Maier, T.; Chang, H.C.; Stines, A.P.; Georgopoulos, C.; Frishman, D.; Hayer-Hartl, M.; Mann, M.; et al. Proteome-wide analysis of chaperonin-dependent protein folding in *Escherichia coli*. *Cell* **2005**, *122*, 209–220. [CrossRef]
43. Arora, G.; Sajid, A.; Virmani, R.; Singhal, A.; Kumar, C.M.S.; Dhasmana, N.; Khanna, T.; Maji, A.; Misra, R.; Molle, V.; et al. Ser/Thr protein kinase PrkC-mediated regulation of GroEL is critical for biofilm formation in *Bacillus anthracis*. *NPJ. Biofilms. Microbiomes* **2017**, *3*, 7. [CrossRef] [PubMed]
44. Virmani, R.; Singh, Y.; Hasija, Y. *Bacillus anthracis* GroEL Mediates Folding of Serine/Threonine Protein Kinase, PrkC. *Indian. J. Microbiol* **2018**, *58*, 520–524. [CrossRef]
45. Calloni, G.; Chen, T.; Schermann, S.M.; Chang, H.C.; Genevaux, P.; Agostini, F.; Tartaglia, G.G.; Hayer-Hartl, M.; Hartl, F.U. DnaK functions as a central hub in the *E. coli* chaperone network. *Cell Rep.* **2012**, *1*, 251–264. [CrossRef] [PubMed]
46. Jain, S.; Smyth, D.; O'Hagan, B.M.; Heap, J.T.; McMullan, G.; Minton, N.P.; Ternan, N.G. Inactivation of the dnaK gene in *Clostridium difficile* 630 Δerm yields a temperature-sensitive phenotype and increases biofilm-forming ability. *Sci. Rep.* **2017**, *7*, 17522. [CrossRef]
47. Fujiwara, K.; Taguchi, H. Filamentous morphology in GroE-Depleted *Escherichia coli* induced by impaired folding of FtsE. *J. Bacteriol.* **2007**, *189*, 5860–5866. [CrossRef]

foods

MDPI

Article

The Genotyping Diversity and Hemolytic Activity of *Cronobacter* spp. Isolated from Plant-Based Food Products in Poland

Monika Garbowska [1,*], Anna Berthold-Pluta [1], Lidia Stasiak-Różańska [1], Antoni Pluta [1], Stephen Forsythe [2] and Ilona Stefańska [3]

[1] Department of Technology and Food Assessment, Division of Milk Technology, Institute of Food Sciences, Warsaw University of Life Sciences—SGGW, 02-787 Warsaw, Poland; anna_berthold@sggw.edu.pl (A.B.-P.); lidia_stasiak_rozanska@sggw.edu.pl (L.S.-R.); antoni_pluta@sggw.edu.pl (A.P.)

[2] FoodMicrobe.com, Keyworth, Nottinghamshire NG12 5GY, UK; sforsythe4j@gmail.com

[3] Department of Preclinical Sciences, Institute of Veterinary Medicine, Warsaw University of Life Sciences—SGGW, 02-787 Warsaw, Poland; ilona_stefanska@sggw.edu.pl

* Correspondence: monika_garbowska@sggw.edu.pl

Abstract: The present study aimed to determine the genotyping diversity and hemolytic properties of 24 strains of *Cronobacter* spp. (15 *Cronobacter sakazakii*, 6 *Cronobacter malonaticus*, 2 *Cronobacter turicensis*, and 1 *Cronobacter condimenti*) isolated from commercial ready-to-eat leaf vegetables, sprouts, nuts, and dried fruits. The multilocus sequence typing (MLST) method was used to determine the sequence types (ST) and clonal complexes (CC) of these strains. The study demonstrated the high genotypic diversity of the *Cronobacter* genus bacteria isolated from plant-based foods. Five novel sequence types (804, 805, 806, 807, and 808) and the presence of novel alleles in the *ppsA*, *gltB*, *gyrB*, and *infB* loci were detected. In total, 16 of the 24 strains were assigned to the sequence types ST99, ST258, ST17, ST648, ST21, ST494, and ST98. One *C. sakazakii* strain (s12) isolated from alfalfa sprouts was assigned to the clonal complex CC4, which encompasses strains often associated with severe infections leading to meningitis in infants. In addition, 87.5% and 16.7% of the *Cronobacter* spp. strains showed β-hemolysis of equine and sheep red blood cells, respectively. The presence of the pathogenic species *C. sakazakii*, *C. malonaticus*, and *C. turicensis* in ready-to-eat plant-derived food products shows they are potential sources of infection, especially to those with compromised immunity, which substantiates their further multi-faceted characterization. The significance of this study may prove useful not only in epidemiological investigations, but also in assessing the risk of infections caused by the presence of *Cronobacter*.

Keywords: *Cronobacter* spp.; MLST; hemolysis; sequence types; clonal complexes; sprouts; nuts; ready-to-eat leaf vegetables

check for updates

Citation: Garbowska, M.; Berthold-Pluta, A.; Stasiak-Różańska, L.; Pluta, A.; Forsythe, S.; Stefańska, I. The Genotyping Diversity and Hemolytic Activity of *Cronobacter* spp. Isolated from Plant-Based Food Products in Poland. *Foods* **2023**, *12*, 3873. https://doi.org/10.3390/foods12203873

Academic Editor: Xinjun Du

Received: 21 September 2023
Revised: 15 October 2023
Accepted: 19 October 2023
Published: 23 October 2023

1. Introduction

Cronobacter bacteria are motile, facultatively anaerobic, Gram-negative rods of the family *Enterobacteriaceae*. The history of this genus is relatively short, as it was only in 2007 that organisms previously classified on the basis of their phenotypic features as *Enterobacter sakazakii* (and even earlier as yellow-pigment-producing *Enterobacter cloacae*) were assigned to the new genus *Cronobacter*, then composed of four species. However, as knowledge of the diversity of the genus improved, it was expanded to the currently agreed composition of seven species. Of the seven *Cronobacter* species, *C. sakazakii*, *C. malonaticus*, and *C. turicensis* have been recognized as important human pathogens resulting in opportunistic infections [1,2]. These bacteria can cause severe, invasive infections in preterm infants, low-birth-weight infants, neonates hospitalized in intensive care, and immunocompromised infants, as well as immunocompromised adults and the elderly [2,3]. Infections caused by

Cronobacter spp. in infants are associated with a high mortality rate, ranging from 40 to 80% [4,5], with frequent neurological complications observed in 20% of convalescents [6,7]. *Cronobacter* bacteria cause bacteremia, necrotizing enterocolitis, meningitis, and sepsis in neonates and infants, and wound and urinary tract infections in adults [7,8]. In 2002, the International Commission for Microbiological Specification for Foods (ICMSF) placed *Cronobacter* (then known as *Enterobacter sakazakii*) in group I.B., which covers pathogenic bacteria that are a "severe hazard for restricted populations, life threatening or substantial chronic sequelae or long duration" [9]. The number of *Cronobacter* infections reported in the literature is relatively low and is most likely underestimated, as it is generally not a mandated notifiable disease [7,10].

The prevalence of *Cronobacter* spp. in various food products has been confirmed in investigations conducted across many countries [11–26], with *C. sakazakii* being the most frequently isolated species. The presence of pathogenic *C. sakazakii*, *C. malonaticus*, and *C. turicensis* in ready-to-eat food products (RTE) makes them putative sources of infections [16,19–23,27]. *Cronobacter* spp. have been isolated from a range of plant-origin foods, including ready-to-eat vegetables, cereals, and nuts [19,23,28]. Productive sources of *Cronobacter* strains are fresh or dried herbs and spices [29]. Although the bacterium has been isolated from various plant-based food products, no foodborne infections have been reported. The majority of *Cronobacter* infections occur in the adult population, but are less severe. Cases of *Cronobacter* infection in all age groups are probably under-reported for a number of reasons, such as misidentification as *Enterobacter cloacae* [29]. To date, very little is known about the genotyping diversity and hemolytic activity possessed by such plant-origin *Cronobacter*. The characterization of strains from plant-based foods with genomic features similar to clinically relevant strains of different *Cronobacter* spp. sequence types (STs) suggests that these foods may serve as potential vehicles for the transmission of opportunistic pathogens. Therefore, it is necessary to understand the genotyping diversity of *Cronobacter* spp. associated with plant-based foods to improve food safety.

As an intestinal pathogen, the main route of *Cronobacter* entry into the human body is through the consumption of contaminated food. These pathogens show a high tolerance to stressful environments, being one of the most heat-tolerant members of the *Enterobacteriaceae* family [5], extremely resistant to low water activity of the environment [30,31], able to tolerate acidic conditions as low as pH 4.2 [7] and produce biofilms that increase their survival under food production conditions [32]. After entering the host, the infection strategy consists of the following stages: (i) the colonization of the mucosa (intestinal, respiratory, or the urinary tract epithelia); (ii) the circumvention, subversion, and exploitation of host defenses; (iii) systemic spread and multiplication (within the blood or phagocytes); and (iv) host damage (through the expression of toxins and/or damage due to the proinflammatory modulation of the human immune system) [8].

Various hemolysins and hemolysis-related genes have been reported in *Cronobacter* genomes by Joseph et al. [4]. The hemolysin gene (*hly*) was present in all the genomes, with the only exceptions being *C. sakazakii* strain 701 and *C. malonaticus* strain 507. Most of the strains had two copies of the hemolysin gene and the hemolysin activator protein precursor gene [4]. Cruz et al. [33] identified the hemolysin gene (*hly*) as a hemolysin III homolog (COG1272), and Jang et al. [34], studying 390 strains, showed that all seven species tested possessed the hemolysin III COG1272 gene homolog. Additionally, three other hemolysin genes were identified, including genes encoding the cystathionine β-synthase (CBS) domain containing hemolysin, putative hemolysin, and 21 kDa hemolysin [34]. Umeda et al. [35] reported that all analyzed *Cronobacter* strains exhibited β-hemolytic activity against guinea pig, horse, and rabbit erythrocytes and that 92.9% of the strains were capable of the α-hemolysis of sheep erythrocytes. The further characterization of individual *Cronobacter* species, including strains occurring in food, is, however, needed to establish their hemolytic capacity. Results related to the hemolytic activity of *Cronobacter* spp. could extend the knowledge in this field, enabling the assessment of whether other closely related species commonly misidentified as *Cronobacter* (like *Franconibacter helveticus*, *Franconibacter pulveris*,

Siccibacter colletis, and *Siccibacter turicensis*) differ in their hemolytic phenotypes from the *Cronobacter* genus species [36].

Multi-locus sequence typing (MLST) is a genetic method recommended for the characterization, differentiation, and typing of many microorganisms. Conventional MLST genotypes bacterial strains according to loci (commonly ~500 nucleotides) for seven housekeeping genes, and it enables the recognition of the sequence types (ST) and clonal complexes (CC) of the tested *Cronobacter* strains and thus the assessment of their genetic diversity and clinical significance [4,37].

The genetic basis of the virulence of *Cronobacter* spp. strains has not been fully elucidated, but some sequence types have been found to be associated with specific types of infection [37]. Infant meningitis due to *Cronobacter* spp. is more frequently caused by strains of *C. sakazakii* belonging to the CC4 clonal complex (especially ST4), whereas *C. sakazakii* ST12 is strongly associated with cases of necrotizing enterocolitis in infants. Infections in children and adults are mainly caused by the clonal complexes of *C. sakazakii* CC4 and *C. malonaticus* CC7, respectively [2,37].

Since the severity of infections due to *Cronobacter* spp. may be related to the genotype of the strain that caused the infection, it is important to determine the STs of isolates from both clinical and food sources. Therefore, the present study aimed to identify the sequence types of *Cronobacter* spp. strains isolated from commercial ready-to-eat leaf vegetables, sprouts, nuts, and dried fruits. The genotypic analysis of the strains could prove useful not only in epidemiological investigations, but also in their risk assessments.

2. Materials and Methods

2.1. Materials

The study was conducted with 24 strains of *Cronobacter* spp., including 15 strains of *C. sakazakii* (9n, 10n, 11m, s12, s14, s21, s22, s41, s42, s44, s45, s47, s48, lv25, and lv27), 6 *C. malonaticus* (5n, 6n, 7n, 8n, 12m, and lv31), 2 *C. turicensis* (1n and lv54), and 1 *C. condimenti* (s37). These isolates were from the bacterial collection of the Division of Milk Technology, Warsaw University of Life Sciences, Poland. The isolation of these strains from plant-derived food products, including ready-to-eat leaf vegetables, sprouts, nuts, and dried fruits, was described in our previous studies (Table 1) [19,23].

Table 1. Strains of *Cronobacter* spp. used in the study and their origin.

Isolate	Species	Origin
s12	*C. sakazakii*	Alfalfa sprouts
9n	*C. sakazakii*	Brazilian nuts
10n	*C. sakazakii*	Brazilian nuts
14	*C. sakazakii*	Alfalfa sprouts
s44	*C. sakazakii*	Mix of sprouts
s45	*C. sakazakii*	Mix of sprouts
s21	*C. sakazakii*	Leek sprouts
s47	*C. sakazakii*	Mix of sprouts
s48	*C. sakazakii*	Mix of sprouts
s22	*C. sakazakii*	Leek sprouts
11m	*C. sakazakii*	Mixes of dried fruits, seeds, and nuts
lv25	*C. sakazakii*	Rucola
lv27	*C. sakazakii*	Endive escarola
s41	*C. sakazakii*	Sunflower sprouts
s42	*C. sakazakii*	Sunflower sprouts
5n	*C. malonaticus*	Hazelnuts
6n	*C. malonaticus*	Cashew nuts
7n	*C. malonaticus*	Pini nuts
8n	*C. malonaticus*	Macadamia nuts

Table 1. *Cont.*

Isolate	Species	Origin
12m	*C. malonaticus*	Mixes of dried fruits, seeds, and nuts
lv31	*C. malonaticus*	Lambs lettuce
1n	*C. turicensis*	Almonds
lv54	*C. turicensis*	Mix of leaf vegetables
s37	*C. condimenti*	Small radish sprouts

The *Cronobacter* spp. strains were stored frozen on Tryptone Soy Broth (TSB) (Oxoid Argenta, Poznań, Poland) with a 10% glycerol addition at a temperature of −40 °C. They were recovered from the frozen state by transferring 0.1 mL of a defrosted culture onto TSB of a given strain to 10 mL of sterile TSB, with incubation at a temperature of 35 °C for 24 h. Afterwards, each strain was inoculated onto Tryptone Soy Agar (TSA) medium (Oxoid Argenta, Poznań, Poland).

2.2. MLST Analysis

The MLST typing followed the methodology of Baldwin et al. [38]. The seven loci analyzed were glutaminyl tRNA synthetase gene (*glnS*), glutamate synthase large subunit gene (*gltB*), ATP synthase beta chain (*atpD*), DNA gyrase beta subunit (*gyrB*), phosphoenolpyruvate synthase A (*ppsA*), the gene encoding the translation initiation factor IF-2 (*infB*), and the gene encoding the translocase protein of the elongation factor EF-G (*fusA*). The gene fragments were amplified using primers and PCR conditions according to the protocol available in the *Cronobacter* MLST database https://pubmlst.org/organisms/cronobacter-spp/primers (accessed on 25 May 2020). The genomic DNA was isolated using the GenElute Bacterial Genomic DNA Kit (Sigma Aldrich, Poznań, Poland), according to the manufacturer's instructions. The PCR and sequencing primers were synthesized at Eurofins Genomics (Ebersberg, Germany). The PCR was performed using the Phusion High-Fidelity PCR Master Mix with the HF buffer kit (ThermoFisher Scientific, Poland), in a total volume of 50 μL containing 20–40 ng of template DNA and 10 pmol of each primer. The polymerase used in the study was characterized by a 50-fold lower error insertion frequency compared to the standard Taq polymerase. The amplification products were purified using the GenElute PCR Clean-Up Kit (Sigma-Aldrich) or EXOSAP (ThermoFisher Scientific). The purified amplicons, obtained using standard and alternative primers (https://pubmlst.org/organisms/cronobacter-spp/primers and https://pubmlst.org/static/organisms/cronobacter-spp/Cronobacter_alternative_primers.pdf) (accessed on 20 June 2020), were sequenced in both directions (Eurofins Genomics). Afterwards, the obtained nucleotide sequences were compared with the sequences deposited in the *Cronobacter* MLST database (https://pubmlst.org/cronobacter/) (accessed on 17 October 2020). Alleles were determined for the seven loci, which identified their sequence type and clonal complex. Novel alleles and sequence types were assigned by the MLST database curator, Prof. Stephen Forsythe.

2.3. Hemolytic activity of Cronobacter spp. Strains

Hemolytic capability was determined on Columbia agar plates with a 5% addition of horse blood (COH) or sheep blood (COS) (Biomerieux, Warsaw Poland). Various hemolytic types were identified using the following reference strains: *Streptococcus pneumoniae* ATCC 6305, *Bacillus cereus* ATCC 14579, and *Listeria innocua* ATCC 33090 (Oxoid Argenta, Poznań Poland).

Single colonies of *Cronobacter* bacteria grown on TSA medium were transferred using a sterile, disposable loop to a blood agar plates and incubated at 37 °C. The hemolysis zones on the plates were measured after 96 h of incubation. The appearance of a green zone around the colony was recorded as α-hemolysis, a transparent zone around the colony was recorded as β-hemolysis, and no changes on the plate was defined as γ-hemolysis [39].

3. Results and Discussion

3.1. MLST Analysis

The MLST analysis results confirmed the earlier identification of all the tested strains obtained using RFLP-PCR: 15 strains belonged to the *C. sakazakii* species, 6 strains to the *C. malonaticus* species, 2 strains to the *C. turicensis* species, and one strain to the *C. condimenti* species. Interestingly, this profiling was consistent with the results of earlier intraspecies strain differentiation conducted with the RAPD-PCR method, indicating that the strains sharing the same RAPD pattern concurred with the same sequence type [19,23].

The present study showed that the 24 strains of *Cronobacter* spp. isolated from plant-derived foods (ready-to-eat leaf vegetables, sprouts, nuts, and dried fruits) belonged to 14 different sequence types (ST) (Table 2).

Table 2. Comparison of 7-loci MLST for 24 strains from 4 *Cronobacter* species.

PubMLST ID	Isolate	Species	*atpD*	*fusA*	*glnS*	*gltB*	*gyrB*	*infB*	*ppsA*	ST	CC
4062	s12	*C. sakazakii*	5	1	3	3	5	5	4	4	4
4063	9n	*C. sakazakii*	3	12	16	5	16	20	14	17	17
4064	10n	*C. sakazakii*	3	12	16	5	16	20	14	17	17
4065	s14	*C. sakazakii*	3	11	13	18	11	17	13	21	21
4066	s44	*C. sakazakii*	3	8	52	54	21	65	73	99	99
4067	s45	*C. sakazakii*	3	8	52	54	21	65	73	99	99
4068	s21	*C. sakazakii*	3	8	52	54	21	65	73	99	99
4069	s47	*C. sakazakii*	3	8	52	54	21	65	73	99	99
4070	s48	*C. sakazakii*	3	8	52	54	21	65	73	99	99
4071	s22	*C. sakazakii*	3	8	52	54	21	65	73	99	99
4072	11m	*C. sakazakii*	11	8	24	220	15	56	261	494	-
4073	lv25	*C. sakazakii*	175	1	120	275	21	234	333	648	-
4074	lv27	*C. sakazakii*	175	1	120	275	21	234	333	648	-
3574	s41	*C. sakazakii*	3	1	120	94	270	1	368	**804**	-
3575	s42	*C. sakazakii*	3	1	120	94	270	1	368	**804**	-
4075	5n	*C. malonaticus*	89	13	107	8	10	35	160	258	-
4076	6n	*C. malonaticus*	89	13	107	8	10	35	160	258	-
4077	7n	*C. malonaticus*	89	13	107	8	10	35	160	258	-
4078	8n	*C. malonaticus*	89	13	107	8	10	35	160	258	-
3139	12m	*C. malonaticus*	64	7	64	7	10	16	**381**	**805**	-
3576	lv31	*C. malonaticus*	3	8	10	94	5	93	74	**807**	-
3573	1n	*C. turicensis*	46	147	42	21	237	193	318	**806**	-
3140	lv54	*C. turicensis*	46	5	4	**314**	**279**	**265**	**382**	**808**	-
1896	s37	*C. condimenti*	24	86	96	28	63	42	147	98	-

New assigned allele number and STs are underlined and in bold.

The main STs identified were *C. sakazakii* ST99 (40%, 6/15), which was isolated from various sprouts, and *C. malonaticus* ST258 (66.7%, 4/6), which was isolated from various nuts. Interestingly, out of the 24 strains, six novel sequence types were defined: ST806 and ST808 in the *C. turicensis* 1n and lv54 strains; ST805 and ST807 in the *C. malonaticus* 12m and lv31 strains; and ST804 in the *C. sakazakii* s41 and s42 strains. In the case of the *C. malonaticus* 12m strain, the analyses showed the presence of a novel allele in the *ppsA* locus and the closest match to the 355 allele with four differences (A231G, G291A, T387G, and T438C). The *C. turicensis* lv54 strain showed novel alleles in as many as four loci, i.e., *gltB*, *gyrB*, *infB*, and *ppsA*. The sequencing chromatograms indicates as the closest match *pps*159 (C48T, C57T, C405G, and T459C), *gltB*64 (A252G, T351C, and T480C), *infB*80 (C21G), and *gyrB*31 (C96T and T138C). The remaining detected sequence types included: ST4, ST17, ST648, ST21, ST494 (*C. sakazakii*), and ST98 (*C. condimenti*). Among the 15 *C. sakazakii* isolates, 4 (s12, 9n, 10m, and 11m,) had STs: ST4, ST17, and ST494, assigned to isolates from clinical sources in the PubMLST *Cronobacter* database (accessed: 10 May 2023).

It is notable that *C. sakazakii* strains belonging to the sequence type ST4 and the clonal complex CC4 are often associated with severe infections leading to meningitis in neonates, children, and adults. These strains have been isolated from milk powder, food products, ice creams, and powdered infant formulas [2,37]. In the present study, the only ST4 strain was *C. sakazakii* s12 (6.6%, 1/15), which was isolated from alfalfa sprouts. Wang et al. [40] analyzed 84 *Cronobacter* spp. isolates obtained from foods imported to Beijing in 2006–2015 and identified mainly the pathogenic sequence types ST1 and ST4 in the case of 31.67% (19/60) and 21.67% (13/60) of *C. sakazakii* strains, respectively, as well as the ST7 type in 70.59% (12/17) of *C. malonaticus* strains. The sequence type ST1 belongs to the clonal complex 1 of *C. sakazakii* (CC1), which is the second major ST in the PubMLST database after ST4 (CC4). ST4 (*C. sakazakii*) and ST7 (*C. malonaticus*) were the predominant STs identified by Li et al. [41] for the *Cronobacter* spp. isolated from wet rice and flour products in China. Similarly, Fei et al. [42] determined the sequence types ST4 (19/56, 33.9%), ST1 (12/56, 21.4%), and ST64 (11/56, 16.1%) in *C. sakazakii* as dominant for *Cronobacter* spp. isolated from powdered infant formula collected from Chinese retail markets. The results of this study may provide a theoretical basis for investigating the transmission routes and genotyping diversity of *Cronobacter* spp. and developing more effective methods for preventing this organism.

3.2. *Hemolytic Activity of Cronobacter spp.*

The hemolytic activity of the *Cronobacter* strains was determined (Table 3). In total, 87.5% and 16.7% of the *Cronobacter* spp. strains were capable of β-hemolysis on the culture medium with the addition of horse and sheep red blood cells, respectively.

Table 3. Hemolysis of horse blood and sheep blood agars by *Cronobacter* spp. strains.

Isolate	Species	Hemolysis (Zone in mm)	
		Horse Blood Agar	Sheep Blood Agar
s12	*C. sakazakii*	β (2.9)	α
9n	*C. sakazakii*	β (2.4)	α
10n	*C. sakazakii*	β (2.8)	α
s14	*C. sakazakii*	β (2.7)	α
s44	*C. sakazakii*	β (3.3)	α
s45	*C. sakazakii*	β (2.8)	α
s21	*C. sakazakii*	β (2.4)	α
s47	*C. sakazakii*	β (2.5)	α
s48	*C. sakazakii*	β (2.8)	α
s22	*C. sakazakii*	β (3.2)	α
11m	*C. sakazakii*	β (1.0)	α
lv25	*C. sakazakii*	β (2.5)	α
lv27	*C. sakazakii*	β (1.1)	α
s41	*C. sakazakii*	β (3.5)	β (1.2)
s42	*C. sakazakii*	β (3.1)	β (1.0)
5n	*C. malonaticus*	β (1.9)	α
6n	*C. malonaticus*	β (1.0)	α
7n	*C. malonaticus*	β (2.6)	α
8n	*C. malonaticus*	β (2.4)	α
12m	*C. malonaticus*	β (1.0)	α
lv31	*C. malonaticus*	α	α
1n	*C. turicensis*	α	β (1.0)
lv54	*C. turicensis*	α	β (1.0)
s37	*C. condimenti*	β (1.0)	α

A greater capability for β-hemolysis was determined on the horse blood agar, as indicated by a zone width ranging from 1.0 to 3.5 mm, than on the sheep blood agar—a zone width ranging from 1.0 to 1.2 mm. Regardless of ST and CC, all the *C. sakazakii* strains produced β-hemolysis on the horse blood agar and 86.7% of the strains caused α-hemolysis on the sheep blood agar. Both *C. sakazakii* strains (s41 and s42) belonging to ST 804 were capable of causing β–hemolysis on both blood agars. In turn, both analyzed *C. turicensis* (1n and lv54) strains assigned to different STs (806 and 808) were α-hemolytic on the horse blood agar and weakly β-hemolytic on the sheep blood agar. All the analyzed strains of *C. malonaticus* were capable of α-hemolysis on the sheep blood agar, while one strain assigned to ST 807 was α-hemolytic on both blood agar media. In the case of *C. condimenti* s37, β-hemolysis was shown on the horse blood agar, whereas it caused α-hemolysis on the sheep blood agar. All the *Cronobacter* strains assigned to a given ST or CC showed the same hemolytic activity profile on both blood agar media. The obtained results of the hemolytic activity of the *Cronobacter* strains indicated that the type of hemolysis was not species-specific, but was a strain-dependent feature.

The presence of various hemolysins and hemolysin-related genes in *Cronobacter* spp. has been described by many authors [33,34,43]. Fakruddin et al. [44] showed hemolytic activity on human blood agar with two out of six *C. sakazakii* strains isolated from food samples, whereas Rajani et al. [45] demonstrated β-hemolysis on bovine blood agar after a 4-day incubation at a temperature of 37 °C in the case of all 11 *C. sakazakii* isolates tested. Furthermore, Cui et al. [46] found that 27 of 31 (87%) *Cronobacter* isolates were not capable of hemolysis, except for *C. sakazakii* SC26, *C. malonaticus* SD16, *C. malonaticus* SD26, and *C. muytjensii* SD83, as indicated by the cleared zones produced around them. Finally, Umeda et al. [35] reported that 57 (100%) *Cronobacter* strains exhibited β-hemolytic activity against guinea pig, horse, and rabbit erythrocytes, and that 92,9% of the strains were capable of the α-hemolysis of sheep erythrocytes. It is known that a gene encoding a hemolysin is present in *Cronobacter* spp. [33], however, whether this hemolysin is active and associated with cytotoxicity has not yet been clarified. This requires conducting more in-depth genetic studies to assign the functionality of these various hemolysin genes to the appropriate phenotype.

4. Conclusions

The present study demonstrated the considerable genotypic diversity of *Cronobacter* strains isolated from plant-based ready-to-eat foods. Five novel sequence types (804, 805, 806, 807, and 808) were detected and novel alleles were found in four loci: *ppsA*, *gltB*, *gyrB*, and *infB*. Therefore, future studies should aim to detect more new STs, which, in the study of the genotypic diversity of *Cronobacter* spp., are beneficial for monitoring some sources and will enable the development of more effective methods to control these bacteria. The sequence type of one *C. sakazakii* strain (s12) was found to be ST4, which is strongly associated with meningitis in newborns. It was also found that 87.5% of the *Cronobacter* spp. strains were capable of β-hemolysis on the culture medium with horse red blood cells and 16.7% on the medium with sheep red blood cells. The presence of the pathogenic species *C. sakazakii*, *C. malonaticus*, and *C. turicensis* in ready-to-eat plant-derived food products shows they are potential sources of infection, especially to those with compromised immunity, which substantiates their further multi-faceted characterization [47]. The fact that active hemolysins appear in isolates from plant-based food seems to be disturbing, as this feature is usually associated with pathogens originating from clinical sources.

Author Contributions: Conceptualization, M.G.; methodology, M.G. and I.S.; investigation, M.G. and I.S.; writing—original draft preparation, M.G.; writing—review and editing, M.G., A.B.-P., I.S., S.F., A.P., L.S.-R.; visualization, M.G.; supervision, M.G.; project administration, M.G. All authors have read and agreed to the published version of the manuscript.

Funding: This research was funded by National Science Centre, research project No. DEC-2019/03/X/NZ1/01583 and the APC was funded by Warsaw University of Life Sciences.

Data Availability Statement: Data are contained within the article.

Conflicts of Interest: The authors declare no conflict of interest. Author Stephen Forsythe was employed by the company FoodMicrobe.com Ltd., Adams Hill, Keyworth, Nottinghamshire NG12 5GY, UK. The remaining authors declare that the research was conducted in the absence of any commercial or financial relationships that could be construed as a potential conflict of interest. The authors declare that this study received funding from National Science Centre (Poland). The funder was not involved in the study design, collection, analysis, interpretation of data, the writing of this article or the decision to submit it for publication.

References

1. FAO/WHO. Enterobacter sakazakii (*Cronobacter* spp.) in Powdered Follow-up Formulae. In *Microbiological Risk Assessment Series No. 15. Food and Agriculture Organization of the United Nations*; World Health Organization: Rome, Italy, 2008.
2. Forsythe, S.J. Updates on the Cronobacter genus. *Annu. Rev. Food Sci. Technol.* **2018**, *9*, 23–44. [CrossRef] [PubMed]
3. Carvalho, G.G.; Calarga, A.P.; Zorgi, N.E.; Astudillo-Trujillo, C.A.; Pardini Gontijo, M.T.; Brocchi, M.; Giorgio, S.; Kabuki, D.Y. Virulence and DNA sequence analysis of *Cronobacter* spp. isolated from infant cereals. *Int. J. Food Microbiol.* **2022**, *376*, 109745. [CrossRef]
4. Joseph, S.; Forsythe, S.J. Insights into the emergent bacterial pathogen Cronobacter spp., generated by multilocus sequence typing and analysis. *Front. Microbiol.* **2012**, *3*, 397. [CrossRef] [PubMed]
5. Yang, H.Y.; Kim, S.K.; Choi, S.Y.; You, D.H.; Lee, S.C.; Bang, W.S.; Yuk, H.G. Effect of acid, desiccation and heat stresses on the viability of *Cronobacter sakazakii* during rehydration of powdered infant formula and in simulated gastric fluid. *Food Control* **2015**, *50*, 336–341. [CrossRef]
6. Shi, C.; Sun, Y.; Liu, Z.; Guo, D.; Sun, H.; Sun, Z.; Chen, S.; Zhnag, W.; Wen, Q.; Peng, X.; et al. Inhibition of *Cronobacter sakazakii* virulence factors by citral. *Sci. Rep.* **2017**, *7*, 43243. [CrossRef]
7. Phair, K.; Pereira, S.G.; Kealey, C.; Fanning, S.; Brady, D.B. Insights into the mechanisms of *Cronobacter sakazakii* virulence. *Microb. Pathog.* **2022**, *169*, 105643. [CrossRef]
8. Jang, H.; Gopinath, G.R.; Eshwar, A.; Srikumar, S.; Scott, N.; Gangiredla, J.; Isha, R.; Isha, P.R.; Finkelstein, S.B.; Negrete, F.; et al. The secretion of toxins and other exoproteins of Cronobacter: Role in virulence, adaption, and persistence. *Microorganisms* **2020**, *8*, 229. [CrossRef]
9. International Commission on Microbiological Specifications for Foods (ICMSF). Microorganisms in Food 7. In *Microbiological Testing in Food Safety Management*; Kluwer Academic/Plenum Publishers: New York, NY, USA, 2002; pp. 168–169.
10. Henry, M.; Fouladkhah, A. Outbreak history, biofilm formation, and preventive measures for control of *Cronobacter sakazakii* in infant formula and infant care settings. *Microorganisms* **2019**, *7*, 77. [CrossRef]
11. Iversen, C.; Forsythe, S.J. Isolation of Enterobacter sakazakii and other Enterobacteriaceae from powdered infant formula milk and related products. *Food Microbiol.* **2004**, *21*, 771–776. [CrossRef]
12. Kandhai, M.C.; Heuvelink, A.E.; Reij, M.W.; Beumer, R.R.; Dijk, R.; van Tilburg, J.J.H.C.; van Schothorst, M.; Gorris, L.G.M. A study into the occurrence of *Cronobacter* spp. in The Netherlands between 2001 and 2005. *Food Control* **2010**, *21*, 1127–1136. [CrossRef]
13. Kim, J.-B.; Park, Y.-B.; Kang, S.-H.; Lee, M.-J.; Kim, K.-C.; Jeong, H.-R.; Kim, D.-H.; Yoon, M.-H.; Lee, J.-B.; Oh, D.-H. Prevalence, genetic diversity, and antibiotic susceptibility of Cronobacter spp. (Enterobacter sakazakii) isolated from Sunshik, its ingredients and soils. *Food Sci. Biotechnol.* **2011**, *20*, 941–948. [CrossRef]
14. Lee, Y.-D.; Park, J.-H.; Chang, H. Detection, antibiotic susceptibility and biofilm formation of *Cronobacter* spp. from various foods in Korea. *Food Control* **2012**, *24*, 225–230. [CrossRef]
15. Ogihara, H.; Kiribe, N.; Fukuda, N.; Furukawa, S.; Morinaga, Y.; Igimi, S. *Cronobacter* spp. in commercially available dried food in Japan. *Biocontrol. Sci.* **2014**, *19*, 209–213. [CrossRef]
16. Garbowska, M.; Berthold-Pluta, A.; Stasiak-Różańska, L. Microbiological quality of selected spices and herbs including the presence of *Cronobacter* spp. *Food Microbiol.* **2015**, *49*, 1–5. [CrossRef]
17. Mohammed, M.A.; Sallam, K.I.; Tamura, T. Prevalence, identification and molecular characterization of *Cronobacter sakazakii* isolated from retail meat products. *Food Control* **2015**, *53*, 206–211. [CrossRef]
18. Vojkovska, H.; Karpiskova, R.; Orieskova, M.; Drahovska, H. Characterization of *Cronobacter* spp. isolated from food of plant origin and environmental samples collected from farms and from supermarkets in the Czech Republic. *Int. J. Food Microbiol.* **2016**, *217*, 130–136. [CrossRef]
19. Berthold-Pluta, A.; Garbowska, M.; Stefańska, I.; Pluta, A. Microbiological quality of selected ready-to-eat leaf vegetables, sprouts and non-pasteurized fresh fruit-vegetable juices including the presence of *Cronobacter* spp. *Food Microbiol.* **2017**, *65*, 221–230. [CrossRef]
20. Li, C.; Zeng, H.; Zhang, J.; Luo, D.; Chen, M.; Lei, T.; Yang, X.; Wu, H.; Cai, S.; Ye, Y.; et al. *Cronobacter* spp. isolated from aquatic products in China: Incidence, antibiotic resistance, molecular characteristic and CRISPR diversity. *Int. J. Food Microbiol.* **2020**, *335*, 108857. [CrossRef]
21. Carvalho, G.G.; Calarga, A.P.; Teodoro, J.R.; Queiroz, M.M.; Astudillo-Trujillo, A.C.; Levy, C.E.; Brocchi, M.; Kabuki, D.Y. Isolation, comparison of identification methods and antibiotic resistance of *Cronobacter* spp. *in infant foods. Food Res. Int.* **2020**, *137*, 109643. [CrossRef]

22. Zeng, H.; Li, C.; Ling, N.; Zhang, J.; Chen, M.; Lei, T.; Wu, S.; Yang, X.; Luo, D.; Ding, Y.; et al. Prevalence, genetic analysis and CRISPR typing of *Cronobacter* spp. isolated from meat and meat products in China. *Int. J. Food Microbiol.* **2020**, *321*, 108549. [CrossRef]
23. Berthold-Pluta, A.; Garbowska, M.; Stefańska, I.; Stasiak-Różańska, L.; Aleksandrzak-Piekarczyk, T.; Pluta, A. Microbiological quality of nuts, dried and candied fruits, including the prevalence of *Cronobacter* spp. *Pathogens* **2021**, *10*, 900. [CrossRef] [PubMed]
24. Parra-Flores, J.; Maury-Sintjago, E.; Rodriguez-Fernández, A.; Acuña, S.; Cerda, F.; Aguirre, J.; Holy, O. Microbiological Quality of Powdered Infant Formula in Latin America. *J. Food Prot.* **2020**, *83*, 534–541. [CrossRef] [PubMed]
25. Pakbin, B.; Brück, W.M.; Allahyari, S.; Rossen, J.W.A.; Mahmoudi, R. Antibiotic resistance and molecular characterization of *Cronobacter sakazakii* strains isolated from powdered infant formula milk. *Foods* **2022**, *11*, 1093. [CrossRef] [PubMed]
26. Fei, P.; Xing, M.; Feng, Y.; Liu, S.; Chang, Y.; Wang, Y.; Yu, Y.; Shi, E.; Zhang, Y.; Bian, X.; et al. *Cronobacter sakazakii* in goat milk-based infant formula from shaanxi province, China. *Foodborne Pathog. Dis.* **2022**, *19*, 304–310. [CrossRef]
27. Xu, X.; Li, C.; Wu, Q.; Zhang, J.; Huang, J.; Yang, G. Prevalence, molecular characterization, and antibiotic susceptibility of *Cronobacter* spp. in Chinese ready-to-eat foods. *Int. J. Food Microbiol.* **2015**, *204*, 17–23. [CrossRef]
28. Lou, X.; Liu, T.; Zhang, W.; Yu, H.; Wang, H.; Song, S.; Chen, Q.; Fang, Z. The occurrence and distribution characteristics of Cronobacter in diverse cereal kernels, flour, and flour-based products. *Food Microbiol.* **2019**, *84*, 103269. [CrossRef]
29. Holý, O.; Forsythe, S. *Cronobacter* spp. as emerging causes of healthcare-associated infection. *J. Hosp. Infect.* **2014**, *86*, 169–177. [CrossRef]
30. Burgess, C.M.; Gianotti, A.; Gruzdev, N.; Holah, J.; Knøchel, S.; Lehner, A.; Margas, E.; Schmitz Esser, S.; Saldinger, S.S.; Tresse, O. The response of foodborne pathogens to osmotic and desiccation stresses in the food chain. *Int. J. Food Microbiol.* **2016**, *221*, 37–53. [CrossRef]
31. Srikumar, S.; Cao, Y.; Yan, Q.; Van Hoorde, K.; Nguyen, S.; Cooney, S.; Gopinath, G.R.; Tall, B.D.; Sivasankaran, S.K.; Lehner, A.; et al. RNA sequencing-based transcriptional overview of xerotolerance in *Cronobacter sakazakii* SP291. *Appl. Environ. Microbiol.* **2019**, *85*, e01993-18. [CrossRef]
32. Ling, N.; Forsythe, S.; Wu, Q.; Ding, Y.; Zhang, J.; Zeng, H. Insights into *Cronobacter sakazakii* biofilm formation and control strategies in the food industry. *Engineering* **2020**, *6*, 393–405. [CrossRef]
33. Cruz, A.; Xicohtencatl-Cortes, J.; Gonzalez-Pedrajo, B.; Bobadilla, M.; Eslava, C.; Rosas, I. Virulence traits in Cronobacter species isolated from different sources. *Can. J. Microbiol.* **2011**, *57*, 735–744. [CrossRef]
34. Jang, H.; Chase, H.R.; Gangiredla, J.; Grim, C.J.; Patel, I.R.; Kothary, M.H.; Jackson, S.A.; Mammel, M.K.; Carter, L.; Negrete, F.; et al. Analysis of the molecular diversity among Cronobacter species isolated from filth flies using targeted PCR, pan genomic DNA microarray, and whole genome sequencing analyses. *Front. Microbiol.* **2020**, *11*, 561204. [CrossRef] [PubMed]
35. Umeda, N.S.; De Filippis, I.; Forsythe, S.J.; Brandão, M.L.L. Phenotypic characterization of *Cronobacter* spp. strains isolated from foods and clinical specimens in Brazil. *Food Res. Int.* **2017**, *102*, 61–67. [CrossRef] [PubMed]
36. Svobodová, B.; Vlach, J.; Junková, P.; Karamonová, L.; Blažková, M.; Fukal, L. Novel method for reliable identification of Siccibacter and Franconibacter strains: From "Pseudo-Cronobacter" to new Enterobacteriaceae genera. *Appl. Environ. Microbiol.* **2017**, *83*, e00234-17. [CrossRef] [PubMed]
37. Forsythe, S.J.; Dickins, B.; Jolley, K.A. Cronobacter, the emergent bacterial pathogen Enterobacter sakazakii comes of age; MLST and whole genome sequence analysis. *BMC Genom.* **2014**, *15*, 1121–1134. [CrossRef]
38. Baldwin, A.; Loughlin, M.; Caubilla-Barron, J.; Kucerova, E.; Manning, G.; Dowson, C.; Forsythe, S. Multilocus sequence typing of *Cronobacter sakazakii* and *Cronobacter malonaticus* reveals stable clonal structures with clinical significance which do not correlate with biotypes. *BMC Microbiol.* **2009**, *9*, 223. [CrossRef]
39. Buxton, R. Blood Agar Plates and Hemolysis Protocols. *Am. Soc. Microbiol.* **2005**, *15*, 1–9. Available online: https://asm.org/getattachment/7ec0de2b-bb16-4f6e-ba07-2aea25a43e76/protocol-2885.pdf (accessed on 10 April 2020).
40. Wang, Q.; Forsythe, S.J.; Zhao, X.-J.; Wanga, Z.-W.; Li, D.; Ma, D.; Cao, J.-Y.; Zeng, J. Species identification and molecular characterization of *Cronobacter* spp. isolated from food imported over nine years into Beijing, China. *Food Microbiol.* **2019**, *82*, 11–19. [CrossRef]
41. Li, Q.; Li, C.; Ye, Q.; Gu, Q.; Wu, S.; Zhang, Y.; Wei, X.; Xue, L.; Chen, M.; Zeng, H.; et al. Occurrence, molecular characterization and antibiotic resistance of *Cronobacter* spp. isolated from wet rice and flour products in Guangdong, China. *Curr. Res. Food Sci.* **2023**, *7*, 100554. [CrossRef]
42. Fei, P.; Jiang, Y.; Jiang, Y.; Yuan, X.; Yang, T.; Chen, J.; Wang, Z.; Kang, H.; Forsythe, S.J. Prevalence, Molecular Characterization, and Antibiotic Susceptibility of *Cronobacter sakazakii* Isolates from Powdered Infant Formula Collected from Chinese Retail Markets. *Front. Microbiol.* **2017**, *8*, 2026. [CrossRef]
43. Joseph, S.; Desai, P.; Ji, Y.; Cummings, C.A.; Shih, R.; Degoricija, L.; Rico, A.; Brzoska, P.; Hamby, S.E.; Masood, N.; et al. Comparative analysis of genome sequences covering the seven Cronobacter species. *PLoS ONE* **2012**, *7*, e49455. [CrossRef] [PubMed]
44. Fakruddin, M.; Rahaman, M.; Ahmed, M.M.; Hoque, M.M. Stress tolerant virulent strains of *Cronobacter sakazakii* from food. *Biol. Res.* **2014**, *47*, 63. [CrossRef] [PubMed]
45. Rajani, C.S.R.; Chaudhary, A.; Swarna, A.; Puniya, A.K. Identification and virulence of Enterobacter sakazakii. *J. Food Ind. Microbiol.* **2016**, *2*, 108. [CrossRef]

46. Cui, J.; Hu, J.; Du, X.; Yan, C.; Xue, G.; Li, S.; Cui, Z.; Huang, H.; Yuan, J. Genomic analysis of putative virulence factors affecting cytotoxicity of Cronobacter. *Front. Microbiol.* **2020**, *10*, 3104. [CrossRef] [PubMed]
47. Yong, W.; Guo, B.; Shi, X.; Cheng, T.; Chen, M.; Jiang, X.; Ye, Y.; Wang, J.; Xie, G.; Ding, J. An investigation of an acute gastroenteritis outbreak: *Cronobacter sakazakii*, a potential cause of food-borne illness. *Front. Microbiol.* **2018**, *9*, 2549. [CrossRef] [PubMed]

foods

MDPI

Article

Detection of *Vibrio parahaemolyticus* Based on Magnetic and Upconversion Nanoparticles Combined with Aptamers

Xinjie Song [1,†], Wei Li [2,†], Li Wu [1], Tianfeng Lv [1], Yao Zhang [1], Juan Sun [1], Xuping Shentu [3], Xiaoping Yu [3] and Yuanfeng Wu [1,*]

[1] Zhejiang Provincial Key Laboratory of Chemical and Biological Processing Technology for Agricultural Products, School of Biological and Chemical Engineering, Zhejiang University of Science and Technology, Liuxia Street Number 318, Hangzhou 310023, China; xjsong@zust.edu.cn (X.S.); wuli@zust.edu.cn (L.W.); LTF107@163.com (T.L.); zhangyao@zust.edu.cn (Y.Z.); sunjuan18@zust.edu.cn (J.S.)

[2] Korean Medicine (KM) Application Center, Korea Institute of Oriental Medicine, Daegu 41062, Republic of Korea; liwei1986@kiom.re.kr

[3] Zhejiang Provincial Key Laboratory of Biometrology and Inspection and Quarantine, College of Life Science, China Jiliang University, Hangzhou 314423, China; stxp@cjlu.edu.cn (X.S.); yxp@cjlu.edu.cn (X.Y.)

* Correspondence: wuyuanfeng@zju.edu.cn

† These authors contributed equally to this work.

Abstract: *Vibrio parahaemolyticus* is a halophilic and heat-labile gram-negative bacterium and is the most prevalent foodborne bacterium in seafood. In order to develop a rapid and sensitive method for detecting the foodborne pathogenic bacterium *Vibrio parahaemolyticus*, an aptamer-modified magnetic nanoparticle and an aptamer-modified upconversion nanoparticle were synthesised and used as a capture probe and a signal probe, respectively. The aptamer-modified magnetic nanoparticle, *V. parahaemolyticus* cell, and aptamer-modified upconversion nanoparticle formed a sandwich-like complex, which was rapidly separated from a complex matrix using a magnetic force, and the bacterial concentration was determined by fluorescence intensity analysis. The results showed that the fluorescence intensity signal correlated positively with the concentration of *V. parahaemolyticus* in the range of 3.2×10^2 to 3.2×10^5 CFU/mL, with a linear equation of y = 296.40x − 217.67 and a correlation coefficient of $R^2 = 0.9610$. The detection limit of the developed method was 4.4 CFU/mL. There was no cross-reactivity with other tested foodborne pathogens. This method is highly specific and sensitive for the detection of *V. parahaemolyticus*, and can achieve the qualitative detection of this bacterium in a complex matrix.

Keywords: *Vibrio parahaemolyticus*; aptamer; magnetic nanoparticles; upconversion nanoparticles; detection

check for updates

Citation: Song, X.; Li, W.; Wu, L.; Lv, T.; Zhang, Y.; Sun, J.; Shentu, X.; Yu, X.; Wu, Y. Detection of *Vibrio parahaemolyticus* Based on Magnetic and Upconversion Nanoparticles Combined with Aptamers. *Foods* **2023**, *12*, 4433. https://doi.org/10.3390/foods12244433

Academic Editor: Xinjun Du

Received: 14 October 2023
Revised: 14 November 2023
Accepted: 7 December 2023
Published: 11 December 2023

1. Introduction

Foodborne illnesses caused by pathogenic microorganisms are common and result in many food safety challenges [1]. Seafood is a nutrient-rich food source, and its consumption has been increasing worldwide in recent decades. The main safety concern surrounding this increase in consumption is foodborne pathogen-induced infections [2]. Common pathogenic bacteria in seafood include *Vibrio*, *Salmonella*, *Listeria*, *Shigella*, *Staphylococcus*, *Clostridium*, and *Escherichia coli* [3,4]. *Vibrio parahaemolyticus*, a halophilic and heat-labile gram-negative bacterium, is the most prevalent foodborne bacterium in seafood. *V. parahaemolyticus* contamination can lead to gastroenteritis, infection, and sepsis [5]. *V. parahaemolyticus*-caused diseases have been frequently reported all over the world in recent years, making *V. parahaemolyticus* a major concern in seafood safety. A study showed that 4256 patients were infected by *V. parahaemolyticus* between 2003 and 2016 in Korea [6]. Chen et al. reported that, among 410 outbreaks of bacterial foodborne illnesses occurring from 2015 to 2020 in Zhejiang province, China, 56.69% (232 cases) were caused by *V. parahaemolyticus* [7]. Another report mentioned that vibriosis was responsible for approximately

80,000 illnesses and 100 deaths annually in the United States [8]. Recently, *Vibrio* spp. infections have attracted more concern because climate warming and global trade increases might increase human *Vibrio* infections worldwide [9]. Therefore, rapid and reliable *V. parahaemolyticus* detection is essential in the control of food safety.

Current *V. parahaemolyticus* detection methods include conventional microbiological culturing [10], molecular biological detection [11], and immunological detection [12]. Hu et al. [13] reported a nanogold-assisted HRM-qPCR method for detecting *V. parahaemolyticus*, in combination with self-screening and traditional targeting, with a detection limit of 5×10^1 CFU/mL. Feng et al. [14] developed a recombinase-aided amplification assay for detecting *V. parahaemolyticus*. This assay could detect concentrations of 7×10^3 CFU/mL directly and could detect levels as low as 0.1 CFU/mL when a 4 h enrichment was performed. Zeng et al. [15] reported the development of a PCR-based lateral flow test strip for detecting *V. parahaemolyticus* in codfish, with a detection limit of 50 CFU/mL. Similarly, Yang [16] developed a fluorescence immunoassay based on the inner filter effect and BSA-gold nanoclusters. Each of these methods has advantages and disadvantages, and developing new methods for the rapid and sensitive detection of *V. parahaemolyticus* is crucial for the surveillance of foodborne diseases.

Upconversion nanoparticles (UCNPs) are novel types of fluorescent materials that exhibit a special luminescence phenomenon: the wavelength of the emitted light is less than the wavelength of the excited light, that is, upconversion luminescence [17]. In the 1960s, Bloembergen first proposed the phenomenon of upconversion, and Porter proved that the upconversion process is a nonlinear optical process that converts two or more low-energy photons into high-energy photons [18]. With the development of research, rare-earth-doped upconversion nanoparticles have been found to have remarkable optical properties, including high resistance to light bleaching, a sharp emission bandwidth, a long lifetime, excellent spectral characteristics, and high photostability. Upconversion luminescence nanoparticles can emit visible or ultraviolet light upon excitation with near-infrared light and are widely applied in biosensing for food safety. Compared to conventional organic fluorescent dyes and quantum dots, the unique anti-Stokes shift luminescence signal of UCNPs can be used to identify autofluorescence in complex real matrices, such as in food, the environment, and biological tissues, thereby reducing background interference and improving sensitivity. Aptamers are synthetic short single-stranded oligonucleotides (DNA or RNA) screened from random DNA libraries in vitro by SELEX [19]. Aptamers have attracted much attention in the field of food safety detection due to their affinity and specificity for targets [20]. Detection methods based on aptamers were studied for the analysis of foodborne pathogens including *V. parahaemolyticus* [21]. Furthermore, magnetic immobilisation separation using Fe_3O_4 magnetic nanoparticles as carriers to enrich pathogens in microbiology can enhance the analysis selectivity and sensitivity [22]. Therefore, in this study, we developed a novel *V. parahaemolyticus* detection method using aptamer-based magnetic fluorescent nanoprobes. Magnetic and upconversion nanoparticles were combined with amino-modified aptamers to capture target pathogenic bacteria based on aptamer specificity. *V. parahaemolyticus* levels were quantitatively analysed based on the fluorescence intensity.

2. Materials and Methods

2.1. Materials

$FeCl_3 \cdot 6H_2O$ and 1,6-hexanediamine were purchased from Shanghai Macklin Biochemical Co., Ltd. (Shanghai, China). Ethylene glycol, ethanol, methanol, isopropanol, and sodium hydroxide were purchased from Shanghai Lingfeng Chemical Reagent Co., Ltd. (Shanghai, China), and 25% glutaraldehyde was purchased from Yonghua Chemical Co., Ltd. (Shanghai, China). YCl_3, $YbCl_3$, $ErCl_3$, 1-octadecene (ODE), and ammonium fluoride (NH_4F) were purchased from Aladdin Reagent Co., Ltd. (Shanghai, China). The 3-aminopropyltriethoxysilane (APTES), 0.01 mol/L phosphate-buffered saline (PBS, pH 7.4), oleic acid (OA), sodium acetate, tetraethyl orthosilicate (TEOS), and *V. parahaemolyti-*

cus aptamer (5′-NH_2-TCTAAAAAATGGGCAAAAAAACATGACTCGTTGAGATACT-3′ used herein were purchased from the Shanghai Sangon Biological Science & Technology Company. Alkaline peptone water medium and thiosulfate citrate bile salt sucrose agar medium were purchased from Beijing Landbridge Technology Co., Ltd. (Beijing, China). *V. parahaemolyticus* (CGMCC 1.1997), *Escherichia coli* (CGMCC 1.12883), *Staphylococcus aureus* (CGMCC 1.6750), and *Salmonella* (CGMCC 1.10603) were purchased from the China General Microbial Strain Preservation and Management Centre. Enterotoxigenic *Escherichia coli* (ATCC 35401) was purchased from the American Type Culture Collection.

2.2. Apparatus

A JEM 2100F high-resolution field emission transmission electron microscope (TEM JEOL, Tokyo, Japan), X-ray diffractometer (Malvern Panalytical, Malvern, UK), high resolution transmission electron microscope (TEM, HITACHI H-7600; HITACHI, Tokyo Japan), fluorescence spectrophotometer (F-4500 HITACHI, Tokyo, Japan), and Fourier transform infrared spectrometer (FTIR, BRKER, Saarbrucken, Germany) were used.

2.3. Synthesis of Amine-Functionalised Fe_3O_4 Magnetic Nanoparticles

Amine-functionalised Fe_3O_4 magnetic nanoparticles were prepared as previously described [23,24]. Briefly, anhydrous sodium acetate (2.0 g), 1,6-hexanediamine (6.5 g), and $FeCl_3{\cdot}6H_2O$ (1.0 g) in glycol (30.0 mL) were mixed at 50 °C. The mixture was allowed to react for 10 h at 200 °C. After being cooled to room temperature, the products were separated by a magnetic stirrer before being collected, washed with ultrapure water (3 times) and ethanol (3 times), and dried at 60 °C for 8 h. After drying, a black powder consisting of amino-functionalised Fe_3O_4 magnetic nanoparticles was collected. TEM imaging, XRD patterns, FTIR spectra, and VSM magnetisation curves were used for the property analysis of the Fe_3O_4 magnetic nanoparticles.

2.4. UCNP Synthesis and Surface Modifications

To synthesise UNCP, firstly, 1 mmol of $ReCl_3$ (Y:Yb:Er = 78:20:2), 6 mL of OA, and 15 mL of ODE were concurrently added to a 100 mL three-necked flask; the mixed solution was heated for 30 min at 160 °C and allowed to cool to 50 °C [25]. Subsequently, 10 mL of a methanol solution, containing 2.5 mmol of NaOH and 3.9 mmol of NH_4F, was added to the mixed solution and agitated for 30 min. Afterwards, the mixture was heated for 1 h at 100 °C and then 300 °C under the protection of argon gas, and kept at the latter temperature for 30 min. After cooling to room temperature, the product was precipitated using ethanol and was washed and centrifuged thrice. The final product was dried for 12 h at 60 °C [26].

Surface modification of the UCNPs was performed as previously described [27]. Firstly, 20 mg of the synthesised UCNPs was ultrasonicated for 30 min in 60 mL of isopropanol. Subsequently, 20 mL of distilled water and 2.5 mL of 25% ammonia were rapidly added to the mixture, and it was magnetically stirred. Afterwards, 20 mL of isopropanol and 60 μL of tetraethoxysilane (TEOS) were added to the solution and allowed to react for 3 h, before 30 mL of isopropanol and 200 μL of APTES were added. The product was left to stand for 2 h at room temperature after 1 h of reaction, before being washed and centrifuged several times. Finally, the product was vacuum-dried for 12 h at 60 °C. The amino-functionalised UCNPs were collected and kept at 4 °C until subsequent use. The dried powders of UCNPs and UCNPs@SiO_2 were used to perform the characterisation of TEM imaging, XRD patterns, and FTIR spectra. To prevent the influence of aggregation on the result, both the UCNPs and UCNPs@SiO_2 nanoparticles were uniformly suspended in water, and the phase and fluorescence spectra were measured with excitation and emission spectra of 980 nm and in the 400–800 nm region directly.

2.5. Synthesis of Aptamer-Fe_3O_4 Magnetic Nanoparticles and Aptamer-UCNPs

Amino-functionalised Fe_3O_4 magnetic nanoparticles (2 mg) were ultrasonicated in 1 mL of PBS (pH 7.4) for 30 min. Subsequently, 250 μL of 25% glutaraldehyde was added to the mixture, and the solution was shaken for 2 h at 25 °C. The product was magnetically separated and washed three times with PBS and then dispersed in 1 mL of PBS. Afterwards, 20 μL of the 25 μM aptamer was added to the uniform amino-functionalised Fe_3O_4 magnetic nanoparticle solution, and the mixture was shaken for 12 h at 37 °C, magnetically separated, and washed thrice with PBS. The final product was dispersed in 1 mL PBS and stored at 4 °C [28]. The aptamer in the supernatant before and after the reaction was measured using UV-vis spectroscopy to confirm the immobilisation of the aptamer on the Fe_3O_4 magnetic nanoparticles.

The aptamer-UCNP preparation was similar to the preparation of the aptamer-Fe_3O_4 magnetic nanoparticles. First, the amino-functionalised UCNPs were ultrasonicated in 1 mL of PBS (pH 7.4) for 30 min. Subsequently, 250 μL of 25% glutaraldehyde was added, and the mixture was shaken for 2 h at 25 °C. The product was separated by centrifugation (12,000 rpm, 10 min) and washed thrice with PBS before being dispersed in 1 mL of PBS. Afterwards, 40 μL of the 25 μM aptamer was added to the uniform amino-functionalised UCNP solution, and the mixture was shaken for 12 h at 37 °C. The mixture was separated via centrifugation and washed four times with PBS. The final product was dispersed in 1 mL of PBS and stored at 4 °C. The aptamers in the supernatant before and after the reaction were measured using UV-vis spectroscopy to confirm the immobilisation of the aptamer on the UNCPs. Different concentrations of UNCPs (1, 2, 4, and 6 mg) were tested to optimise the synthesis of the aptamer-UCNPs.

2.6. Optimization of Reacted Amount of Aptamer-Fe_3O_4 Magnetic Nanoparticles and Aptamer-UCNPs

To optimise the amount of reacting aptamer-Fe_3O_4 magnetic nanoparticles, 40, 80, 120, 160, and 200 μL of aptamer-MNPs were added to a fixed amount (150 μL) of aptamer-UCNP solution. Secondly, 100 μL of *V. parahaemolyticus* (at 10^5 CFU/mL) was added to the aptamer-UCNP solutions simultaneously, and the mixture was replenished to 500 μL with PBS buffer and shaken slowly for 2 h at 37 °C. After reacting, the product was magnetically separated and washed thrice with PBS to remove unbound components. Finally, the product was resuspended in 400 μL of PBS buffer, and the fluorescence intensity of the suspension was measured at 542 nm using a fluorescence spectrophotometer with an excitation spectrum of 980 nm and emission spectra in the 400–800 nm region.

Afterwards, to optimise the amount of reacting aptamer-UCNPs, 50, 100, 150, 200, and 250 μL of aptamer-UCNPs (at 6 mg/mL) were added and reacted with the aptamer-Fe_3O_4 magnetic nanoparticles following the aforementioned procedure. The optimal amount of aptamer-UCNPs was determined based on the fluorescence intensity of the suspension, which was measured at 542 nm using a fluorescence spectrophotometer with an excitation spectrum of 980 nm and emission spectra in the 400–800 nm region.

2.7. Detection of V. parahaemolyticus Using the Developed Fluorescent Method

To detect *V. parahaemolyticus* using our developed method, 100 μL of *V. parahaemolyticus* was added to 500 μL of the aptamer-Fe_3O_4 magnetic nanoparticles suspension and gently vibrated for 2 h at 37 °C. The reactant was magnetically separated and washed thrice with PBS buffer. The aptamer-UCNP suspension was then added, and the mixture was gently vibrated for 1 h at 37 °C. Subsequently, the product was magnetically separated, washed thrice with PBS buffer, and resuspended in 400 μL of PBS buffer. The fluorescence intensity of the suspension was measured in the 400–800 nm region using a fluorescence spectrophotometer with an excitation spectrum of 980 nm. *V. parahaemolyticus* concentrations from 3.2×10^2 CFU/mL to 3.2×10^5 CFU/mL were tested to determine the detection limit of our method.

2.8. Detection Specificity

Four foodborne pathogenic bacteria, *Escherichia coli*, *Enterotoxigenic Escherichia coli*, *Salmonella*, and *Staphylococcus aureus*, were used to evaluate the specificity of the developed method. The tested foodborne pathogenic bacteria at a concentration of 10^5 CFU/mL were mixed with the aptamer-Fe_3O_4 magnetic nanoparticles, reacted under gentle shaking, and separated using a magnetic force. Next, the separated mixture was mixed with aptamer-UNCPs and separated using a magnetic force after 30 min of reaction under gentle shaking. The separation was uniformly resuspended in PBS buffer by ultrasonication. The fluorescence intensities of the suspensions were measured using excitation and emission spectra of 980 and in the 400–800 nm region, respectively.

3. Results and Discussion

3.1. Mechanism of *V. parahaemolyticus* Detection

In our study, a *V. parahaemolyticus* detection method was developed using aminated nanoparticles and amino-modified aptamers. As the aptamer specifically binds to the target, aptamer-modified nanoparticles can identify and capture *V. parahaemolyticus*. Aptamer UCNPs and aptamer-Fe_3O_4 magnetic nanoparticles were used as fluorescent signals and a separation medium, respectively, in this study. First, the tested sample was mixed with aptamer-Fe_3O_4 magnetic nanoparticles and allowed to react under gentle shaking for 30 min. Aptamer-Fe_3O_4 magnetic nanoparticles could capture *V. parahaemolyticus* using aptamers when *V. parahaemolyticus* was present in the tested sample. The aptamer-Fe_3O_4-*V. parahaemolyticus* can be separated using a magnetic force. The separated aptamer-Fe_3O_4-*V. parahaemolyticus* was mixed with aptamer-UNCPs to create sandwich-like aggregates, aptamer-Fe_3O_4-*V. parahaemolyticus*-UNCPs, which can be formed and separated using a magnetic force (Figure 1). Due to the presence of UCNPs, the fluorescence intensity of the sandwich-like aggregate can be measured via fluorescence spectrophotometry with an excitation spectrum of 980 nm and emission spectra in the 400–800 nm region, and the intensity of this signal can be increased by increasing the concentration of *V. parahaemolyticus* for detection.

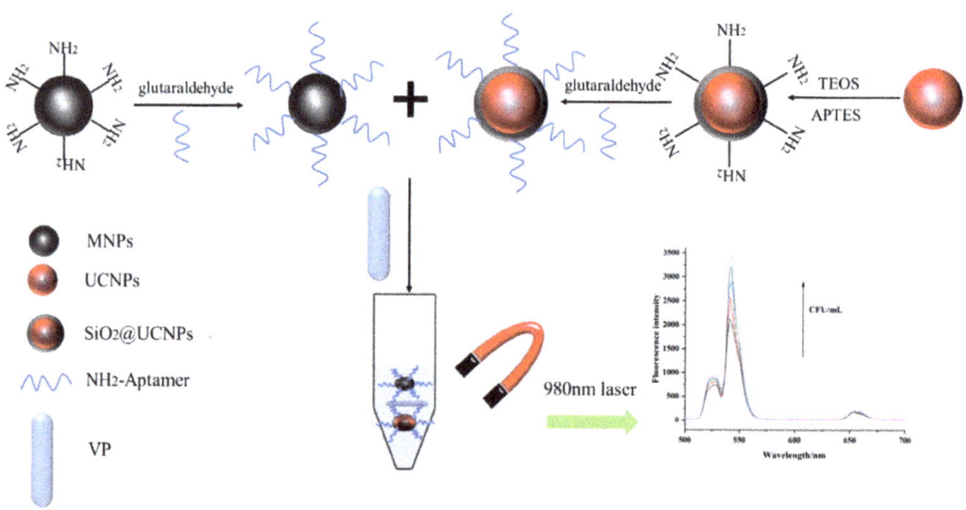

Figure 1. A schematic description of the proposed fluorescence detection platform.

3.2. Characterisation of Amine-Functionalised Magnetic Fe_3O_4 Nanoparticle

Bare Fe_3O_4 magnetic nanoparticles must undergo surface modifications because of their high activity levels and susceptibility to oxidation. The TEM images (Figure 2A) indicated that the Fe_3O_4 magnetic nanoparticles were uniformly spherical with a size of 50 nm. The XRD results (Figure 2B) indicated that the Fe_3O_4 magnetic nanoparticles have typical X-ray diffraction lines at $2\theta = 18.6°, 35.8°, 43.7°, 53.9°, 556.8°$, and $63°$, which can be indexed as (111), (311), (400), (422), (511), and (440); this is consistent with the standard card of Fe_3O_4 (JCPDS Card no. 19-0629). No diffraction peaks corresponding to other impurities were observed, indicating that the product was Fe_3O_4 with high purity. The FTIR spectra (Figure 2C) showed sharp peaks at 580, 1047, and 1629 cm^{-1}, representing characteristic absorption peaks of the Fe–O bond, the stretching vibration of the CH_2 bond, and the bending vibration of the NH bond, respectively [29]. This result indicated that the Fe_3O_4 surface was successfully modified with amino groups. The magnetisation curves revealed that the saturation magnetisation of Fe_3O_4 increased with the applied magnetic field strength and rapidly reached a saturated state (Figure 2D). The magnetisation curve appeared as an "S" shape (Figure 2D), indicating that the synthesised Fe_3O_4 exhibits superparamagnetism at room temperature.

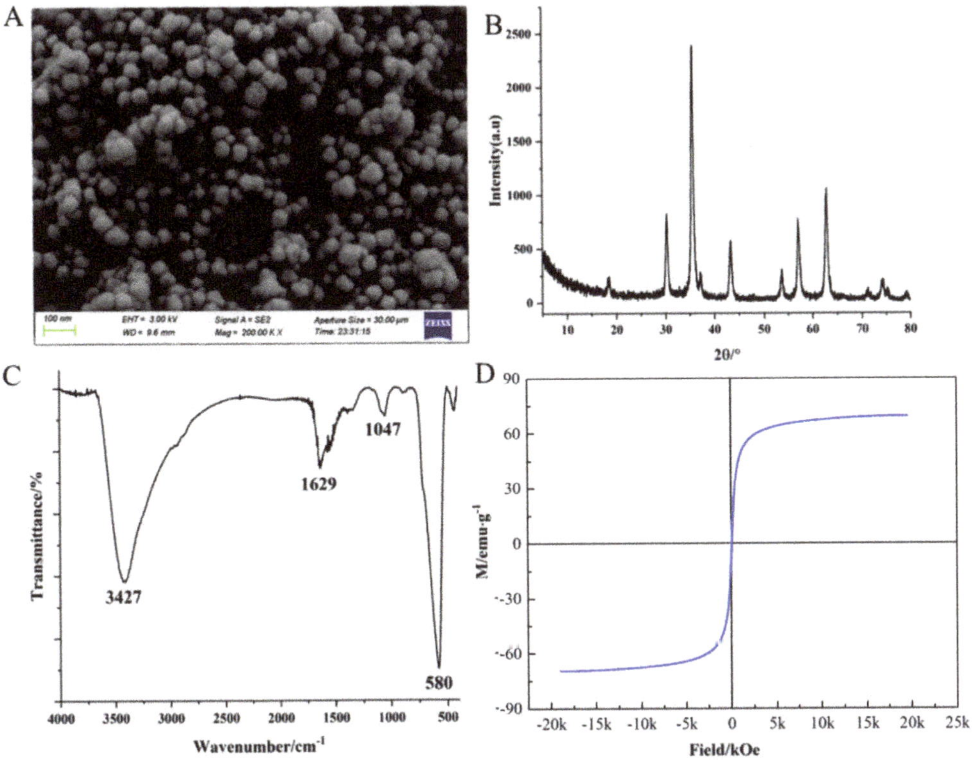

Figure 2. Characterisation of Fe_3O_4 magnetic nanoparticles modified with amino groups (TEM image of MNPs (**A**), XRD patterns of MNPs (**B**), FTIR spectra of MNPs (**C**), and VSM magnetisation curves of MNPs (**D**)).

3.3. UCNP Characterisation

To increase the solubility and stability of UCNPs in the water phase and to provide more modifiable active groups on the surface of the UCNPs, TEOS was used to form SiO_2 on the surface, and the compound was modified with amino groups using APTES. The TEM images of the UCNPs showed that the particles were hexagonal in shape and approximately 100 nm in size (Figure 3A). After the UCNPs were coated with SiO_2, a shell formed on their surface, which was clearly visible in the TEM images (Figure 3B). Figure 3C indicates that the UCNPs exhibited typical X-ray diffraction lines at 2θ = 29.95°, 30.79°, 34.69°, 39.61°, 43.42°, 46.84°, 53.56°, and 53.59° corresponding to the (110), (101), (200), (111), (210), (102), and (300) planes of $NaYF_4$. According to the FTIR spectra, the stretching vibration peak of the Si-O bond at 1097 cm^{-1} and the bending vibration peak of the amino group at 1631 cm^{-1} appeared in the UCNPs@SiO_2 and were absent in the untreated UCNPs (Figure 3D). This further supports the successful surface modification of the UCNPs. The fluorescence intensity of the two materials was measured using a fluorescence spectrophotometer with excitation and emission spectra of 980 nm and in the 400–800 nm region, respectively. The results showed that the fluorescence intensity of the modified UCNPs was much lower than that of the unmodified UCNPs, indicating that SiO_2 affected their fluorescence intensity. Upon excitation with the 980 nm laser, the nanomaterials displayed three sets of fluorescence emission bands, of which the most obvious emission peak was at 542 nm (Figure 3E); therefore, this fluorescence value at 542 nm was chosen as the detection signal for the subsequent experiments.

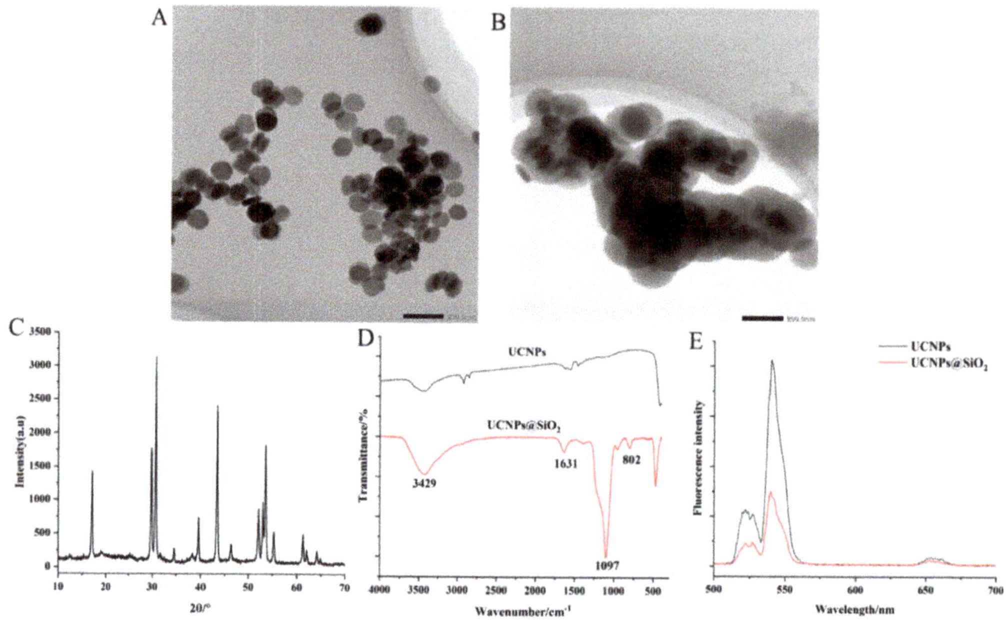

Figure 3. Characterisation of UCNPs and UCNPs@SiO_2 (TEM image of UCNPs (**A**), TEM image of UCNPs@SiO_2 (**B**), XRD patterns of UCNPs (**C**), FTIR spectra of UCNPs and UCNPs@SiO_2 (**D**), and fluorescence spectrum of UCNPs, and UCNPs@SiO_2 (**E**)).

3.4. Aptamer-Fe_3O_4 Magnetic Nanoparticles and Aptamer-UCNP Characterisation

The synthesised nanoparticles have a surface amino group, which can be cross-linked with the 5′-terminal-modified amino group aptamer using glutaraldehyde to enable functional modifications [30]. Since the aptamer is an oligonucleotide chain with a characteristic absorption peak at 260 nm [31,32], aptamer–nanoparticle binding was indirectly characterised by measuring the absorbance of the aptamer stock solution and the absorbance of the supernatant during binding. The UV-vis spectra results of the aptamer and aptamer-Fe_3O_4 magnetic nanoparticles (Figure 4A) show that the absorbance of the supernatant at 260 nm decreased significantly after the binding of different concentrations of aptamers, indicating that the aptamers successfully combined with the nanomaterials [33]. The capture capacity of aptamer-UCNPs was highly affected by the aptamers immobilised on the surface of the nanomaterials. Different reaction concentrations of UCNPs (1, 2, 4, and 6 mg/mL) were investigated to characterise the capture capacity and immobilisation efficiency of the aptamer-modified nanomaterials. The reduction in the absorbance of the supernatant at 260 nm increased as the concentration of UCNPs increased, indicating that more aptamer molecules were immobilised on the nanomaterial (Figure 4B). When the reaction concentration of the UCNPs was 6 mg/mL, the absorbance of the supernatant after the reaction was the lowest, indicating that a high initial reaction concentration could improve the immobilisation of the aptamer on the nanoparticles. Considering economy and efficiency, 6 mg/mL of UCNPs was selected for all subsequent experiments to synthesise the aptamer-UNCPs.

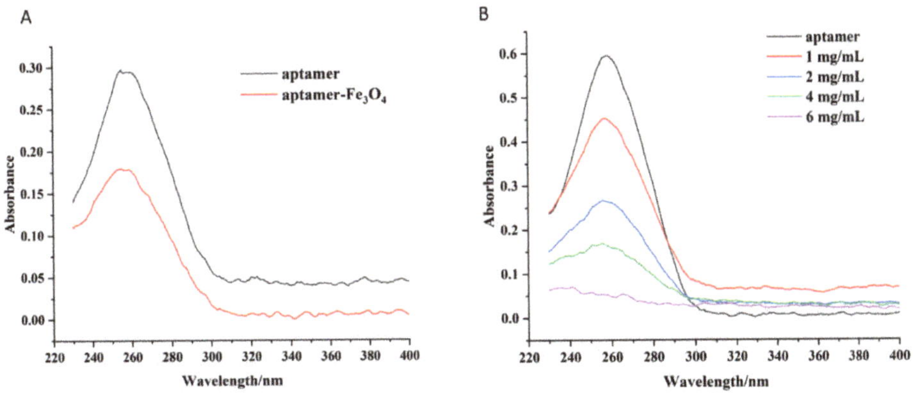

Figure 4. Confirmation of aptamer immobilisation on the Fe_3O_4 magnetic nanoparticles (**A**) and UCNPs (**B**) using UV-vis spectra.

3.5. Optimal Amount of Aptamer-Fe_3O_4 Magnetic Nanoparticles and Aptamer-UCNPs

Different amounts of aptamer-Fe_3O_4 magnetic nanoparticles and aptamer-UCNPS were tested in the reaction system to optimise the detection method. Different amounts of Fe_3O_4 magnetic nanoparticles (40, 80, 120, 160, or 200 μL) were added to the reaction system. The fluorescence intensity gradually increased with the increased amount of added aptamer-Fe_3O_4 magnetic nanoparticles and peaked at 120 μL of aptamer-Fe_3O_4 magnetic nanoparticles (Figure 5A,B). Furthermore, the intensity decreased when more aptamer-Fe_3O_4 magnetic nanoparticles were added (Figure 5B). Thus, 120 μL of the aptamer-Fe_3O_4 magnetic nanoparticles was used for subsequent experiments. Similarly, when different amounts of aptamer-UCNPs (50, 100, 150, 200, or 250 μL) were added to the reaction system, the fluorescence intensity of the reaction system gradually increased as the amount of aptamer-UCNPs increased (Figure 5C). When more than 200 μL of aptamer-UCNPs was added, the fluorescence intensity of the system no longer increased (Figure 5D), indicating the aptamer-UCNPs were exceeded. Thus, 200 μL of the aptamer-UCNPs was used for the subsequent experiments.

3.6. Detection Method Sensitivity

The developed method was used to detect different concentrations of *V. parahaemolyticus* (with 120 μL of aptamer-Fe_3O_4 magnetic nanoparticles and 200 μL of aptamer-UCNPs used in the reaction system). The fluorescence intensities of the supernatants were measured at the end of the reaction. The results showed that the fluorescence intensity increased as the concentration of *V. parahaemolyticus* increased (Figure 6A) in the concentration range of 3.2×10^2–3.2×10^5 CFU/mL (Figure 6B). The linear equation was $y = 296.40x - 217.67$ (where x is log CFU mL^{-1} and y is the fluorescence intensity), with $R^2 = 0.9610$, indicating that there was a positive correlation between the *V. parahaemolyticus* concentration and fluorescence intensity. The detection limit was 4.4 CFU/mL (3 ε/S, where ε is the standard deviation of the blank sample and S is the slope of the linear equation). Yu et al. reported a universal pathogen-sensing platform based on a smart hydrogel aptasensor embedded with gold nanoclusters to detect live *V. paraheamolyticus* in water with a detection limit of 10 CFU/mL [34]. Zhai et al. have developed an immunomagnetic separation and quantum-dot-based immunofluorescence method to detect *V. paraheamolyticus* with a detection limit of 10^2 CFU/mL [35]. Similarly, Ren et al. used a fluorescence resonance energy transfer (FRET)-based paper sensor to detect *V. parahaemolyticus*; the detection limit was 8.9 CFU/mL [36]. Our results demonstrate that the fluorescence method we developed, based on magnetic and upconversion nanoparticles combined with aptamers, has sufficient sensitivity to detect *V. parahaemolyticus*.

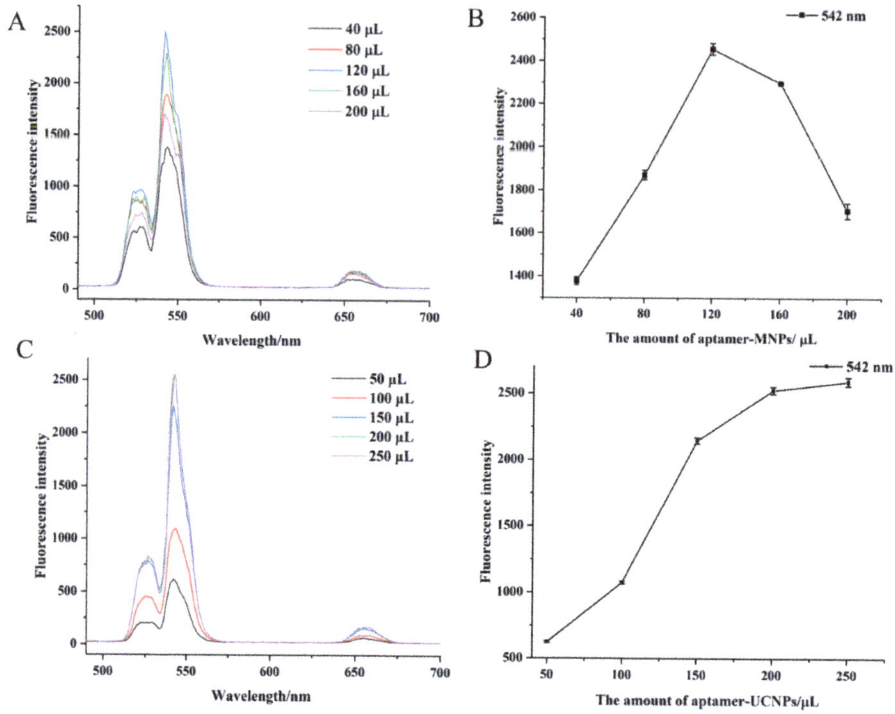

Figure 5. The fluorescence spectra of different additions of aptamer-MNPs (**A**), absorption peaks at 542 nm of different additions of aptamer-MNPs (**B**), the fluorescence spectra of different additions of aptamer-UCNPs (**C**), absorption peaks at 542 nm of different additions of aptamer-UCNPs (**D**).

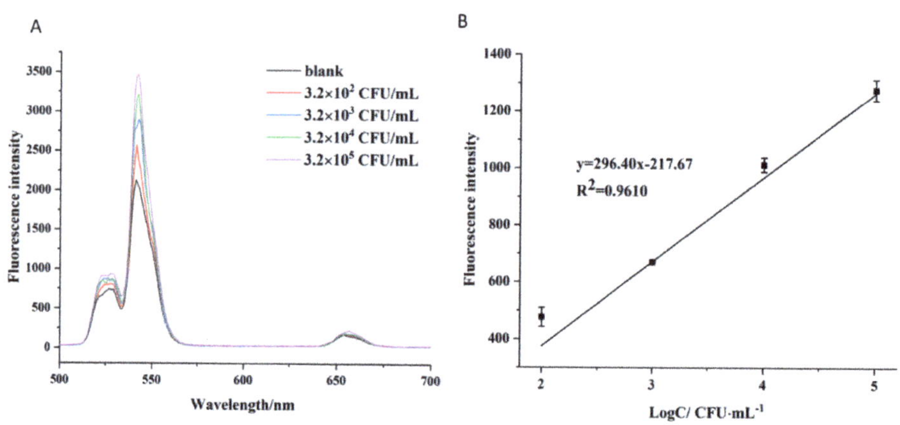

Figure 6. Fluorescence spectra of compounds in the presence of different *V. parahaemolyticus* concentrations (**A**), the linear relationship between the fluorescence intensity and concentration of *V. parahaemolyticus* (**B**), where F is the sample fluorescence intensity and F_0 is the blank control.

3.7. Detection Method Specificity

To evaluate the specificity of the developed method for *V. parahaemolyticus*, four foodborne pathogenic bacteria, including *E. coli*, *enterotoxigenic E. coli*, *S. aureus*, and *Salmonella*, were added to the reaction system, and the fluorescence intensity of the supernatant after magnetic separation was measured. The tested *E. coli*, *Enterotoxigenic E. coli*, *S. aureus*, and *Salmonella* concentrations were 10^5 CFU/mL; 10^5 CFU/mL of *V. parahaemolyticus* was used as a positive control. The results revealed that only the separated *V. parahaemolyticus* suspension generated high fluorescence intensities, whereas the *E. coli*, *enterotoxigenic E. coli*, *S. aureus*, and *Salmonella* suspensions generated low fluorescence intensities using the developed method, proving that the detection method was highly specific for *V. parahaemolyticus* (Figure 7). Recently, Li et al. presented a colorimetric and SERS dual-mode detection method targeting *V. parahaemolyticus* using aptamers and multifunctional composite magnetic material, which exhibited no cross-reactivity with *E. coli*, *S. typhimurium*, *E. coli*, *S. aureus*, or *L. monocytogenes* [37]. Similarly, Parsaeimehr and Ozbay reported on a colorimetric detection method based on PCR and the utilisation of DNAzyme; their method also showed no cross-reactivity with *E. coli* or *S. aureus* [38]. Based on the results of the current study, the developed fluorescence method has good specificity for *V. parahaemolyticus*, is simple and sensitive, and can be adapted and applied to detect other foodborne pathogens.

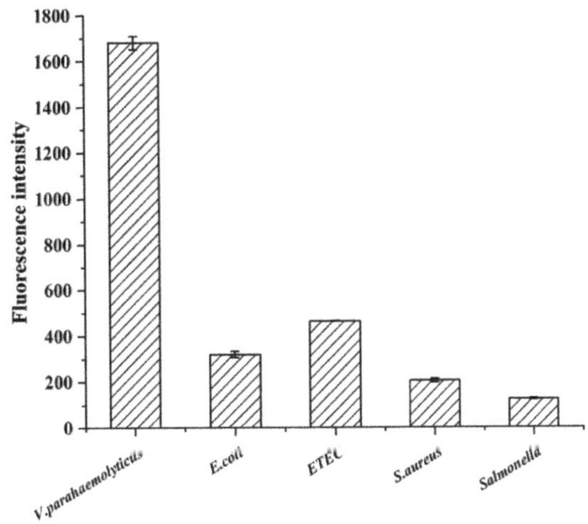

Figure 7. The specificity of the developed method with other bacteria.

4. Conclusions

In the present study, Fe_3O_4 magnetic nanoparticles and UCNPs were successfully synthesised. The Fe_3O_4 magnetic nanoparticles have a uniform size of 50 nm and exhibit superparamagnetism at room temperature. The TEM images of the UCNPs showed that the particles were hexagonal in shape and approximately 100 nm in size. The SiO_2 synthesised UCNPs had three fluorescence emission bands, of which the most obvious emission peak was at 542 nm. After the immobilisation of the aptamer against *V. parahaemolyticus*, aptamer-Fe_3O_4 magnetic nanoparticles and aptamer-UCNPs were used as a capture medium and fluorescence signal source in detection, respectively. A fluorescence method for *V. parahaemolyticus* was designed using the compatibility of the aptamer and the target *V. parahaemolyticus*, which can rapidly isolate and detect *V. parahaemolyticus* from a complex matrix using magnetic forces. The developed fluorescence method-based aptamer-modified magnetic and upconversion nanoparticles had high specificity for *V. parahaemolyticus* and had no cross-reactivity with tested foodborne pathogenic bacteria, including *E. coli*, *enterotoxigenic E. coli*, *S. aureus*, and *Salmonella*. This method had a detection limit of 4.4 CFU/mL, the fluorescence signal had a clear linear relationship with the concentration of *V. parahaemolyticus* in the range of 3.2×10^2–3.2×10^5 CFU/mL, with a linear equation of y = 296.40x − 217.67 and a correlation coefficient of R^2 = 0.9610. These results indicate that the developed fluorescence method can achieve the qualitative detection of *V. parahaemolyticus* in a complex matrix. Furthermore, this method can be adapted to detect other bacteria by changing the aptamers used and has broad application prospects in the field of pathogenic bacterial detection.

Author Contributions: X.S. (Xinjie Song) and W.L.: Experimental design, Methodology, Writing—original draft, Supervision. L.W. and T.L.: Formal analysis, Data analysis. Y.Z. and J.S.: Validation Formal analysis. X.S. (Xuping Shentu) and X.Y.: Investigation, Formal analysis, Supervision. Y.W. Project administration, Funding acquisition. All authors have read and agreed to the published version of the manuscript.

Funding: This research was supported by the Key Research and Development Program of Zhejiang Province (2021C02061) and Chunhui Project Foundation of Education Department of China (202200217).

Data Availability Statement: Data is contained within the article.

Conflicts of Interest: The authors declare no conflict of interest.

References

1. Bavisetty, S.C.B.; Vu, H.T.K.; Benjakul, S.; Vongkamjan, K. Rapid pathogen detection tools in seafood safety. *Curr. Opin. Food Sci.* **2018**, *20*, 92–99. [CrossRef]
2. Afreen, M.; Bağdatli, İ. Food-borne pathogens in seafood. *EJAR* **2021**, *5*, 44–58.
3. Cai, G.; Zheng, L.; Liao, M.; Li, Y.; Wang, M.; Liu, N.; Lin, J. A microfluidic immunosensor for visual detection of foodborne bacteria using immunomagnetic separation, enzymatic catalysis and distance indication. *Microchim. Acta* **2019**, *186*, 757. [CrossRef] [PubMed]
4. Asgari, S.; Dhital, R.; Aghvami, S.A.; Mustapha, A.; Zhang, Y.; Lin, M. Separation and detection of *E. coli* O157:H7 using a SERS-based microfluidic immunosensor. *Microchim. Acta* **2022**, *189*, 111. [CrossRef]
5. Nair, G.B.; Ramamurthy, T.; Bhattacharya, S.K.; Dutta, B.; Takeda, Y.; Sack, D.A. Global dissemination of *Vibrio parahaemolyticus* Serotype O3:K6 and its serovariants. *Clin. Microbiol. Rev.* **2007**, *20*, 39–48. [CrossRef]
6. Park, K.; Mok, J.S.; Kwon, J.Y.; Ryu, A.R.; Kim, S.H.; Lee, H.J. Food-borne outbreaks, distributions, virulence, and antibiotic resistance profiles of Vibrio parahaemolyticus in Korea from 2003 to 2016: A review. *Fish. Aquat. Sci.* **2018**, *21*, 3. [CrossRef]
7. Chen, L.; Sun, L.; Zhang, R.; Liao, N.; Qi, X.; Chen, J. Surveillance for foodborne disease outbreaks in Zhejiang Province, China, 2015–2020. *BMC Public Health* **2022**, *22*, 135. [CrossRef]
8. Wu, Q.; Liu, J.; Malakar, P.K.; Pan, Y.; Zhao, Y.; Zhang, Z. Modeling naturally-occurring Vibrio parahaemolyticus in post-harvest raw shrimps. *Food Res. Int.* **2023**, *173*, 113462. [CrossRef]
9. Ndraha, N.; Hsiao, H. A climate-driven model for predicting the level of Vibrio parahaemolyticus in oysters harvested from Taiwanese farms using elastic net regularized regression. *Microb. Risk Anal.* **2022**, *21*, 100201. [CrossRef]
10. Su, Y.C.; Duan, J.; Wu, W.-H. Selectivity and specificity of a chromogenic medium for detecting Vibrio parahaemolyticus. *J. Food Prot.* **2005**, *68*, 1454–1456. [CrossRef]
11. Zhai, Y.; Zhao, C.; Li, L.; Xu, K.; Wang, J.; Song, X.; Li, H. Production of phage display-derived peptide and the application for detecting Vibrio parahaemolyticus by combined PCR technology. *Food Anal. Methods* **2020**, *13*, 1906–1917. [CrossRef]
12. Wang, L.; Zhang, J.; Bai, H.; Li, X.; Lv, P.; Guo, A. Specific Detection of Vibrio Parahaemolyticus by Fluorescence Quenching Immunoassay Based on Quantum Dots. *Appl. Biochem. Biotechnol.* **2014**, *173*, 1073–1082. [CrossRef] [PubMed]
13. Hu, A.; Kong, L.; Lu, Z.; Qiao, J.; Lv, F.; Meng, F.; Bie, X. Research on nanogold-assisted HRM-qPCR technology for highly sensitive and accurate detection of Vibrio parahaemolyticus. *LWT Food Sci. Technol.* **2022**, *162*, 113488. [CrossRef]
14. Feng, Z.-S.; Li, J.-Y.; Zhang, J.-Y.; Li, F.-Y.; Guan, H.-X.; Zhang, R.-Q.; Liu, H.; Guo, Q.; Shen, X.-X.; Kan, B.; et al. Development and evaluation of a sensitive recombinase aided amplification assay for rapid detection of Vibrio parahaemolyticus. *J. Microbiol. Methods* **2022**, *193*, 106404. [CrossRef] [PubMed]
15. Zeng, D.; Chen, S.; Jiang, L.; Ren, J.; Ling, N.; Su, J.; Zhao, Y.; Jiang, Y.; Xue, F.; Tang, F.; et al. A polymerase chain reaction based lateral flow test strip with propidium monoazide for detection of viable Vibrio parahaemolyticus in codfish. *Microchem. J.* **2020**, *159*, 105418. [CrossRef]
16. Yang, G.; Wei, C.; Tang, Y.; Zhang, J.; Zhang, X.; Chi, H.; Tao, L.; Kong, C. A novel fluorescence immunoassay based on inner filter effect and gold nanoclusters for Vibrio parahaemolyticus determination. *Results Chem.* **2021**, *3*, 100208. [CrossRef]
17. Lee, J.H.; Oh, M.; Kim, B. Phage biocontrol of zoonotic food-borne pathogen Vibrio parahaemolyticus for seafood safety. *Food Control* **2023**, *144*, 109334. [CrossRef]
18. Ndraha, N.; Huang, L.; Wu, V.C.H.; Hsiao, H.-I. Vibrio parahaemolyticus in seafood: Recent progress in understanding influential factors at harvest and food-safety intervention approaches. *Curr. Opin. Food Sci.* **2023**, *48*, 100927. [CrossRef]

19. Schaumburg, F.; Carrell, C.S.; Henry, C.S. Rapid bacteria detection at low concentrations using sequential immunomagnetic separation and paper-based isotachophoresis. *Anal. Chem.* **2019**, *91*, 9623–9630. [CrossRef]
20. Hou, Y.; Zhu, L.; Hao, H.; Zhang, Z.; Ding, C.; Zhang, G.; Bi, J.; Yan, S.; Liu, G.; Hou, H. A novel photoelectrochemical aptamer sensor based on rare-earth doped Bi_2WO_6 and Ag_2S for the rapid detection of Vibrio parahaemolyticus. *Microchem. J.* **2021**, *165*, 106132. [CrossRef]
21. Zhang, Y.; Lai, B.S.; Juhas, M. Recent advances in aptamer discovery and applications. *Molecules* **2019**, *24*, 941. [CrossRef] [PubMed]
22. Duan, N.; Wu, S.; Chen, X.; Huang, Y.; Wang, Z. Selection and Identification of a DNA Aptamer Targeted to Vibrio parahemolyticus. *J. Agric. Food Chem.* **2012**, *60*, 4034–4038. [CrossRef] [PubMed]
23. Rong, Y.; Hassan, M.; Ouyang, Q.; Chen, Q. Lanthanide ion (Ln^{3+})-based upconversion sensor for quantification of food contaminants: A review. *Compr. Rev. Food Sci.* **2021**, *20*, 3531–3578. [CrossRef] [PubMed]
24. Wang, L.; Bao, J.; Wang, L.; Zhang, F.; Li, Y. One-pot synthesis and bioapplication of amine-functionalized magnetite nanoparticles and hollow nanospheres. *Chem. Eur. J.* **2006**, *12*, 6341–6347. [CrossRef]
25. Nodehi, R.; Shayesteh, H.; Rahbar-Kelishami, A. Fe_3O_4@NiO core–shell magnetic nanoparticle for highly efficient removal of Alizarin red S anionic dye. *Int. J. Environ. Sci. Technol.* **2022**, *19*, 2899–2912. [CrossRef]
26. Li, Z.; Zhang, Y. An efficient and user-friendly method for the synthesis of hexagonal-phase $NaYF_4$:Yb, Er/Tm nanocrystals with controllable shape and upconversion fluorescence. *Nanotechnology* **2008**, *19*, 345606. [CrossRef]
27. Li, H.; Ahmad, W.; Rong, Y.; Chen, Q.; Zuo, M.; Ouyang, Q.; Guo, Z. Designing an aptamer based magnetic and upconversion nanoparticles conjugated fluorescence sensor for screening *Escherichia coli* in food. *Food Control* **2020**, *107*, 106761. [CrossRef]
28. Sheng, W.; Shi, Y.; Ma, J.; Wang, L.; Zhang, B.; Chang, Q.; Duan, W.; Wang, S. Highly sensitive atrazine fluorescence immunoassay by using magnetic separation and upconversion nanoparticles as labels. *Microchim. Acta* **2019**, *186*, 564. [CrossRef]
29. Wu, S.; Zhang, H.; Shi, Z.; Duan, N.; Fang, C.C.; Dai, S.; Wang, Z. Aptamer-based fluorescence biosensor for chloramphenicol determination using upconversion nanoparticles. *Food Control* **2015**, *50*, 597–604. [CrossRef]
30. Liu, R.; Zhang, Y.; Ali, S.; Haruna, S.A.; He, P.; Li, H.; Ouyang, Q.; Chen, Q. Development of a fluorescence aptasensor for rapid and sensitive detection of Listeria monocytogenes in food. *Food Control* **2021**, *122*, 107808. [CrossRef]
31. Wang, T.; Chen, C.; Larcher, L.M.; Barrero, R.A.; Veedu, R.N. Three decades of nucleic acid aptamer technologies: Lessons learned, progress and opportunities on aptamer development. *Biotechnol. Adv.* **2019**, *37*, 28–50. [CrossRef] [PubMed]
32. Ying, N.; Wang, Y.; Song, X.; Yang, L.; Qin, B.; Wu, Y.; Fang, W. Lateral flow colorimetric biosensor for detection of Vibrio parahaemolyticus based on hybridization chain reaction and aptamer. *Microchim. Acta* **2021**, *188*, 381. [CrossRef] [PubMed]
33. Chen, M.; Song, Y.; Han, L.; Zhou, D.; Wang, Y.; Pan, L.; Tu, K. An Ultrasensitive Upconversion Fluorescence Aptasensor Based on Graphene Oxide Release and Magnetic Separation for Staphylococcus aureus Detection. *Food Anal. Methods* **2022**, *15*, 2791–2800. [CrossRef]
34. Yu, J.; Xiao, S.; Yu, Z.; Hui, Y.; Li, T.; Wu, D.; Bi, W.; Gan, N.; Jia, Z. On-site and dual-mode detection of live Vibrio parahaemolyticus in waters: A universal pathogen sensing platform based on a smart hydrogel aptasensor imbedded with gold nanoclusters. *Sens. Actuators B Chem.* **2022**, *366*, 131947. [CrossRef]
35. Zhai, Y.; Meng, X.; Li, L.; Liu, Y.; Xu, K.; Zhao, C.; Wang, J.; Song, X.; Li, J.; Jin, M. Rapid detection of *Vibrio parahaemolyticus* using magnetic nanobead-based immunoseparation and quantum dot-based immunofluorescence. *RSC Adv.* **2021**, *11*, 38638–38647. [CrossRef]
36. Ren, Y.; Cao, L.; Zhang, X.; Jiao, R.; Ou, D.; Wang, Y.; Zhang, D.; Shen, Y.; Ling, N.; Ye, Y. A novel fluorescence resonance energy transfer (FRET)-based paper sensor with smartphone for quantitative detection of Vibrio parahaemolyticus. *Food Control* **2023**, *145*, 109412. [CrossRef]
37. Li, J.; Lin, X.; Wu, J.; Ying, D.; Duan, N.; Wang, Z.; Wu, S. Multifunctional magnetic composite nanomaterial for colorimetric-SERS dual-Mode detection and photothermal sterilization of Vibrio parahaemolyticus. *Chem. Eng. J.* **2023**, *477*, 147113. [CrossRef]
38. Parsaeimehr, A.; Ozbay, G. Utilizing HRPzyme, a cost-effective Vibrio parahaemolyticus detection method. *LWT* **2023**, *189*, 115461. [CrossRef]

foods

MDPI

Article

Spatial Distribution and Enrichment Dynamics of Foodborne Norovirus in Oyster Tissues

Mao Mao [1,2], Zilei Zhang [1,3], Xuchong Zhao [4], Haoran Geng [5], Liang Xue [1,*] and Danlei Liu [2,5,*]

1 State Key Laboratory of Applied Microbiology Southern China, Institute of Microbiology, Guangdong Academy of Sciences, Guangzhou 510070, China; mimi12770415@163.com (M.M.); zhangzilei@shcc.edu.cn (Z.Z.)
2 Shanghai International Travel Healthcare Center, Shanghai Customs District P. R. China, Shanghai 200335, China
3 Inspection and Quarantine Technology Communication Department, Shanghai Customs College, Shanghai 201204, China
4 Jinan Center for Disease Control and Prevention, Jinan 250021, China; zhaoxuchong@163.com
5 Shanghai-MOST Key Laboratory of Health and Disease Genomics, NHC Key Lab of Reproduction Regulation, Shanghai Institute for Biomedical and Pharmaceutical Technologies, Fudan University, Shanghai 200237, China; hrgeng20@fudan.edu.cn
* Correspondence: xueliang@gdim.cn (L.X.); liudanlei@fudan.edu.cn (D.L.)

Abstract: The prevalence of norovirus in oysters poses a significant threat to food safety, necessitating a comprehensive understanding of contamination patterns. This study explores the temporal dynamics of norovirus distribution in various oyster tissues over a contamination period ranging from 6 to 96 h. Four tissues—the gill, palp, digestive gland, and stomach—were subjected to systematic monitoring using RT-qPCR for absolute quantification. Results revealed rapid norovirus detection in all tissues six hours post-contamination, with subsequent variations in detection rates. Gill and digestive gland tissues exhibited a peak in detection at 12–24 h, aligning with the oyster's gastrointestinal circulatory system. The digestive gland, distinguished by specific enrichment and adsorption capabilities, demonstrated the highest virus concentration at 48 h. In contrast, the stomach displayed a reemergence of norovirus. Beyond 72 h, detection remained exclusive to the digestive gland, with Ct values comparable to earlier time points. At 96 h, a limited amount of norovirus was detected in the digestive gland, emphasizing the importance for timely monitoring. In addition to providing critical insights into optimal detection strategies, these findings highlight the time-related characteristics of norovirus contamination in oysters. The study identifies the digestive gland as a key target for reliable monitoring, providing valuable data to improve protocols for reducing hazards associated with oyster consumption and foodborne norovirus infections. This research contributes to the understanding of norovirus dynamics in oyster tissues and reinforces current efforts aimed at ensuring food safety and public health.

Keywords: norovirus contamination; oyster tissues; temporal dynamics; detection strategies; foodborne infections

Citation: Mao, M.; Zhang, Z.; Zhao, X.; Geng, H.; Xue, L.; Liu, D. Spatial Distribution and Enrichment Dynamics of Foodborne Norovirus in Oyster Tissues. *Foods* **2024**, *13*, 128. https://doi.org/10.3390/foods13010128

Academic Editor: Efstathios Giaouris

Received: 30 November 2023
Revised: 23 December 2023
Accepted: 26 December 2023
Published: 29 December 2023

1. Introduction

Foodborne noroviruses, commonly referred to as noroviruses (NoV), are a significant global public health concern due to their association with foodborne illnesses and outbreaks. These highly contagious viruses are a leading cause of acute gastroenteritis, affecting individuals of all ages. Large-scale outbreaks in schools and communities can lead to the closure of growing areas and the recall of shellfish products, resulting in serious life threats and huge economic losses [1]. The global impact of NoV is growing, with an estimated 684 million cases and 212,000 deaths occurring each year [2]. NoVs cause acute gastroenteritis in humans for the following reasons: NoVs are usually associated with foodborne and waterborne transmission; NoVs are highly stable in the environment, and only 18 viral particles could cause disease; there are currently no antiviral drugs specific for NoVs [3]. The primary mode of transmission is the fecal–oral route, which can occur through contaminated water, person-to-person contact, and, importantly for this study, consumption of contaminated food, particularly seafood. Based on genetic differences found in the capsid protein, the NoV is currently classified into at least 49 genotypes within 10 genogroups (GI to GX) [4]. Acute gastroenteritis cases in humans are associated with two epidemiologically dominant genogroups, GI and GII, that are further subdivided into nine genotypes and 27 genotypes, respectively. According to global surveillance data, several NoV genotypes are responsible for outbreaks, with GII.4 constituting the majority [5,6].

In recent years, significant attention has been given to NoV outbreaks that are linked to the consumption of raw or undercooked oysters. Oysters, filter-feeding bivalve mollusks, can bioaccumulate NoV from contaminated water sources [7]. These sources include effluent from wastewater treatment facilities, combined sewer/stormwater overflows, malfunctioning septic tanks, and recreational and commercial fishing vessels [8]. There are varieties of oysters all over the world, such as *Crossostrea gigas*, *Crassostrea gigas angulate*, *Crassostrea hongkongensis*, *Ostrea rivularis Gould*, and *Crassostrea sikamea*. The Pacific oyster (*Crassotrea gigas*), naturally distributed around Japan, China, and Korea, is widely now an important cultivated oyster species worldwide [7]. It is becoming increasingly evident that viruses interact with oysters, and the dynamics in oysters after infection have been strongly linked to massive oyster mortality and human illnesses, such as gastroenteritis and NoV outbreaks. In outbreaks caused by shellfish, genotyping of norovirus in stool and oyster samples may reveal discrepancies. This raises the question of what role the oyster played and could be overcome by analyzing viral diversity in greater detail [9].

Currently, there are numerous issues linked to viral contamination. Over the past 10 years, more than 40% of RASFF (Rapid Alert System for Food and Feed) notifications were related to NoV detection in oysters [10]. Studies from various countries have consistently shown a high prevalence of NoV in commercially harvested oyster samples, raising significant concerns regarding the contamination of oysters with NoV [11]. Chassaing et al. conducted a meta-analysis of publications on NoV-related foodborne outbreaks from 1993 to 2019 and found that bivalve molluscan shellfish accounted for 53% of the implicated food sources [12]. For perspective, the prevalence of NoV in oysters from other regions is reported in the range of <2% in Australia, 9% in France, 3.9 to 20% in the USA, 16.9% in China, 32.1% in Spain, and up to 71.6% in the United Kingdom [13]. Our previous research entailed a year-long investigation into oyster contamination in aquaculture farms and revealed that 16.9% (60/356) of the collected oyster samples tested positive for NoV [14]. A systematic review and meta-analysis indicated a substantial burden of NoV in shellfish worldwide, with GII.4 being the predominant genotype [15]. These findings emphasize the widespread prevalence of NoV contamination in oyster samples and highlight the need for further research to elucidate the dynamics of norovirus distribution in oyster populations.

The detection of NoV in both food and environmental samples remains a significant challenge, primarily due to the complex composition of sample tissues and the low concentration of the virus present. Current human NoV cell culture systems are complicated and do not allow infectious NoV detection at levels found in contaminated foods [16]. Similarly to other viruses, detecting and monitoring NoV RNA encompasses several steps, including sample collection, concentration and enrichment, laboratory assay, and data normalization and interpretation. In the absence of methods to determine the viral infectivity of NoVs in food samples, foodborne viruses are detected mostly by molecular methods. Molecular techniques are currently the gold standard for discovering NoVs, including reverse transcription-polymerase chain reaction (RT-PCR), droplet digital RT-PCR (RT-ddPCR), and quantitative real-time RT-PCR. Among those methods, RT-qPCR has emerged as a valuable tool in the detection and quantification of NoVs due to its high sensitivity and specificity, as well as its low risk of carry-over contamination [17]. As of now, various commercially available RT-qPCR assays demonstrate the sensitivity of real-time RT-PCR at approximately 10–50 genome copies per reaction for NoV GI and 1–300 genome copies per reaction for NoV GII [18]. For accurate genotype identification of positive samples, RT-PCR and nested RT-PCR were used, followed by amplicon sequencing.

The adoption of RT-qPCR for NoV detection aligns with the international standard ISO 15216 [19], therefore enhancing the credibility and applicability of the detection method Moreover, while the absence of genomes is necessarily a sign of that the corresponding infectious viruses are absent, ISO 15216, when strictly applied, imposes a theoretical limit of detection (LOD) of greater than 18 particles, usually defined as the minimal NoV infective dose [20]. In addition to using of RT-qPCR, several studies have focused on optimizing and comparing the enrichment processes used in the preprocessing of oyster tissues, aimed at enhancing the efficiency and accuracy of detecting NoV in oyster samples [19,21–23] Meanwhile, differential detection rates of NoV have been shown within distinct oyster tissues, shedding light on the complex dynamics of NoV distribution and persistence within oyster organisms [24,25]. Therefore, understanding the distribution and prevalence of NoV in oysters is crucial for improving detection rate and reducing the false-positive rate.

There have been numerous studies investigating shellfish virus contamination. However, most of the above studies focused on a specific country, a certain period, and limited species types. This is not conducive to a comprehensive understanding of the overall contamination of NoV in shellfish worldwide. In this study, we explored the spatial dynamic distribution of NoV in different oyster tissues by constructing an artificially contaminated oyster model to clarify the in vivo circulation path and excretion time of NoV. Simultaneously, the generated data can support the development of virus detection preprocessing methods for oysters and contribute to the design of depuration strategies.

2. Materials and Methods

2.1. Artificial Contamination Oysters with NoV

All experiments were performed in a biosafety level 2 (BSL-2) containment laboratory using standard BSL-2 work practices. In this study, viral titers were measured using the RT-qPCR method. The stool sample with the highest copy number for NoV was about 1.42×10^7 genome copies per mL.

The oysters (*Crassotrea gigas*) were obtained from Zhejiang Province, China. After surface cleaning, oysters were transferred into 20 L of clean water with 1.8% sea salt (Yier Biological Engineering Co., Ltd., Guangzhou, China), dissolved to simulate a marine environment. An air pump (Sunsun group co., Ltd., Zhoushan, China) supplied oxygen continuously. Before the experiment, the oysters were immersed in salt water for 24 h. To ensure the cleanliness of the oysters used in the experiments, three oysters from each group were shucked, homogenized, and tested for NoV before being artificially contaminated. After 24 h of acclimation, all fresh oysters were randomly divided into two groups (16–18 oysters per group). Each group was exposed to 4 L of water with 1.8% sea salt under oxygenation with 1.0 mL NoV-positive GII.4 stool samples (final titer around 10^6 genome copies/liter) at room temperature to simulate a NoV-polluted environment. The water was circulated for 96 h continuously, and the experiment lasted for 5 days. During the whole experiment, oyster mortality was checked daily, and dead oysters were promptly removed. Two biological replicates were taken randomly at each time point from each group to ensure four parallel oyster samples for each time point.

2.2. Sample Collection and Processing Procedures

Oyster samples were collected at specific time points (6 h, 12 h, 24 h, 48 h, 72 h, 96 h) for further analysis. Samples were tested for NoV under ISO 15216 [19]. Briefly, the oysters' external surfaces were thoroughly washed and sterilized with alcohol before being opening. And then the oysters were dissected, and the digestive glands, gills, stomach, and labial palps were isolated. The latest sterile disposable plasticware and razor blades were used to process each sample. A tissue weighing 1.0 ± 0.1 g was measured and transferred to a 1.5 mL centrifuge tube. Proteinase K (1.0 mL, 3000 U/L) was added, and the samples were incubated for 1 h at 37 °C with stirring at 200 rpm. As a follow-up step, a 15 min incubation was conducted in a water bath at 60 °C. After centrifugation at 13,000× *g* for 5 min at 4 °C, the supernatant was carefully transferred to a clean tube. To extract viral RNA, the HiPure Viral RNA Kit (Magen, Guangzhou, China) was used following the manufacturer's guidelines. Viral RNA extraction was performed with 140.0 μL of supernatant, and elution was collected in 20.0 μL of DNase/RNase-free water. The extracted RNA was tested immediately, and the remaining RNA was stored at −80 °C for further molecular analysis.

2.3. Quantitative Real-Time PCR (RT-qPCR) for NoV

According to our previous report, duplicate one-step RT-qPCR assays for GI and GII NoV were conducted in a single tube using the Bio-Rad CFX96 qPCR machine [26]. Each 20.0 μL reaction mixture consisted of 6.0 μL of viral RNA samples. The cycling conditions were as follows: 5 min at 42 °C, and then 15 s at 95 °C, followed by 40 cycles consisting of 5 s at 95 °C and 10 s at 60 °C. Fluorescence was read at the end of each 60 °C extension step. The PCR cycle threshold (Ct) values were determined with Biorad CFX96 Manager Software version 3.1, which utilizes amplification-based threshold determination with default settings. A cycle threshold Ct value of ≤40 and a signal demonstrating a substantial exponential increase were considered positive samples. NoV GII genome copy numbers per reaction were calculated using an RT-qPCR standard curve, as previously described, and then converted to genome copies per gram of the tissues. Statistical evaluation was performed using Microsoft Excel and GraphPad Prism 7 (GraphPad Software, USA, www.graphpad.com).

3. Results

3.1. Dynamics of NoV Contamination in Oyster Tissues

In this comprehensive exploration of NoV contamination dynamics within oyster tissues, our focus was on the intricate interplay between the oyster's anatomical structure and its gastrointestinal circulatory system. A strategic selection of four tissues—the gill, palp, digestive gland, and stomach—was chosen to elucidate the internal circulation pathway and timeframe of NoV excretion within contaminated oysters (Figure 1) [27]. To achieve this, we employed RT-qPCR technology for precise quantitative analysis across tissues, with the original data available in Supporting Information Table S1. Results revealed a notable progression in NoV detection, with positive signals identified in all four tissues after the initial contamination stage at 6 h. After 12 to 24 h, only the gill and digestive gland tissues showed evidence of infection, with the gill tissue exhibiting a greater viral quantity than the digestive gland. The 48 h mark witnessed the virus reaching its zenith concentration in the digestive gland, coinciding with the absence of detectable virus in the gill tissue but reappearing in the stomach. A fascinating feature of this pattern is that it aligns with that of oyster tissues' digestive circulation, suggesting a specific enrichment and adsorption role for NoV in the gastrointestinal system. Beyond 72 h, only the digestive gland tissue displayed detectable NoV levels, with Ct values comparable to those at 12 and 24 h post-contamination, differing by a modest 2.3 Ct value from the peak at 48 h. At 96 h, only a small amount of NoV was detected in the digestive gland tissue. It is, however, insufficient to ascertain the virus's infectivity at this time due to the limited cultivation system for NoV.

3.2. Relationship between Sampling Sites and the Detection Rate

The study delved into the relationship between various sampling sites and the detection efficiency of NoV, revealing noteworthy variations in detection rates across different tissues (Figure 2). The average Ct values demonstrated distinct trends, registering 33.92 for the gill, 34.31 for the palp, 35.28 for the stomach, and 33.82 for the digestive gland. Standard deviations (Ct STDEV) indicated larger levels of variability in the gill and digestive gland tissues, with values of 1.49 compared to the palp (0.58) and stomach (1.87). However, the digestive gland exhibited outstanding performance with a detection rate of 95.83% (23/24). In contrast, the gill, palp, and stomach showed lower detection rates of 50.00% (12/24),

16.67% (4/24), and 33.33% (8/24), respectively. This underscores the digestive gland as the most reliable site for NoV detection, as detection rates in all tissues, excluding the digestive gland, were below the 50% threshold. The observed variability in detection rates underscores the importance of strategic tissue selection for NoV detection in oysters.

3.3. Relationship between Contamination Time and Detection Rate

The investigation explored the nuanced interplay between contamination time and NoV detection rate in oysters, revealing distinctive patterns through statistical analysis (Figure 3). At an early contamination time of 6 h, a robust detection rate of 93.75% (15/16) was observed, with an average Ct value of 35.33 and a standard deviation (Ct STDEV) of 1.51. As contamination progressed for 12 and 24 h, the detection rate remained stable at 50.00% (8/16), accompanied by average Ct values of 33.19 and 33.57, respectively, indicating a sustained level of contamination. However, at 48 h, a slight decrease in detection rate to 50.00% (8/16) was coupled with an increased Ct STDEV of 2.04, indicating increased variability in detection levels. Subsequently, at 72 and 96 h, detection rates declined to 25.00% (4/16), with corresponding average Ct values of 33.75 and 34.72 and notably lower Ct STDEV values of 0.37 and 0.09, respectively. Based on these findings, NoV detection in oysters is highly dependent on the duration of contamination, with a high detection rate initially descending over time. Considering the variability in Ct values and detection rates, timely monitoring is essential to accurately assess the level of NoV contamination in oysters.

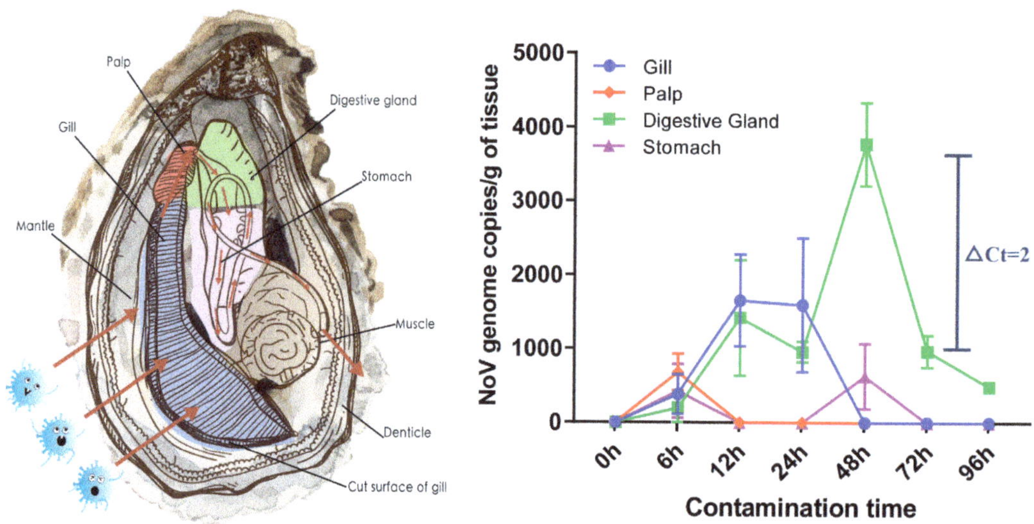

Figure 1. Schematic illustration of oyster tissue structure (**left**) and dynamics in the line chart of norovirus contamination in oyster tissues (**right**). Each tissue is colored from left to right in the same order. The bar graph presents the mean and STDEV from four parallel samples, with superimposed symbols at the mean and a connecting line. Oyster's anatomical structure and its gastrointestinal circulatory system were adapted with permission from Ref. [27]. 2021, Elsevier. The arrow indicates the direction of the virus's movement.

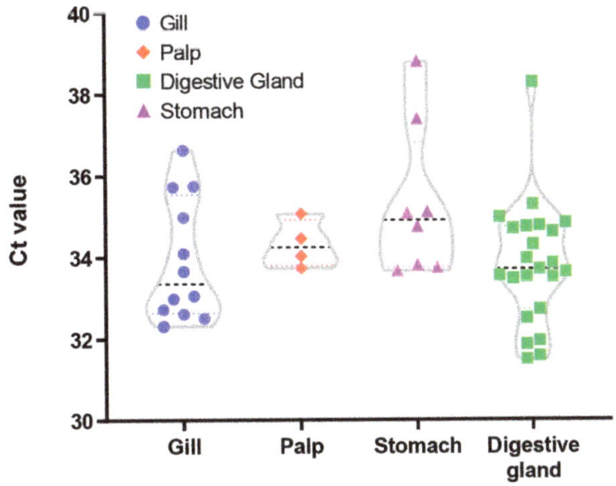

	Gill	Palp	Stomach	Digestive gland
Ct average	33.92	34.31	35.28	33.82
Ct STDEV	1.49	0.58	1.87	1.49
Detection rate	50.00% (12/24)	16.67% (4/24)	33.33% (8/24)	95.83% (23/24)

Figure 2. Violin plot of the relationship between sampling sites and Ct value. Each dot represents an organ from an oyster. Each compartment of the tissue is represented by a distinct color. Below are the density plots, and the corresponding tables highlight the calculated mean and standard deviation (STDEV) for each distribution. The mean and STDEV were calculated for positive samples from each issue group.

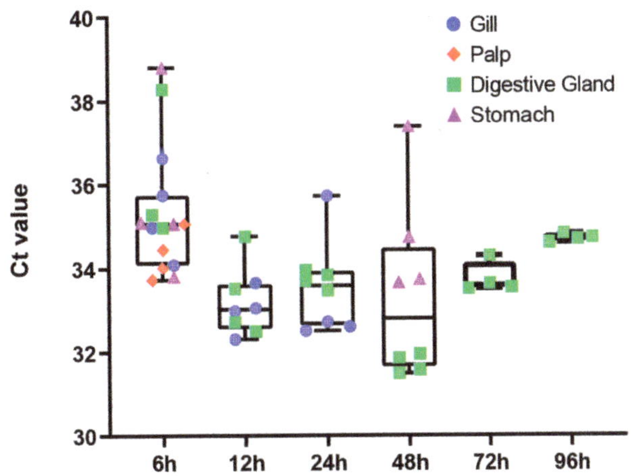

	6h	12h	24h	48h	72h	96h
Ct average	35.33	33.19	33.57	33.30	33.75	34.72
Ct STDEV	1.51	0.79	1.05	2.04	0.37	0.09
Detection rate	93.75% (15/16)	50.00% (8/16)	50.00% (8/16)	50.00% (8/16)	25.00% (4/16)	25.00% (4/16)

Figure 3. Boxplot diagram of the relationship between contamination time and detection rate. Each of the dots represents an organ from an oyster. Each tissue compartment is represented by a distinct color. Below are the density plots, and the corresponding tables highlight the calculated mean and standard deviation (STDEV) for each distribution. The mean and STDEV were calculated for positive samples of each time point group.

4. Discussion

The periodic monitoring of viral pathogens, such as NoV, is essential for determining infection status in the general population [15]. A gold standard for NoV detection has thus emerged in the form of RT-qPCR assays, recognized for their unparalleled sensitivity and ability to identify genetically diverse NoV strains [4]. In view of the fact that culture methods for NoV are not applicable to food matrices, the ISO 15216 approach relies instead on the method of RT-qPCR for detecting viral genomes, rather than the quantification of infectious viruses [19]. It should be noted that although the methods outlined in ISO 15216 are widely accepted for virus recovery and quantification from food surfaces, their applicability to the complex pretreatment matrices of food requires further assessment.

In order to mitigate the impact of food matrices on NoV detection, we investigated the spatial dynamic distribution of NoV in different oyster tissues. NoV contamination in oyster tissues was discussed in a nuanced manner based on our findings, including the impact of tissue-specific enrichment and the possibility of long-term persistence of NoV in digestive gland tissues. It can be concluded from our results that the digestive gland is the optimal target for NoV detection. Due to its consistency and high detection rate, the digestive gland makes an ideal target for NoV surveillance; it provides valuable insights for fine-tuning sampling strategies and improving surveillance progress accuracy. Timely and consistent monitoring plays a crucial role in accurately measuring the extent of NoV contamination in oysters, given the observed variability in Ct values and detection rates Additionally, the preprocessing steps outlined in the existing ISO methods demonstrate stability, and prompt testing upon sample collection holds the potential to enhance overall detection efficiency. This temporal perspective enhances the understanding of NoV dynamics in oyster populations and aids in the development of effective monitoring and mitigation strategies.

Detection methods for NoV in food samples continue to face challenges in effectively separating the virus from complex food matrices. A critical step to resolving the bottleneck previously mentioned is to identify where NoV is distributed in oyster tissues, what binding sites it uses, and what mechanisms it uses for adsorption. NoVs persist in oyster tissues for extended periods through binding to specific antigens in the gill and digestive gland tissue. Immunohistochemical experiments have further illustrated the enrichment of NoV viral particles and constructed virus-like particles (VLPs) in various oyster tissues [28]. Research has also identified substances in oyster tissues that can be recognized by monoclonal antibodies targeting histo-blood group antigens (HBGAs) [29,30], with A-type HBGAs particularly implicated in the adsorption of NoV [31,32]. In prior investigations, our team employed a GII.4-type NoV bacterial cell surface display system to capture, identify, and validate protein ligands specific to NoV binding in oyster tissues. We observed that oyster heat shock protein 70 (HSP70) is a candidate vital ligand for the specific binding of NoV in oyster tissues, providing a better understanding of NoV attachment and transmission in oysters [33]. Through the comprehensive investigation of the spatial distribution and enrichment dynamics of NoV within various oyster tissues in this project, we provide valuable support and direction for subsequent studies focused on specific adsorption sites and target selection. This research contributes essential insights to advance the understanding of NoV interactions within oyster tissues, addressing a pivotal aspect in the refinement of NoV detection strategies.

The enrichment of viruses in oyster tissues is primarily attributed to physical filtration during bivalve feeding. However, the accumulation of NoV in oyster tissues may involve more specific interactions, significantly complicating the purification process [34]. Heat treatment, while effective in inactivating the virus, compromises shellfish texture and flavor, thereby diminishing their sensory qualities and consumer value [35]. Purification procedures entail placing harvested shellfish in clean water to naturally reduce bacterial and viral content through the digestion and excretion processes. This method is widely employed in the global shellfish industry to eliminate or diminish the contamination of pathogens. Nevertheless, these purification methods exhibit limited efficacy in addressing contamination concerns. The depuration process can eliminate certain viruses such as feline calicivirus, but others, such as NoV, can remain persistent in oyster tissues for long periods [36].

Consequently, understanding the dynamics of NoV enrichment in certain tissues, including the digestive gland, will help target purification efforts and contribute to the development of more efficient decontamination processes for oysters. There may also be persistent latent infections that cannot be detected using PCR methods due to low viral loads. A 10- to 100-fold increase may occur due to the small volume analyzed by RT-qPCR, the low rate of recovery during genome extraction, and the presence of RT-qPCR inhibitors. Thus, except in some specific cases (i.e., in cases with very high levels of pollution or artificial contamination), the number of genome copies of NoV in oysters is usually close to the limit of detection. All these considerations highlight the difficulties in interpreting a positive or negative result when using such an approach [37,38]. Based on the findings of this study, it is recommended to extend the purification duration for an effective reduction of NoV contamination in oysters. The shellfish harvest area will be closed for a minimum of 30 days in Western Canada if the Canadian Food Inspection Agency (CFIA) detects NoV in any single oyster sample. Prolonged purification allows natural shellfish digestion and excretion to reduce viral content more comprehensively. Meanwhile, regular monitoring of water quality levels is essential during the purification process to ensure a controlled and safe environment. The Canadian Shellfish Sanitation Programme (CSSP) manual is a reference document for monitoring, classifying, and controlling areas where bivalve molluscan shellfish are harvested. It is necessary to detect oyster quality during transportation, storage, and processing. Standardization for prolonged purification in the oyster aquaculture industry is of paramount significance. This proactive approach aims to optimize the purification process and enhance its efficiency in mitigating NoV contamination risks in oysters.

Furthermore, the research has certain limitations due to time, energy, and constraints. The NoV extraction method, according to ISO 15216, was designed for digestive gland tissues. The reliability and validity of the gills and palps may not be equally reliable for different oyster tissues. Results should also be interpreted with caution due to the methodological limitations of primary studies, which are reflected in the low certainty of the evidence rating since only oyster tissue changes were considered and not shifts in temperature regimes, oyster species, or NoV genotypes. Some studies show that low temperatures weaken shellfish metabolism [39]. There were significantly higher viral loads among oysters from cold waters (5 °C) than those from warmer waters (>10 °C) [40]. Based on that, it sheds some light on the traditional oyster purification procedures and suggests that temperature might be considered a regulating method. Our study was conducted at room temperature. A further step towards understanding oyster spatial distribution and enrichment dynamics with NoV is to consider temperature as a variable. But again, more artificially contaminated experiments and actual field data are needed to confirm this.

5. Conclusions

To conclude, this study focused on the spatial distribution and enrichment dynamics of NoV within different oyster tissues, shedding light on crucial factors influencing the purification process. Various tissues, including the gills, palps, stomach, and digestive glands, were monitored to discern NoV contamination kinetics. The results offer a thorough assessment of NoV contamination in oysters, providing valuable data for the development of efficient purification strategies and detection methods. The findings contribute to ongoing efforts to improve food safety, particularly concerning oyster consumption, and emphasize the significance of refined purification techniques to reduce the dangers associated with NoV.

Supplementary Materials: The following supporting information can be downloaded at: https://www.mdpi.com/article/10.3390/foods13010128/s1, Table S1: The original data of quantitative real-time PCR (RT-qPCR) for Norovirus.

Author Contributions: Conceptualization, M.M., L.X. and D.L.; data curation, Z.Z., X.Z. and H.G.; funding acquisition, M.M., Z.Z. and D.L.; methodology, Z.Z., H.G., L.X. and D.L.; project administration, L.X.; resources, X.Z.; visualization, M.M. and D.L.; writing—original draft, M.M. and D.L.; writing—review and editing, Z.Z. and L.X. All authors have read and agreed to the published version of the manuscript.

Funding: This research was funded by the National Natural Science Foundation of China (32302250), the State Key Laboratory of Applied Microbiology Southern China (Grant No. SKLAM007-2022 and PKLMSH005-2021) and the Shanghai Sailing Program (22YF1415600).

Institutional Review Board Statement: Not applicable.

Informed Consent Statement: Not applicable.

Data Availability Statement: Data is contained within the article and supplementary material.

Conflicts of Interest: The authors declare no conflicts of interest.

References

1. Rupnik, A.; Keaveney, S.; Devilly, L.; Butler, F.; Doré, W. The Impact of Winter Relocation and Depuration on Norovirus Concentrations in Pacific Oysters Harvested from a Commercial Production Site. *Food Environ. Virol.* **2018**, *10*, 288–296. [CrossRef] [PubMed]
2. Havelaar, A.H.; Kirk, M.D.; Torgerson, P.R.; Gibb, H.J.; Hald, T.; Lake, R.J.; Praet, N.; Bellinger, D.C.; de Silva, N.R.; Gargouri, N.; et al. World health organization global estimates and regional comparisons of the burden of foodborne disease in 2010. *PLoS Med.* **2015**, *12*, e1001923. [CrossRef] [PubMed]
3. Atmar, R.L.; Opekun, A.R.; Gilger, M.A.; Estes, M.K.; Crawford, S.E.; Neill, F.H.; Graham, D.Y. Norwalk Virus Shedding after Experimental Human Infection. *Emerg. Infect. Dis.* **2008**, *14*, 1553–1557. [CrossRef] [PubMed]
4. Chhabra, P.; de Graaf, M.; Parra, G.I.; Chan, M.C.; Green, K.; Martella, V.; Wang, Q.; White, P.A.; Katayama, K.; Vennema, H.; et al. Updated classification of norovirus genogroups and genotypes. *J. Gen. Virol.* **2019**, *100*, 1393–1406. [CrossRef] [PubMed]
5. Jin, M.; Wu, S.; Kong, X.; Xie, H.; Fu, J.; He, Y.; Feng, W.; Liu, N.; Li, J.; Rainey, J.J.; et al. Norovirus Outbreak Surveillance, China, 2016–2018. *Emerg. Infect. Dis.* **2020**, *26*, 437–445. [CrossRef]
6. Cannon, J.L.; Barclay, L.; Collins, N.R.; Wikswo, M.E.; Castro, C.J.; Magaña, L.C.; Gregoricus, N.; Marine, R.L.; Chhabra, P.; Vinjé, J. Genetic and Epidemiologic Trends of Norovirus Outbreaks in the United States from 2013 to 2016 Demonstrated Emergence of Novel GII.4 Recombinant Viruses. *J. Clin. Microbiol.* **2017**, *55*, 2208–2221. [CrossRef] [PubMed]
7. Ludwig-Begall, L.F.; Mauroy, A.; Thiry, E. Noroviruses—The state of the art, nearly fifty years after their initial discovery. *Viruses* **2021**, *13*, 1541. [CrossRef] [PubMed]
8. Campos, C.J.A.; Lees, D.N. Environmental transmission of human noroviruses in shellfish waters. *Appl. Environ. Microb.* **2014**, *80*, 3552–3561. [CrossRef]
9. Desdouits, M.; Wacrenier, C.; Ollivier, J.; Schaeffer, J.; Le Guyader, F.S. A targeted metagenomics approach to study the diversity of norovirus GII in shellfish implicated in outbreaks. *Viruses* **2020**, *12*, 978. [CrossRef]
10. Scientific Opinion on an update on the present knowledge on the occurrence and control of foodborne viruses. *EFSA J.* **2011**, *9*, 2190.
11. Lowther, J.; Gustar, N.; Powell, A.; Brien, S.O.; Lees, D.N. A one-year survey of norovirus in UK oysters collected at the point of sale. *Food Environ. Virol.* **2018**, *10*, 278–287. [CrossRef] [PubMed]
12. Chassaing, M.; Boudaud, N.; Belliot, G.; Estienney, M.; Majou, D.; de Rougemont, A.; Gantzer, C. Interaction between norovirus and Histo-Blood Group Antigens: A key to understanding virus transmission and inactivation through treatments? *Food Microbiol.* **2020**, *92*, 103594. [CrossRef] [PubMed]
13. Green, T.J.; Yin Walker, C.; Leduc, S.; Michalchuk, T.; Mcallister, J.; Roth, M.; Janes, J.K.; Krogh, E.T. Spatial and temporal pattern of norovirus dispersal in an oyster growing region in the northeast pacific. *Viruses* **2022**, *14*, 762. [CrossRef] [PubMed]
14. Zhang, H.; Liu, D.; Zhang, Z.; Hewitt, J.; Li, X.; Hou, P.; Wang, D.; Wu, Q. Surveillance of human norovirus in oysters collected from production area in Shandong Province, China during 2017–2018. *Food Control* **2021**, *121*, 107649. [CrossRef]
15. Li, Y.; Xue, L.; Gao, J.; Cai, W.; Zhang, Z.; Meng, L.; Miao, S.; Hong, X.; Xu, M.; Wu, Q.; et al. A systematic review and meta-analysis indicates a substantial burden of human noroviruses in shellfish worldwide, with GII.4 and GII.2 being the predominant genotypes. *Food Microbiol.* **2023**, *109*, 104140. [CrossRef] [PubMed]
16. Raymond, P.; Paul, S.; Guy, R.A. Impact of capsid and genomic integrity tests on norovirus extraction recovery rates. *Foods* **2023**, *12*, 826. [CrossRef] [PubMed]
17. Qian, W.; Huang, J.; Wang, T.; Fan, C.; Kang, J.; Zhang, Q.; Li, Y.; Chen, S. Ultrasensitive and visual detection of human norovirus genotype GII.4 or GII.17 using CRISPR-Cas12a assay. *Virol. J.* **2022**, *19*, 150. [CrossRef]

18. Rupprom, K.; Chavalitshewinkoon-Petmitr, P.; Diraphat, P.; Kittigul, L. Evaluation of real-time RT PCR assays for detection and quantification of norovirus genogroups I and II. *Virol. Sin.* **2017**, *32*, 139–146. [CrossRef]
19. *ISO 15216-1:2017/Amd 1:2021*; Microbiology of the Food Chain—Horizontal Method for Determination of Hepatitis a Virus and Norovirus Using Real-Time RT-PCR—Part 1: Method for Quantification. International Organization for Standardization (ISO): Geneva, Switzerland, 2021.
20. Teunis, P.F.M.; Moe, C.L.; Liu, P.E.; Miller, S.; Lindesmith, L.; Baric, R.S.; Le Pendu, J.; Calderon, R.L. Norwalk virus: How infectious is it? *J. Med. Virol.* **2008**, *80*, 1468–1476. [CrossRef]
21. Tunyakittaveeward, T.; Rupprom, K.; Pombubpa, K.; Howteerakul, N.; Kittigul, L. Norovirus monitoring in oysters using two different extraction methods. *Food Environ. Virol.* **2019**, *11*, 374–382. [CrossRef]
22. Quang Le, H.; Suffredini, E.; Tien Pham, D.; Kim To, A.; De Medici, D. Development of a method for direct extraction of viral RNA from bivalve molluscs. *Lett. Appl. Microbiol.* **2018**, *67*, 426–434. [CrossRef] [PubMed]
23. Plante, D.; Bran Barrera, J.A.; Lord, M.; Iugovaz, I.; Nasheri, N. Development of an RNA extraction protocol for norovirus from raw oysters and detection by qRT-PCR and Droplet-Digital RT-PCR. *Foods* **2021**, *10*, 1804. [CrossRef] [PubMed]
24. Lowmoung, T.; Pombubpa, K.; Duangdee, T.; Tipayamongkholgul, M.; Kittigul, L. Distribution of naturally occurring norovirus genogroups i, II, and IV in oyster tissues. *Food Environ. Virol.* **2017**, *9*, 415–422. [CrossRef] [PubMed]
25. Pilotto, M.R.; Souza, D.S.M.; Barardi, C.R.M. Viral uptake and stability in Crassostrea gigas oysters during depuration, storage and steaming. *Mar. Pollut. Bull.* **2019**, *149*, 110524. [CrossRef] [PubMed]
26. Liu, D.; Zhang, Z.; Wu, Q.; Tian, P.; Geng, H.; Xu, T.; Wang, D. Redesigned duplex RT-qPCR for the detection of GI and GII human noroviruses. *Engineering* **2020**, *6*, 442–448. [CrossRef]
27. Kibenge, F.S.B.; Strange, R.J. Chapter 1—Introduction to the anatomy and physiology of the major aquatic animal species in aquaculture. In *Aquaculture Pharmacology*; Kibenge, F.S.B., Baldisserotto, B., Chong, R.S., Eds.; Academic Press: Cambridge, MA, USA, 2021; pp. 1–111.
28. Maalouf, H.; Zakhour, M.; Le Pendu, J.; Le Saux, J.; Atmar, R.L.; Le Guyader, F.S. Distribution in tissue and seasonal variation of norovirus genogroup i and II ligands in oysters. *Appl. Environ. Microb.* **2010**, *76*, 5621–5630. [CrossRef] [PubMed]
29. Le Guyader, F.S.; Loisy, F.; Atmar, R.L.; Hutson, A.M.; Estes, M.K.; Ruvoën-Clouet, N.; Pommepuy, M.; Le Pendu, J. Norwalk Virus–specific Binding to Oyster Digestive Tissues. *Emerg. Infect. Dis.* **2006**, *12*, 931–936. [CrossRef] [PubMed]
30. Tian, P.; Engelbrektson, A.L.; Jiang, X.; Zhong, W.; Mandrell, R.E. Norovirus recognizes histo-blood group antigens on gastrointestinal cells of clams, mussels, and oysters: A possible mechanism of bioaccumulation. *J. Food Prot.* **2007**, *70*, 2140–2147. [CrossRef]
31. Tan, M.; Xia, M.; Cao, S.; Huang, P.; Farkas, T.; Meller, J.; Hegde, R.S.; Li, X.; Rao, Z.; Jiang, X. Elucidation of strain-specific interaction of a GII-4 norovirus with HBGA receptors by site-directed mutagenesis study. *Virology* **2008**, *379*, 324–334. [CrossRef]
32. Morozov, V.; Hanisch, F.; Wegner, K.M.; Schroten, H. Pandemic GII.4 sydney and epidemic GII.17 kawasaki308 noroviruses display distinct specificities for Histo-Blood group antigens leading to different transmission vector dynamics in pacific oysters. *Front. Microbiol.* **2018**, *9*, 2826. [CrossRef]
33. Zhang, Z.; Liu, D.; Wu, Q.; Wang, D.; Johnson, K.N.; Johnson, K.N. Oyster heat shock protein 70 plays a role in binding of human noroviruses. *Appl. Environ. Microb.* **2021**, *87*, e79021. [CrossRef] [PubMed]
34. Torok, V.; Hodgson, K.; Mcleod, C.; Tan, J.; Malhi, N.; Turnbull, A. National survey of foodborne viruses in Australian oysters at production. *Food Microbiol.* **2018**, *69*, 196–203. [CrossRef] [PubMed]
35. Manuel, C.S.; Moore, M.D.; Jaykus, L. Predicting human norovirus infectivity—Recent advances and continued challenges. *Food Microbiol.* **2018**, *76*, 337–345. [CrossRef] [PubMed]
36. Qin, Q.; Shen, J.; Reece, K.S. A deterministic model for understanding nonlinear viral dynamics in oysters. *Appl. Environ. Microb.* **2022**, *88*, e236021. [CrossRef] [PubMed]
37. Hartard, C.; Leclerc, M.; Rivet, R.; Maul, A.; Loutreul, J.; Banas, S.; Boudaud, N.; Gantzer, C. F-Specific RNA bacteriophages, especially members of subgroup II, should be reconsidered as good indicators of viral pollution of oysters. *Appl. Environ. Microb.* **2018**, *84*, e01866-17. [CrossRef]
38. Lowther, J.A.; Gustar, N.E.; Hartnell, R.E.; Lees, D.N. Comparison of norovirus RNA levels in Outbreak-Related oysters with background environmental levels. *J. Food Protect.* **2012**, *75*, 389–393. [CrossRef]
39. Dore, B.; Keaveney, S.; Flannery, J.; Rajko-Nenow, P. Management of health risks associated with oysters harvested from a norovirus contaminated area, Ireland, February–March 2010. *Euro Surveill* **2010**, *15*, 19567. [CrossRef]
40. Campos, C.J.A.; Kershaw, S.; Morgan, O.C.; Lees, D.N. Risk factors for norovirus contamination of shellfish water catchments in England and Wales. *Int. J. Food Microbiol.* **2017**, *241*, 318–324. [CrossRef]

foods MDPI

Article

Development of a Sensitive and Specific Quantitative RT-qPCR Method for the Detection of Hepatitis E Virus Genotype 3 in Porcine Liver and Foodstuff

Jan Bernd Hinrichs [1], Antonia Kreitlow [1], Madeleine Plötz [1], Ulrich Schotte [2], Paul Becher [3], Nele Gremmel [3], Roger Stephan [4], Nicole Kemper [5] and Amir Abdulmawjood [1,*]

1 Institute of Food Quality and Food Safety, University of Veterinary Medicine Hannover, 30173 Hannover, Germany; jan.bernd.hinrichs@tiho-hannover.de (J.B.H.); antonia.kreitlow@tiho-hannover.de (A.K.); madeleine.ploetz@tiho-hannover.de (M.P.)
2 Department C Animal Health and Zoonoses, Central Institute of the Bundeswehr Medical Service Kiel, 24119 Kronshagen, Germany; ulrichschotte@bundeswehr.org
3 Institute of Virology, University of Veterinary Medicine Hannover, 30559 Hannover, Germany; paul.becher@tiho-hannover.de (P.B.); nele.gremmel@tiho-hannover.de (N.G.)
4 Vetsuisse Faculty, Institute for Food Safety and Hygiene, University of Zurich, 8057 Zurich, Switzerland; stephanr@fsafety.uzh.ch
5 Institute for Animal Hygiene, Animal Welfare and Farm Animal Behaviors, University of Veterinary Medicine Hannover, 30173 Hannover, Germany; nicole.kemper@tiho-hannover.de
* Correspondence: amir.abdulmawjood@tiho-hannover.de

Abstract: As an international and zoonotic cause of hepatitis, hepatitis E virus (HEV) poses a significant risk to public health. However, the frequency of occurrence and the degree of contamination of food of animal origin require further research. The aim of this study was to develop and validate a highly sensitive quantitative RT-qPCR assay for the detection and quantification of HEV contamination in porcine liver and food. The focus was on genotype 3, which is most common as a food contaminant in developed countries and Europe. The selected assay has its target sequence in the open reading frame 1 (ORF1) of the HEV genome and showed good results in inclusivity testing, especially for HEV genotype 3. The developed assay seems to show high efficiency and a low intercept when compared to other assays, while having a comparable limit of detection (LOD). In addition, a standard curve was generated using artificially spiked liver to provide more accurate quantitative results for contamination assessment and tracking in this matrix. Application of the assay to test 67 pig livers from different origins resulted in a positivity rate of 7.5%, which is consistent with the results of numerous other prevalence studies. Quantitative detection of the viral genome in the food chain, particularly in pig livers, is essential for understanding the presence and evolution of HEV contamination and thus ensures consumer safety.

Keywords: hepatitis E virus genotype 3; RT-qPCR; pork products; food chain control; public health

Citation: Hinrichs, J.B.; Kreitlow, A.; Plötz, M.; Schotte, U.; Becher, P.; Gremmel, N.; Stephan, R.; Kemper, N.; Abdulmawjood, A. Development of a Sensitive and Specific Quantitative RT-qPCR Method for the Detection of Hepatitis E Virus Genotype 3 in Porcine Liver and Foodstuff. *Foods* **2024**, *13*, 467. https://doi.org/10.3390/foods13030467

Academic Editor: Xinjun Du

Received: 10 January 2024
Revised: 25 January 2024
Accepted: 31 January 2024
Published: 1 February 2024

1. Introduction

Hepatitis E virus (HEV) is a globally occurring non-enveloped (+)-ssRNA virus that has been identified as the causative agent of human hepatitis E [1]. The virus belongs to the species *Paslahepevirus balayani* in the genus *Paslahepevirus* and within the family *Hepeviridae* and is currently divided into 8 different genotypes [2]. While genotypes 1 and 2 are mainly pathogenic for humans, genotypes 3 to 8 have been shown to infect other mammals with zoonotic potential for genotypes 3, 4, and 7 [3].

Transmission of the virus is mainly fecal–oral, with water being an important vector, especially for HEV genotypes 1 and 2, which are most common in developing countries.

In contrast, sporadic outbreaks in industrialized countries are mostly linked to the potentially zoonotic genotypes 3 and 4 [4,5].

Comparatively high anti-HEV-specific IgG prevalences indicate a wide distribution of the virus in Germany and Europe [6]. Although infection with genotype 3, which is the most common in Europe, usually appears to be subclinical, the health risk, especially for immunocompromised patients, cannot be neglected [5,7].

Since human-to-human transmission does not seem to play a major role with this genotype, zoonotic infections are assumed to play a relevant role in the spread of HEV in Europe [8]. The HEV genotype 3 primarily infects pigs in addition to humans, and the European domestic pig population, as well as wild boar, is suspected to act as an important virus reservoir [9,10].

Higher antibody prevalence in occupational groups with frequent contact with pigs suggests that direct transmission of the virus from pigs to humans is possible [11–13]. However, in various outbreaks of hepatitis E, a clear phylogenetic proximity of the virus detected in the patients to the HEV found in the food suspected as the source of infection could also be demonstrated [8,14]. It is therefore assumed that transmission through food of animal origin, especially from pigs, also plays a significant role in the spread of hepatitis E in Europe. In addition to the high antibody and virus prevalence in the European domestic and wild boar population, HEV RNA has been found in a large number of studies at various stages of the food chain, both in pork liver at the slaughterhouse as well as in ready-to-eat sausages in shops [15–18].

Since the virus appears to replicate mainly in hepatocytes, foods containing pork liver are particularly suspected as a source of infection [19]. Several outbreaks in Europe attributable to foodstuff were also linked to the consumption of insufficiently heated pork liver and liver sausages containing pork liver [8,20,21]. Therefore, meat from pigs or wild boar and pork liver sausage have been identified as risk factors for HEV infection in various countries, as well as oysters and offal [22,23]. As a result, there have been increasing proposals in recent years to improve the monitoring of food of animal origin and its raw materials for HEV contamination.

A better understanding of the development of HEV contamination during the processing of foods containing potentially contaminated parts, and in particular food containing pig liver, would also be essential to improve consumer safety [24].

As sufficiently effective cell lines for the cultivation of HEV have only recently become available and there is still no standardized procedure for the multiplication of the virus, as well as serological detection methods not being considered sufficiently sensitive, the detection methods are mostly based on the qualitative detection of the HEV genome [25,26].

With the official testing procedure BVL L 06.17.01-1:2020-11, a standardized and proven method for the extraction and qualitative detection of HEV in food is available in Germany, which uses the assay of Jothikumar et al. 2006 [27] to detect the virus RNA.

However, particularly for the assessment of the contamination level in the food chain and its distribution as well as its development, very sensitive quantitative detection is needed to assess the impact of individual measures, such as improved hygiene. A standard method for the quantification of HEV in food has not yet been established, which makes it difficult to assess the level of contamination in the food chain as accurately as possible.

This problem is partly due to the fact that most of the assays developed to date have focused on a sufficiently inclusive detection of HEV. This is made difficult by the high genetic diversity of the virus, which is particularly pronounced for genotype 3, the most relevant in food in Europe. Most of these assays target ORF2 or the partially overlapping ORF3 of the HEV genome [28]. The ORF3 region is the most homogeneous within the genome and lends itself to very inclusive detection of all genotypes [29]. However, other regions of the HEV genome may also be suitable for more sensitive detection of the food-relevant genotypes, such as a region at the beginning of ORF1 used by Mizuo et al. [30] or another region in ORF2 used by Erker et al. [31]. Most of the detection methods developed so far and applied to food are RT-qPCR assays [28], but some LAMP assays have also been presented in East Asia for monitoring contamination, especially with genotype 4 [32–34]. However, due to the highly heterogeneous HEV genome and the resulting difficulty in selecting a large homologous target area, especially in genotype 3, the RT-PCR assay seems to be more suitable for detection.

Most PCR assays developed to date were originally developed for the detection of HEV in serum from clinically ill patients or the detection in water, but only a few of these assays have been validated or developed for the detection of HEV contamination in food [28].

An example of a commonly used method is the PCR assay for water testing developed by Jothikumar et al., in 2006 [27], which was also tested with HEV-contaminated samples from pigs.

This assay has been widely and successfully used for the qualitative detection of HEV in the food chain and is characterized by its high inclusivity and the associated low probability of false-negative results [35]. It has also been used for quantification after calibration with an RNA-standard. However, especially for the quantitative evaluation of results from food samples, this test does not show optimal results due to the comparatively high ct values for genotype 3 samples.

Highly sensitive detection is required for the optimal quantitative interpretation of results, but this may compromise the inclusivity. For the quantitative evaluation of HEV in the blood of human patients, great progress has been made in recent years with the introduction of an international standard [36] for the calibration of the detection method, so that several quantitative detection methods are already available, e.g., Germer et al. [37] or Frias et al. [38]. For the highly heterogeneous and difficult to process food matrices, there is no such standardized material suitable for calibration, making the development of a quantitative detection method much more difficult.

The lack of such a standard material makes it necessary to test the application of a potentially quantitative method with either natively contaminated liver or appropriately spiked liver. In the case of naturally contaminated liver, sufficient extraction of HEV RNA from the matrix must be ensured, which is usually performed using an internal extraction control. Examples of such controls are the *E. coli* phage (MS2) or the *murine norovirus* (MNV-1), which are usually detected simultaneously with HEV detection in a multiplex assay [35].

One problem for spiking liver was the difficulty of specifically multiplying the HEV in cell culture, making it hard to obtain quantified material for such spiking. A uniform approach in which the sample used for spiking can be used both as an extraction control and to calibrate the quantification of the detection of HEV RNA in complex food matrices would be optimal, so that a calibration covering all steps is possible.

The aim of this study was to develop and validate a highly sensitive quantitative RT-qPCR method for the detection and quantification of HEV genotype 3 contamination in pig liver and food, considering the relevant HEV subtypes and taking into account losses during isolation and transcription in the calibration process.

2. Materials and Methods

2.1. Primer Design

Primers and probes were designed using the Primer-BLAST online design tool [https://www.ncbi.nlm.nih.gov/tools/primer-blast/, accessed on 17 February 2023] provided by the National Center for Biotechnology Information (NCBI) (Bethesda, MD, USA). The primers and probes were designed to be homologous to the reference sequences proposed by Smith et al., (2020) [2] of five different subtypes of HEV genotype 3 frequently found in food of animal origin in Europe [15,17]. Sequence data was obtained from the GenBank provided by NCBI [https://www.ncbi.nlm.nih.gov/genbank/, accessed on 17 February 2023] and included the reference genomes of HEV genotypes 3c, 3f, 3b, 3e, and 3i with GenBank accession numbers FJ70535, AB369687, AP003430, AB248521, and FJ998008, respectively. To prevent cross-reactions with other viruses belonging to the *Norovirus* and *Hepatovirus* genera, potential matches within the amplicon were excluded. This was achieved through computational alignment of the target sequence with the genomes of these viruses, available in the GenBank database.

On this basis, five different primer sets targeting regions in the ORF1 and ORF3 regions of the HEV genome were selected and pre-tested using a SYBR Green Real-time PCR to detect the first international HEV virus panel provided by the Paul Ehrlich Institute (PEI) (Langen, Germany) and a variety of HEV eluates from field samples. For the pre-test, HEV RNA was converted into DNA using a QuantiTect Reverse Transcription Kit™ (Qiagen GmbH, Hilden, Germany) and then analyzed using the FastStart Essential DNA Green I Mastermix™ and a LightCycler® 96 device from Roche Diagnostics GmbH (Mannheim, Germany). A complementary probe was designed for the primers that showed the best inclusivity and detection speed in these pre-tests, and this primer set was then used for further testing in a TaqMan PCR assay. The selected primer sequences (See Table 1) targeted an amplicon in the ORF1 region of the HEV genome located between bp 45 and 131 in the reference strain HEV subtype 3c (FJ705359). All primers and probes were ordered from *Eurofins* Genomics Germany GmbH (Ebersberg, Germany).

Table 1. Primers and probes used for RT-qPCR.

Primer:	Sequence 5′ → 3′	Region	GC%	Tm (°C)
JBH4-HEV		ORF1 (FJ705359)		
ForwardPrimer: Fw_JBH4-HEV	5′-TAAGGCTCCTGGCATTACTACT-3′	45–67	45.45	58.35
ReversePrimer: Rv_JBH4-HEV	5′-GCCGAACCACCACAGCATT-3′	113–131	57.89	61.27
Probe: P_JBH4-HEV	5′-[FAM]-CTGCTCTGGCTGCGGCCAA-[BHQ1]-3′	81–99	68.42	59.96
IPC-ntb2	Source: Anderson et al., (2011) [39]			
ForwardPrimer: IPC-ntb2-fw	5′-ACCACAATGCCAGAGTGACAAC-3′		50	68
ReversePrimer: IPC-ntb2-re	5′-TACCTGGTCTCCAGCTTTCAGTT-3′		47.82	68
Probe: IPC-ntb2	5′-[HEX]-CACGCGCATGAAGTTAGGGGACCA-[BHQ1]-3′		58.3	74

2.2. Implementation of an Internal Amplification Contral (IAC)

An IAC was implemented to assess the possible inhibition of the reaction by background DNA and other substances such as fat. To achieve this, the primers and probes proposed by Anderson et al., (2011) [39], which have been used successfully in other food-optimized detection assays were used (See Table 1). These primers target a sequence in the tobacco (*Nicotiana tabacum*) genome. A DNA-oligonucleotide matching this primer set's target sequence was prepared and quantified by Eurofins Genomics Germany GmbH and added to each reaction. To determine the optimal concentration of the amplification control in each reaction, a dilution series of the oligonucleotide was prepared at concentrations between 1 pmol/µL $\times 10^{-5}$ and 1 pmol/µL $\times 10^{-12}$ and tested in several runs using the PCR settings described in Section 2.3. Potential inhibition of the HEV RNA detection reaction by the IAC or the added primer set was also evaluated during this trial, by adding an HEV positive control to each reaction at a concentration close to the detection limit. The optimum concentration for the IAC, at which the IAC was close to its detection limit and no inhibition of HEV detection was observed during the preliminary testing, was determined to be 1 pmol/µL $\times 10^{-11}$, and 1 µL of this was added as IAC to each reaction.

2.3. Reverse Transcription Quantitative Real-Time PCR (RT-qPCR) Assay

The RT-qPCR was performed using 10 µL 2× master mix (Takyon™ No ROX 2X MasterMix), 0.2 µL reverse transcriptase and 0.2 µL additives provided by Takyon™ to stabilize the reverse transcription. The reagents were produced and provided by Eurogentec (Seraing, Belgium). The concentration of primers for the detection of HEV-RNA was 250 nM and 100 nM for the associated probe. For the implemented IAC, primer and probe concentration was 500 nM and 200 nM, respectively. The primer sets used are found in Table 1. Nuclease-free water was added to bring the total reaction volume to 20 µL. Reverse transcription was performed at a temperature of 48 °C for 20 min, followed by an initial denaturation step at 95 °C for 3 min. Subsequently, forty cycles of a two-step amplification were carried out, consisting of denaturation at 95 °C for 10 s and annealing at 60 °C for 60 s. At the end of each amplification cycle, amplification was detected by measuring an increase of fluorescence using the LightCycler® 96 device (Roche Diagnostics GmbH) The results were recorded and analyzed using the Light Cycler® Application Software 1.1 (Roche Diagnostics GmbH, Mannheim, Germany).

For each reaction, a positive control was performed in one reaction vessel to ensure a successful reaction, while a negative control was performed in another vessel to confirm that no non-specific reactions occurred.

2.4. Reference Method Used for Comparative Analysis of Samples

The official standard method for the qualitative detection of HEV in swine liver in accordance with BVL L 06.17.01-1:2020-11 was selected as reference method. Slight modifications were made to the RNA extraction and the PCR reaction parameters. For RNA extraction, a larger amount of tissue was first minced with a scalpel and mixed, of which 0.25 g of tissue was then used for the first homogenization step. This reduction was conducted in order to achieve a higher cell disruption in accordance with recent studies [40]. The reaction parameters of the PCR were adjusted accordingly in order to achieve optimum adaptation to the PCR device and the reagents used. Reverse transcription was therefore carried out at a temperature of 48 °C as recommended by the manufacturer of the reverse transcriptase (Takyon™). In addition, a shorter initial denaturation of 3 min at 95 °C was chosen, also as recommended by the manufacturer of the master mix, Takyon™. Amplification was performed for 45 cycles as a two-step amplification [Temp. °C, Ramp. °C/s (Duration s); 95 °C, 4.4 °C/s (10 s); 60 °C, 2.2 °C/s (30 s)] as recommended by the manufacturer of the PCR instrument (LightCycler® 96, Roche Diagnostics GmbH, Vienna, Austria).

2.5. Analytical Specificity of the JBH4-HEV RT-qPCR Assay

A total of 46 samples containing HEV RNA from different subtypes of HEV genotypes 3 and 4 were tested to determine the inclusivity of the JBH4-HEV RT-qPCR assay. The sample eluates tested were HEV reference material from the Paul Ehrlich Institute [36], RNA isolated from wild boar liver in previous HEV prevalence studies by Schotte et al. [41], RNA isolated from liver samples from the study by Wist et al. [42], and RNA isolated from field samples of wild boar and domestic pig liver by Gremmel et al. [43]. The selected subtypes were primarily chosen due to their association with foodborne transmission and their significance for human infections, covering HEV subtypes 3c, 3f, 3i, 3e, 3h, 3-i-like, 3b, 3ra, 4c and 4g [44]. Additionally, the exclusivity of the JBH4-HEV RT-qPCR assay was investigated using non-target virus material from closely related species and viruses circulating in the same environment as HEV. The tests included DNA and RNA of various enveloped and non-enveloped viruses provided by the Institute of Virology of the University of Veterinary Medicine Hannover. Potential cross-reactivity was screened with *Porcine circovirus 3* (PCV3), *Porcine parvovirus* (PPV), *Swine influenza virus* (SIV), *Pseudorabies virus* (PRV), *Atypical porcine pestivirus* (APPV), *Transmissible gastroenteritis virus* (TGEV) and the EU and US variants of the *Porcine reproductive and respiratory syndrome virus* (PRRSV). Cross-reactivity with other food-borne viruses such as norovirus or hepatitis A virus was also assessed by digital alignment of the amplicon with known sequences of these viruses available in the GenBank database.

In addition, RNA was isolated from muscle, fat, and liver tissues from pigs previously found to be HEV-negative by both the new assay and the reference method and tested to exclude possible matrix-related cross-reactivity. All eluates were screened using the RT-qPCR presented in this study and the previously described reference RT-qPCR to further validate the results.

2.6. Analytical Sensitivity of the JBH4-HEV RT-qPCR Assay

Analytical sensitivity was first determined using an RNA oligonucleotide designed to be complementary to the target region of the JBH4 HEV primer set in the ORF1 region of HEV subtype 3c (accession number FJ705359). The RNA oligonucleotide was generated and quantified by Eurofins Genomics GmbH. The specified stock concentration of 100 pmol/µL was first adjusted to 1 genome equivalent (GE)/µL $\times 10^{13}$. A dilution series from 1×10^7 GE/µL to 1×10^0 GE/µL was then prepared and tested in 6 replicates. Each dilution step was added directly to the reaction mixture and tested with the newly developed PCR assay as described above. The data obtained during the investigation of this dilution series under minimal external influences was used to create an initial standard curve.

This standard curve was used to assess possible losses during the isolation steps in the spiking experiment and to evaluate these overall.

2.7. Quantification of HEV RNA by the JBH4-HEV RT-qPCR Assay under Field Conditions by a Spiking Experiment

In order to quantify the samples tested as realistically as possible, an additional standard curve was generated based on a spiking experiment. By comparing this curve with the previously generated analytical standard curve based on the pure RNA oligonucleotide, it was possible to account for both isolation losses and interfering background substances from the liver sample.

For this purpose, 6 portions of 0.25 g liver tissue each were taken from a liver that had previously tested negative with both the new PCR and the reference PCR. After initial homogenization, three 25 mg portions were taken from each of these six samples and individually transferred to a new tube containing 600 µL RLT buffer (without ß-mercaptoethanol/DTT) and a GK60 Precellys Lysing Kit™ (Bertin-Technologies, Montigny-le Bretonneux, France). Prior to the second homogenization step, each sample was spiked with 10 µL of the RNA oligonucleotide at a specific dilution. The concentrations used ranged from 1×10^7 GE/µL to 1×10^0 GE/µL and were each tested three times, giving a total of nine samples. After the second homogenization step, the samples were further processed according to the protocol of the Qiagen RNeasy kit™ (Qiagen GmbH) and the isolated RNA was eluted in 50 µL of RNase-free water. If the spiked RNA was completely isolated, the resulting eluates would concentrations are to be expected that are approximately between 1×10^7 GE/µL and 1×10^0 GE/µL. The limit of detection was set at the dilution level at which 2/3 of the samples could be detected.

Treatment was performed on ice to minimize potential losses and isolated RNA eluate was immediately placed on ice and analyzed directly by RT-qPCR.

2.8. Validation of the JBH4-HEV RT-qPCR Assay by Testing Naturally Contaminated Porcine Liver

To evaluate the suitability of the assay for testing field samples, RNA was extracted from 67 livers and tested for HEV RNA using both the reference PCR method [27] and the newly established JBH4-HEV RT-qPCR assay. The livers were randomly collected from slaughterhouses, hunters, and retailers located in the vicinity of Hannover city. Liver samples were processed using a slightly modified extraction protocol according to the official method BVL L 06.17.01-1:2020-11. Modifications were made according to the publication by Zhao et al. [40]. Also longer centrifugation times were chosen.

Briefly, approximately 50 g of liver tissue was collected from four different subserosal sites, pre-mixed, weighed to a total weight of 0.25 g and then placed in a 2 mL reaction tube filled with zirconium beads and 250 µL of 1% phosphate-buffered saline. The sample was then homogenized using the Precellys-Evolution™ homogenizer (Bertin-Technologies) for 5 cycles of 30 s each at a speed of 5 ms^{-1}. The sample was then centrifuged at 13,000× g for 20 min at 5 °C and then 25 mg of the supernatant homogenate was added to a new tube containing 600 µL of buffer RLT (Qiagen GmbH) and homogenized again using the GK60 Precellys Lysing Kit™ (Bertin-Technologies) for 1 cycle of 20 s at a speed of 5 ms^{-1}. The RNA was then extracted and enriched by ultrafiltration using the Qiagen RNeasy Kit™ (Qiagen GmbH) according to the manufacturer's recommendations. All RNA samples were analyzed using both the newly developed assay and the established reference method to verify the results.

Quantitative evaluation of HEV contamination in samples tested positive was estimated using the standard curve generated within the previously described spiking experiment.

3. Results

3.1. Analytical Specificity of the JBH4-HEV RT-qPCR Assay

The JBH4-HEV RT-qPCR assay was able to detect all genotype three samples tested (see Figure 1), while the two genotype four samples showed no measurable amplification (see Table 2). Of the 46 HEV eluates tested, three were negative in the reference PCR, two of which belonged to genotype 3i and one to genotype 4g. The RNA samples isolated from liver and muscle of pigs previously tested negative for HEV in the reference assays also showed no detectable amplification within 40 amplification cycles. None of the samples from a total of eight non-target viruses showed a positive reaction within the 40 amplification cycles, while the IAC was detected in each reaction and showed no evidence of inhibition. There was also no detectable match between PCR primers and genome sequences of the Norovirus and Hepatovirus genera available in NCBI GenBank using the NCBI BLAST algorithm for alignment.

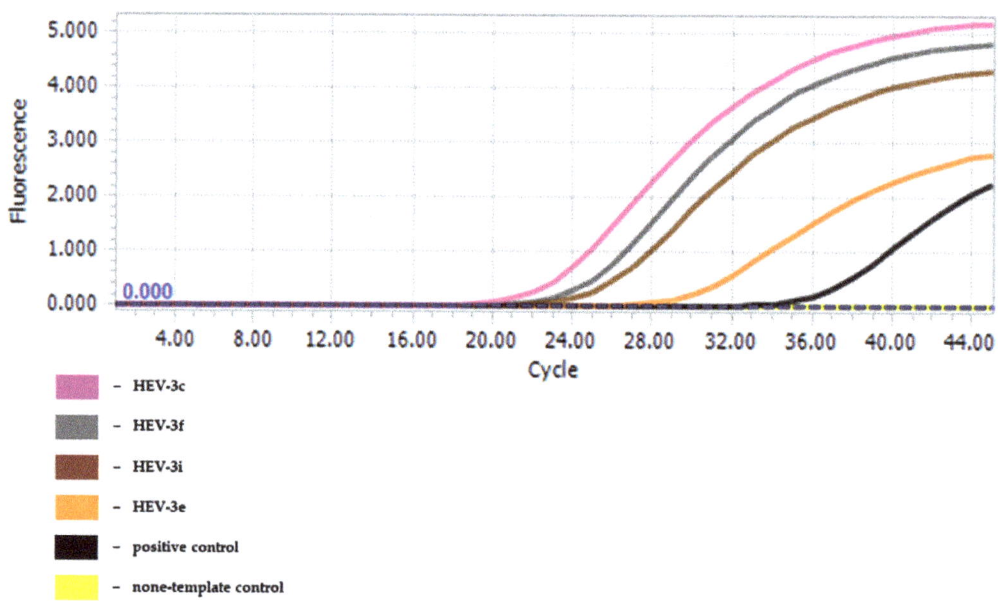

Figure 1. Amplification curves from the analysis of different HEV-positive samples isolated from pig liver and assigned to the HEV genotype 3.

Table 2. Viral eluates used for testing the analytical specificity of the JBH4-HEV RT-qPCR assay

Virus	Origin	Reactivity
Inclusivity		
HEV-Subtype 3b	HEV reference material PEI (Baylis et al., 2012) [45]	+
HEV-Subtype 3c	HEV reference material PEI (Baylis et al., 2012) [45]	+
HEV-Subtype 3c	Eluate isolated from pigliver (Schotte et al., 2022) [41]	+
HEV-Subtype 3c	Eluate isolated from pigliver (Schotte et al., 2022) [41]	+
HEV-Subtype 3c	Eluate isolated from pigliver (Schotte et al., 2022) [41]	+
HEV-Subtype 3c	Eluate isolated from pigliver (Schotte et al., 2022) [41]	+
HEV-Subtype 3c	Eluate isolated from pigliver (Schotte et al., 2022) [41]	+
HEV-Subtype 3c	Eluate isolated from pigliver (Schotte et al., 2022) [41]	+
HEV-Subtype 3c	Eluate isolated from pigliver (Schotte et al., 2022) [41]	+
HEV-Subtype 3c	Eluate isolated from pigliver (Schotte et al., 2022) [41]	+
HEV-Subtype 3c	Eluate isolated from pigliver (Schotte et al., 2022) [41]	+
HEV-Subtype 3c	Eluate isolated from pigliver (Schotte et al., 2022) [41]	+
HEV-Subtype 3c	Eluate isolated from pigliver (Schotte et al., 2022) [41]	+
HEV-Subtype 3c	Eluate isolated from pigliver (Schotte et al., 2022) [41]	+
HEV-Subtype 3c	Eluate isolated from pigliver (Schotte et al., 2022) [41]	+
HEV-Subtype 3c	Eluate isolated from pigliver (Schotte et al., 2022) [41]	+
HEV-Subtype 3c	Eluate isolated from pigliver (Schotte et al., 2022) [41]	+
HEV-Subtype 3c	Eluate isolated from pigliver (Schotte et al., 2022) [41]	+
HEV-Subtype 3c	Eluate isolated from pigliver (Gremmel et al., 2022) [43]	+
HEV-Subtype 3c	Eluate isolated from pigliver (Gremmel et al., 2022) [43]	+
HEV-Subtype 3c	Eluate isolated from pigliver (Gremmel et al., 2022) [43]	+
HEV-Subtype 3c	Eluate isolated from pigliver (Gremmel et al., 2022) [43]	+
HEV-Subtype 3e	HEV reference material PEI (Baylis et al., 2012) [45]	+
HEV-Subtype 3e	Eluate isolated from pigliver (Gremmel et al., 2022) [43]	+
HEV-Subtype 3f	HEV reference material PEI (Baylis et al., 2012) [45]	+
HEV-Subtype 3f	Eluate isolated from pigliver (Schotte et al., 2022) [41]	+
HEV-Subtype 3f	Eluate isolated from pigliver (Schotte et al., 2022) [41]	+
HEV-Subtype 3f	Eluate isolated from pigliver (Schotte et al., 2022) [41]	+
HEV-Subtype 3f	Eluate isolated from pigliver (Schotte et al., 2022) [41]	+
HEV-Subtype 3f	Eluate isolated from pigliver (Schotte et al., 2022) [41]	+
HEV-Subtype 3f	Eluate isolated from pigliver (Schotte et al., 2022) [41]	+
HEV-Subtype 3f	Eluate isolated from pigliver (Schotte et al., 2022) [41]	+
HEV-Subtype 3h	Eluate isolated from pigliver (Wist et al., 2018) [42]	+
HEV-Subtype 3i	Eluate isolated from pigliver (Schotte et al., 2022) [41]	+
HEV-Subtype 3i	Eluate isolated from pigliver (Schotte et al., 2022) [41]	+
HEV-Subtype 3i	Eluate isolated from pigliver (Schotte et al., 2022) [41]	+
HEV-Subtype 3i	Eluate isolated from pigliver (Schotte et al., 2022) [41]	+
HEV-Subtype 3i	Eluate isolated from pigliver (Schotte et al., 2022) [41]	+
HEV-Subtype 3i	Eluate isolated from pigliver (Schotte et al., 2022) [41]	+
HEV-Subtype 3i	Eluate isolated from pigliver (Schotte et al., 2022) [41]	+
HEV-Subtype 3i-like	Eluate isolated from pigliver (Schotte et al., 2022) [41]	+
HEV-Subtype 3i-like	Eluate isolated from pigliver (Schotte et al., 2022) [41]	+
HEV-Subtype 3i-like	Eluate isolated from pigliver (Schotte et al., 2022) [41]	+
HEV-Subtype 3ra	HEV reference material PEI (Baylis et al., 2012) [45]	+
HEV-Subtype 4c	HEV reference material PEI (Baylis et al., 2012) [45]	-
HEV-Subtype 4g	HEV reference material PEI (Baylis et al., 2012) [45]	-
Exclusivity		
PCV3	IoV [1]	-
PPV	IoV [1]	-
SIV	IoV [1]	-
PRV	IoV [1]	-
APPV	IoV [1]	-
TGEV	IoV [1]	-
PRRSV EU	IoV [1]	-
PRRSV US	IoV [1]	-

[1] Institute of Virology, University of Veterinary Medicine Hannover.

3.2. Analytical Sensitivity of the JBH4-HEV RT-qPCR Assay

In the first experiment, the target RNA was detected up to a dilution level of 1×10^1 GE/µL (GenomeEquivalents/µL), corresponding to a number of 20 copies per reaction. Results are found in Table 3. From the six-fold repetition of the measurement of the individual concentration levels, a standard curve was calculated, which could be described with the equation y = −3.295x + 37.811. The value m (−3.295) indicated the slope of the curve and the value b (37.811) indicated the Y-value for log level 0, i.e., the calculated Cq-value for the detection of 1 copy per reaction. Using the formula $E = (10^{(-1/slope)})-1$, the efficiency was then calculated from its correlation with the slope. The equation $E = (10^{(-1/-3.1216)})-1$ thus resulted in a PCR efficiency of 1.011. R^2 was determined at 0.9909.

Table 3. Detection of a dilution series of pure RNA oligonucleotide.

GE/µL	Positive/Tested	Average Cq-Value	Average Deviation
1×10^7	6/6	13.24	0.217
1×10^6	6/6	16.90	0.168
1×10^5	6/6	20.46	0.059
1×10^4	6/6	23.84	1.193
1×10^3	6/6	26.92	0.132
1×10^2	6/6	31.62	1.995
1×10^1	6/6	33.83	0.880
1×10^0	2/6	35.47	0.311

3.3. Analytical Sensitivity of the JBH4-HEV RT-qPCR Assay under Field Conditions in Spiked Liver Samples

Subsequently, the target RNA was also used to spike liver material for generating a matrix-specific standard curve. As described in Section 2.7, 10 µL of each of the dilution levels 10^6 to 10^1 GE/µL were spiked, so that a total of 10^7 to 10^2 GE were present in 25 mg of material. The corresponding results are shown in Table 4.

Table 4. Detection rate of target RNA in artificially contaminated liver tissue at different contamination levels.

GE/25 mg Sample	Tested/Positive	Average Cq-Value	Average Deviation
1×10^7	9/9	20.805	0.285
1×10^6	9/9	24.058	0.220
1×10^5	9/9	27.165	0.415
1×10^4	9/9	30.611	0.353
1×10^3	7/9	35.324	2.072
1×10^2	2/9	36.665	1.831

The resulting standard curve could be described with the equation y = −3.3297x + 38.569. The equation $E = (10^{(-1/-3.3297)})-1$ resulted in a PCR efficiency of 0.9967, and R^2 at 0.9888 (see Figure 2). The detection limit was set at 1×10^3 GE/25 mg liver material. No reaction was affected by inhibition as the IAC was in every sample.

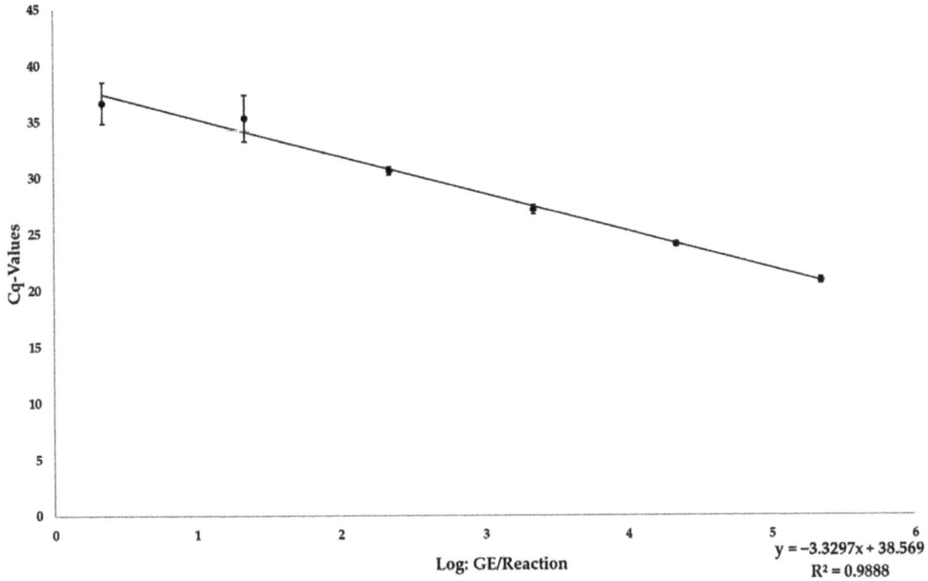

Figure 2. Standard curve from the results of measuring the analytical sensitivity of the JBH4-HEV RT-qPCR assay under field conditions by analyzing liver spiked with different dilution levels of target RNA.

3.4. Validation of the JBH4-HEV RT-qPCR Assay

Of the 67 livers tested, a total of 5 showed a positive reaction in both the new assay and the reference method BVL L 06.17.01-1:2020-11. Of these 5 positive livers, 4/5 were from domestic pigs and 1/5 from wild boar, 3/5 were purchased from local butcher shops, 1/5 was purchased directly at the slaughterhouse and one was from hunting. The contaminations were quantified and fluctuated considerably with values between 972 GE/25 mg and 215,583 GE/25 mg.

4. Discussion

The molecular detection of food contaminants using PCR is challenging due to the high heterogeneity of the matrices and the associated potentially high proportion of interfering substances. The high genetic diversity of the HEV genome, particularly genotype 3, which is one of the predominant food contaminants in Europe, makes it difficult to select a suitable target region and specific primers for detecting the presence of the HEV genome [2,29,46].

Inoue et al. [47] identified three regions in the HEV genome that could be suitable for inclusive molecular detection of HEV due to a sufficiently high sequence identity of over 75%. These include the 5′-untranslated-region (UTR) with the 5′-terminal part of ORF1, the overlapping ORF2/ORF3 region and the central part of ORF2. Most assays focus on the regions in ORF2 and ORF3 to achieve high inclusivity, such as the most commonly used assay for the detection of HEV in food by Jothikumar et al. [27]. In the case of a prioritized consideration of HEV genotype 3 subtypes, and thus the major HEV genotype in pork-containing foods, primer design in ORF1 is sufficiently comprehensive and may offer advantages in terms of sensitivity. In this study, various primer sets were tested in preliminary tests. The selected ORF1-based primer set outperformed the alternative variants in terms of analytical inclusivity and sensitivity.

Since most of the developed assays to date were originally validated for human diagnostics and were only partially used for HEV detection in food, comparable data about their performance in food analysis are lacking. An exception is the HEV assay by

Jothikumar et al. [27], which has also been established for investigating environmental samples. As this assay has been shown to be probably the most sensitive for detecting HEV in pig samples by Ward et al. [35], this assay was used in many prevalence studies including food samples [28]. For this reason, it was selected as reference assay in this study. A total of 46 HEV RNA containing samples were analyzed with both the new JBH4_HEV RT-qPCR assay and the reference assay. The three previously confirmed HEV genotype 3 samples that tested negative using the assay by Jothikumar et al. [27] could be correctly identified by the new JBH4_HEV RT-qPCR suggesting a higher sensitivity. In addition, inclusivity was tested using samples of HEV genotypes 3 and 4 from the World Health Organization's first international reference panel for different hepatitis E virus genotypes for nucleic acid amplification assays [36]. All HEV genotype 3 samples were tested positive whereas the 2 samples of genotype 4, the predominant food-associated HEV variant in East Asia, were not detectable using the JBH4_HEV RT-qPCR. The genotype 4g sample was tested negative in both the new and reference PCR assay while genotype 4c was amplified with a comparatively high cq value of 38 in the reference PCR assay. Insufficient RNA isolation could be considered as the cause for the false negative results, as the selected isolation method was not optimized for serum samples.

The inclusivity test performed in this study focused on subtypes of HEV genotype 3, which are most frequently detected in European foods. As most PCR assays are designed to detect the different HEV genotypes associated with humans, they have been validated for a wide range of genotypes, but with less emphasis on HEV genotype 3 in and from food. For example, Enouf et al. [40] tested eight HEV genotype 3 eluates from patient serum, whereas many other assays determined inclusivity only by comparison with genomic databases. In the study of Jothikumar et al. [27], only 10 HEV genotype 3 eluates were considered. By testing 42 different HEV eluates, the inclusivity of the JBH4_HEV RT-qPCR assay was therefore more thoroughly validated for food-related applications.

Investigating the exclusivity of the JBH4_HEV RT-qPCR included viral contaminants that potentially occur in porcine liver. Some PCR assays developed for detection of HEV showed a similar strategy in their exclusivity testing, e.g., Kaba et al. [48] and Jothikumar et al. [27]. With no false positives, the JBH4_HEV RT-qPCR performed similarly to previously established assays.

The determination of analytical sensitivity and the development of a quantitative HEV-specific RT-qPCR assay is a major challenge, as until recently there were no efficient cell culture models for virus propagation and subsequent quantification of the purified analyte [26]. Although precise quantification of, e.g., liver material is already possible with digital droplet PCR [49], this method is not yet suitable for mass production due to its limited distribution and high costs. Quantified reference material, such as the International Reference Panel [36], which is available for the evaluation of PCR tests for human diagnostics based on serum or fecal samples, does not exist in this form for pig liver and therefore for food matrices. In order to avoid the need for virus replication, artificially produced DNA- and RNA-oligonucleotides, which are identical to the amplicon, can be used for determining analytical sensitivity and detection limit in various food products and raw materials.

This strategy can help to achieve a good comparability between the performance of different assays. For example, compared with the four standard curves obtained by Ward et al. [35] using various HEV-specific TaqMan-RT-qPCR assays, the JBH4_HEV RT-qPCR assay showed a very good efficiency of 101.1% and a comparatively low intercept of 37.81 for the pure analyte. A lower intercept value of 35.96 was only achieved by Ahn et al. [50]. However, their assay showed limitations in terms of inclusivity for the HEV genotype 3 occurring in pigs.

The analytical sensitivity determined for the JBH4_HEV RT-qPCR assay ranged between 2 and 20 GE per reaction and thus, was comparable to PCR assays developed by Gyarmati et al. [51] (1 and 20 GE per reaction), Enouf et al. [52] (10 GE per reaction), Ahn et al. [50] (16.8 GE per reaction) and Jothikumar et al. [27] (4 GE per reaction). In this study, these data were supplemented by a spiking experiment using the amplicon-based RNA oligonucleotide for artificial contamination of porcine liver to assess the detection limit and performance of the assay under the influence of matrix-specific factors.

The achieved detection limit of 20 GE/reaction, an efficiency of 99.67% and a relatively low intercept of 38.57 for the standard curve obtained confirmed that the JBH4_HEV RT-qPCR assay was suitable, highly sensitive and reliable tool for detecting and quantifying HEV in complex samples even taking into account potential RNA losses during sample processing. The cq values measured for each sample were reproducible in all repetitions and showed an overall average deviation of 0.69 Ct. It can be concluded that the RNA yields were stable after each isolation process and that the extraction method used was compatible with the subsequent JBH4_HEV RT-qPCR assay and sufficiently robust. This evaluation of the extraction method including mechanical disruption and subsequent ultrafiltration was consistent with the findings of Hennechart–Collette et al. [53] who recommended the procedure for use on liver sausages.

When using the JBH4_HEV RT-qPCR assay to analyze pig livers from various sources in the Hannover region, 5 out of 67 livers were found to be positive, which corresponds to a prevalence of 7.5%. This result is therefore very close to the results of various prevalence studies on pig livers in Germany and Europe [9], such as the prevalence of 4% found by Wenzel et al. [54] for pig livers from domestic pigs in south-western Germany or the prevalence of 13.5% and 11% found by Baechlein et al. [55] and Boxmann et al. [56] for pig livers from domestic pigs in Germany and the Netherlands, respectively.

Comparably high percentages of positive liver samples were also found in HEV prevalence studies in wild boar, such as the 5.9% positive samples found by Gremmel et al. [43] in northern Germany or the values of 2.8% to 13.3% HEV-positive samples between 2013 and 2017 described by Schotte et al. [41] in various regions of Germany.

The difference in prevalence in the present study may be due to the smaller sample size of 67 livers, considering that Gremmel et al. [43], Baechlein et al. [55] and Boxmann et al. [56] sampled over 200 animals, while Schotte et al., sampled approximately 3500 [41].

In the study by Boxmann et al. [56], the average contamination of the livers analyzed was log(10) 3.2 GE/0.1 g with a range between 1.7–6.2 GE; in the study by Baechlein et al. [55], an average load of log(10) 8 GE/0.1 g is stated.

This range of results is therefore largely consistent with the results of this study, in which an average native viral load of log(10) 4.8 GE/0.1 g was determined in pig liver, with a range of 3.8 to 6.11 GE/0.1 g. In most studies, however, the HEV load of the samples was not quantitatively assessed, so that only very limited comparative data are available [48,57,58].

In general, it is difficult to estimate the risk for consumers from exposure to contaminated pig liver and products thereof. This is due to the fact that no human infectious dose has yet been established for HEV, so reference values from pigs or monkeys have to be used [25]. Furthermore, only limited conclusions can be drawn about the infectivity of the virus from positive PCR results because the method only detects RNA, but not the extent to which it is also a component of an infectious virus or an already partially denatured structure. Previously proposed methods of differentiation, such as the addition of RNase to denature free viral RNA or selective dyes, can only be indicative [59]. A recent study focusing on the isolation of HEV from liver samples showed that 83.3% (15 out of 18) HEV RNA-positive samples contained infectious hepatitis E viral particles and therefore must be considered as a potential source for human infections [43]. Overall, the JBH4_HEV RT-qPCR assay could be adapted as a good tool for use in relevant food matrices with an increased risk of HEV contamination, such as meat products containing pork liver. Specializing and optimizing the assay for these matrices could thus help to track the progression of HEV contamination in the food chain by quantifying the contamination within the different production stages, improve food control, and verify the effectiveness of reduction measures. It can therefore make a significant contribution to protecting consumers from HEV infection by improving food safety through better identification of critical control points in the processing of HEV-contaminated pork liver and through sensitive monitoring of potentially contaminated food. Further research is required to investigate the occurrence and development of the hepatitis E virus in the European food chain [24], including substantial screening of animal food stuffs, particularly pig liver, and more precise recording of the possible food-related spread. Finally, the JBH4_HEV RT-qPCR assay is a suitable tool for generating comprehensive and detailed data to fill these gaps.

Author Contributions: Conceptualization, A.K. and A.A.; Methodology, J.B.H. and A.K.; Validation, J.B.H.; Formal analysis, J.B.H.; Investigation, J.B.H.; Resources, J.B.H., U.S., P.B., N.G., R.S. and A.A.; Data curation, J.B.H.; Writing—original draft, J.B.H.; Writing—review & editing, A.K. M.P., U.S., P.B., N.G., R.S., N.K. and A.A.; Visualization, J.B.H.; Supervision, A.K., M.P. and A.A.; Project administration, A.K. and A.A. All authors have read and agreed to the published version of the manuscript.

Funding: This Open Access publication was funded by the Deutsche Forschungsgemeinschaft (DFG, German Research Foundation)—491094227 "Open Access Publication Funding" and the University of Veterinary Medicine Hannover, Foundation. The study was funded by the Fritz Ahrberg Foundation under the funding number 60070066.

Institutional Review Board Statement: Not applicable.

Informed Consent Statement: Not applicable.

Data Availability Statement: The original contributions presented in the study are included in the article, further inquiries can be directed to the corresponding author.

Acknowledgments: The authors would like to thank Silke Ortaeri and Anke Bertling for their excellent technical assistance.

Conflicts of Interest: The authors declare no conflict of interest.

References

1. Kamar, N.; Bendall, R.; Legrand-Abravanel, F.; Xia, N.-S.; Ijaz, S.; Izopet, J.; Dalton, H.R. Hepatitis E. *Lancet* **2012**, *379*, 2477–2488. [CrossRef]
2. Smith, D.B.; Izopet, J.; Nicot, F.; Simmonds, P.; Jameel, S.; Meng, X.J.; Norder, H.; Okamoto, H.; van der Poel, W.H.M.; Reuter, G.; et al. Update: Proposed reference sequences for subtypes of hepatitis E virus (species *Orthohepevirus A*). *J. Gen. Virol.* **2020**, *101*, 692–698. [CrossRef]
3. Thiry, D.; Mauroy, A.; Pavio, N.; Purdy, M.A.; Rose, N.; Thiry, E.; de Oliveira-Filho, E.F. Hepatitis E Virus and Related Viruses in Animals. *Transbound. Emerg. Dis.* **2017**, *64*, 37–52. [CrossRef] [PubMed]
4. Aslan, A.T.; Balaban, H.Y. Hepatitis E virus: Epidemiology, diagnosis, clinical manifestations, and treatment. *World J. Gastroenterol.* **2020**, *26*, 5543–5560. [CrossRef] [PubMed]
5. Lewis, H.C.; Wichmann, O.; Duizer, E. Transmission routes and risk factors for autochthonous hepatitis E virus infection in Europe: A systematic review. *Epidemiol. Infect.* **2010**, *138*, 145–166. [CrossRef]
6. Izopet, J.; Tremeaux, P.; Marion, O.; Migueres, M.; Capelli, N.; Chapuy-Regaud, S.; Mansuy, J.M.; Abravanel, F.; Kamar, N.; Lhomme, S. Hepatitis E virus infections in Europe. *J. Clin. Virol.* **2019**, *120*, 20–26. [CrossRef]
7. Velavan, T.P.; Pallerla, S.R.; Johne, R.; Todt, D.; Steinmann, E.; Schemmerer, M.; Wenzel, J.J.; Hofmann, J.; Shih, J.W.K.; Wedemeyer, H.; et al. Hepatitis E: An update on One Health and clinical medicine. *Liver Int.* **2021**, *41*, 1462–1473. [CrossRef]
8. Treagus, S.; Wright, C.; Baker-Austin, C.; Longdon, B.; Lowther, J. The Foodborne Transmission of Hepatitis E Virus to Humans. *Food Environ. Virol.* **2021**, *13*, 127–145. [CrossRef] [PubMed]
9. Pavio, N.; Doceul, V.; Bagdassarian, E.; Johne, R. Recent knowledge on hepatitis E virus in Suidae reservoirs and transmission routes to human. *Vet. Res.* **2017**, *48*, 78. [CrossRef]
10. Anheyer-Behmenburg, H.E.; Szabo, K.; Schotte, U.; Binder, A.; Klein, G.; Johne, R. Hepatitis E Virus in Wild Boars and Spillover Infection in Red and Roe Deer, Germany, 2013–2015. *Emerg. Infect. Dis.* **2017**, *23*, 130–133. [CrossRef] [PubMed]
11. Faber, M.; Willrich, N.; Schemmerer, M.; Rauh, C.; Kuhnert, R.; Stark, K.; Wenzel, J.J. Hepatitis E virus seroprevalence, seroincidence and seroreversion in the German adult population. *J. Viral Hepat.* **2018**, *25*, 752–758. [CrossRef]
12. Meng, X.J.; Wiseman, B.; Elvinger, F.; Guenette, D.K.; Toth, T.E.; Engle, R.E.; Emerson, S.U.; Purcell, R.H. Prevalence of antibodies to hepatitis E virus in veterinarians working with swine and in normal blood donors in the United States and other countries. *J. Clin. Microbiol.* **2002**, *40*, 117–122. [CrossRef] [PubMed]
13. Montagnaro, S.; De Martinis, C.; Sasso, S.; Ciarcia, R.; Damiano, S.; Auletta, L.; Iovane, V.; Zottola, T.; Pagnini, U. Viral and Antibody Prevalence of Hepatitis E in European Wild Boars (*Sus scrofa*) and Hunters at Zoonotic Risk in the Latium Region. *J. Comp. Pathol.* **2015**, *153*, 1–8. [CrossRef] [PubMed]

14. Matsuda, H.; Okada, K.; Takahashi, K.; Mishiro, S. Severe hepatitis E virus infection after ingestion of uncooked liver from a wild boar. *J. Infect. Dis.* **2003**, *188*, 944. [CrossRef] [PubMed]
15. Ferri, G.; Vergara, A. Hepatitis E Virus in the Food of Animal Origin: A Review. *Foodborne Pathog. Dis.* **2021**, *18*, 368–377. [CrossRef]
16. Pallerla, S.R.; Schembecker, S.; Meyer, C.G.; Linh, L.T.K.; Johne, R.; Wedemeyer, H.; Bock, C.T.; Kremsner, P.G.; Velavan, T.P. Hepatitis E virus genome detection in commercial pork livers and pork meat products in Germany. *J. Viral Hepat.* **2021**, *28*, 196–204. [CrossRef]
17. Szabo, K.; Trojnar, E.; Anheyer-Behmenburg, H.; Binder, A.; Schotte, U.; Ellerbroek, L.; Klein, G.; Johne, R. Detection of hepatitis E virus RNA in raw sausages and liver sausages from retail in Germany using an optimized method. *Int. J. Food Microbiol.* **2015**, *215*, 149–156. [CrossRef]
18. Harrison, L.; Ramos, T.M.; Wu, X.; DiCaprio, E. Presence of hepatitis E virus in commercially available pork products. *Int. J. Food Microbiol.* **2021**, *339*, 109033. [CrossRef]
19. Choi, C.; Chae, C. Localization of swine hepatitis E virus in liver and extrahepatic tissues from naturally infected pigs by in situ hybridization. *J. Hepatol.* **2003**, *38*, 827–832. [CrossRef]
20. Said, B.; Ijaz, S.; Chand, M.A.; Kafatos, G.; Tedder, R.; Morgan, D. Hepatitis E virus in England and Wales: Indigenous infection is associated with the consumption of processed pork products. *Epidemiol. Infect.* **2014**, *142*, 1467–1475. [CrossRef]
21. Motte, A.; Roquelaure, B.; Galambrun, C.; Bernard, F.; Zandotti, C.; Colson, P. Hepatitis E in three immunocompromized children in southeastern France. *J. Clin. Virol.* **2012**, *53*, 162–166. [CrossRef]
22. Mansuy, J.M.; Gallian, P.; Dimeglio, C.; Saune, K.; Arnaud, C.; Pelletier, B.; Morel, P.; Legrand, D.; Tiberghien, P.; Izopet, J. A nationwide survey of hepatitis E viral infection in French blood donors. *Hepatology* **2016**, *63*, 1145–1154. [CrossRef]
23. La Rosa, G.; Muscillo, M.; Vennarucci, V.S.; Garbuglia, A.R.; La Scala, P.; Capobianchi, M.R. Hepatitis E virus in Italy: Molecular analysis of travel-related and autochthonous cases. *J. Gen. Virol.* **2011**, *92*, 1617–1626. [CrossRef] [PubMed]
24. Ripellino, P.; Pianezzi, E.; Martinetti, G.; Zehnder, C.; Mathis, B.; Giannini, P.; Forrer, N.; Merlani, G.; Dalton, H.R.; Petrini, O.; et al. Control of Raw Pork Liver Sausage Production Can Reduce the Prevalence of HEV Infection. *Pathogens* **2021**, *10*, 107. [CrossRef] [PubMed]
25. Van der Poel, W.H.M.; Dalton, H.R.; Johne, R.; Pavio, N.; Bouwknegt, M.; Wu, T.; Cook, N.; Meng, X.J. Knowledge gaps and research priorities in the prevention and control of hepatitis E virus infection. *Transbound. Emerg. Dis.* **2018**, *65* (Suppl. 1), 22–29. [CrossRef]
26. Fu, R.M.; Decker, C.C.; Dao Thi, V.L. Cell Culture Models for Hepatitis E Virus. *Viruses* **2019**, *11*, 608. [CrossRef]
27. Jothikumar, N.; Cromeans, T.L.; Robertson, B.H.; Meng, X.J.; Hill, V.R. A broadly reactive one-step real-time RT-PCR assay for rapid and sensitive detection of hepatitis E virus. *J. Virol. Methods* **2006**, *131*, 65–71. [CrossRef]
28. Cook, N.; D'Agostino, M.; Wood, A.; Scobie, L. Real-Time PCR-Based Methods for Detection of Hepatitis E Virus in Pork Products: A Critical Review. *Microorganisms* **2022**, *10*, 428. [CrossRef]
29. Vina-Rodriguez, A.; Schlosser, J.; Becher, D.; Kaden, V.; Groschup, M.H.; Eiden, M. Hepatitis E virus genotype 3 diversity: Phylogenetic analysis and presence of subtype 3b in wild boar in Europe. *Viruses* **2015**, *7*, 2704–2726. [CrossRef]
30. Mizuo, H.; Suzuki, K.; Takikawa, Y.; Sugai, Y.; Tokita, H.; Akahane, Y.; Itoh, K.; Gotanda, Y.; Takahashi, M.; Nishizawa, T.; et al. Polyphyletic strains of hepatitis E virus are responsible for sporadic cases of acute hepatitis in Japan. *J. Clin. Microbiol.* **2002**, *40*, 3209–3218. [CrossRef]
31. Erker, J.C.; Desai, S.M.; Mushahwar, I.K. Rapid detection of Hepatitis E virus RNA by reverse transcription-polymerase chain reaction using universal oligonucleotide primers. *J. Virol. Methods* **1999**, *81*, 109–113. [CrossRef]
32. Lan, X.; Yang, B.; Li, B.Y.; Yin, X.P.; Li, X.R.; Liu, J.X. Reverse transcription-loop-mediated isothermal amplification assay for rapid detection of hepatitis E virus. *J. Clin. Microbiol.* **2009**, *47*, 2304–2306. [CrossRef]
33. Zhang, L.Q.; Zhao, F.R.; Liu, Z.G.; Kong, W.L.; Wang, H.; Ouyang, Y.; Liang, H.B.; Zhang, C.Y.; Qi, H.T.; Huang, C.L.; et al. Simple and rapid detection of swine hepatitis E virus by reverse transcription loop-mediated isothermal amplification. *Arch. Virol.* **2012**, *157*, 2383–2388. [CrossRef]
34. Wang, J.; Zhang, H.; Yan, S.; Wang, Y.; Lu, X.; Fu, H. Development of a One Step Reverse Transcription-Loop-Mediated Isothermal Amplification System for a Highly Sensitive Detection of Rabbit Hepatitis E Virus. *Clin. Lab.* **2017**, *63*, 901–905. [CrossRef]
35. Ward, P.; Poitras, E.; Leblanc, D.; Letellier, A.; Brassard, J.; Plante, D.; Houde, A. Comparative analysis of different TaqMan real-time RT-PCR assays for the detection of swine Hepatitis E virus and integration of Feline calicivirus as internal control. *J. Appl. Microbiol.* **2009**, *106*, 1360–1369. [CrossRef]
36. Baylis, S.A.; Hanschmann, K.O.; Matsubayashi, K.; Sakata, H.; Roque-Afonso, A.M.; Kaiser, M.; Corman, V.M.; Kamili, S.; Aggarwal, R.; Trehanpati, N.; et al. Development of a World Health Organization International Reference Panel for different genotypes of hepatitis E virus for nucleic acid amplification testing. *J. Clin. Virol.* **2019**, *119*, 60–67. [CrossRef] [PubMed]
37. Germer, J.J.; Ankoudinova, I.; Belousov, Y.S.; Mahoney, W.; Dong, C.; Meng, J.; Mandrekar, J.N.; Yao, J.D. Hepatitis E Virus (HEV) Detection and Quantification by a Real-Time Reverse Transcription-PCR Assay Calibrated to the World Health Organization Standard for HEV RNA. *J. Clin. Microbiol.* **2017**, *55*, 1478–1487. [CrossRef] [PubMed]
38. Frías, M.; López-López, P.; Zafra, I.; Caballero-Gómez, J.; Machuca, I.; Camacho, Á.; Risalde, M.A.; Rivero-Juárez, A.; Rivero, A. Development and Clinical Validation of a Pangenotypic PCR-Based Assay for the Detection and Quantification of Hepatitis E Virus (*Orthohepevirus* A Genus). *J. Clin. Microbiol.* **2021**, *59*, e02075-20. [CrossRef] [PubMed]

39. Anderson, A.; Pietsch, K.; Zucker, R.; Mayr, A.; Müller-Hohe, E.; Messelhäusser, U.; Sing, A.; Busch, U.; Huber, I. Validation of a Duplex Real-Time PCR for the Detection of *Salmonella* spp. in Different Food Products. *Food Anal. Methods* **2011**, *4*, 259–267. [CrossRef]
40. Zhao, M.Y.; Li, D. Optimization and Implementation of the Virus Extraction Method for Hepatitis E Virus Detection from Raw Pork Liver. *Food Environ. Virol.* **2021**, *13*, 74–83. [CrossRef] [PubMed]
41. Schotte, U.; Martin, A.; Brogden, S.; Schilling-Loeffler, K.; Schemmerer, M.; Anheyer-Behmenburg, H.E.; Szabo, K.; Müller-Graf, C.; Wenzel, J.J.; Kehrenberg, C.; et al. Phylogeny and spatiotemporal dynamics of hepatitis E virus infections in wild boar and deer from six areas of Germany during 2013–2017. *Transbound. Emerg. Dis.* **2022**, *69*, e1992–e2005. [CrossRef]
42. Wist, V.; Kubacki, J.; Lechmann, J.; Steck, M.; Fraefel, C.; Stephan, R.; Bachofen, C. Complete Genome Sequence of a Swiss Hepatitis E Virus Isolate from the Liver of a Fattening Pig. *Genome Announc.* **2018**, *6*, e00113-18. [CrossRef]
43. Gremmel, N.; Keuling, O.; Becher, P.; Baechlein, C. Isolation of 15 hepatitis E virus strains lacking ORF1 rearrangements from wild boar and pig organ samples and efficient replication in cell culture. *Transbound. Emerg. Dis.* **2022**, *69*, e2617–e2628. [CrossRef]
44. Nicot, F.; Dimeglio, C.; Migueres, M.; Jeanne, N.; Latour, J.; Abravanel, F.; Ranger, N.; Harter, A.; Dubois, M.; Lameiras, S.; et al. Classification of the Zoonotic Hepatitis E Virus Genotype 3 into Distinct Subgenotypes. *Front. Microbiol.* **2020**, *11*, 634430. [CrossRef] [PubMed]
45. Baylis, S.A.; Ma, L.; Padley, D.J.; Heath, A.B.; Yu, M.W.; Collaborative Study Group. Collaborative study to establish a World Health Organization International genotype panel for parvovirus B19 DNA nucleic acid amplification technology (NAT)-based assays. *Vox Sang.* **2012**, *102*, 204–211. [CrossRef] [PubMed]
46. Kozyra, I.; Bigoraj, E.; Jabłoński, A.; Politi, K.; Rzeżutka, A. Genetic Diversity and Epidemiological Significance of Wild Boar HEV-3 Strains Circulating in Poland. *Viruses* **2021**, *13*, 1176. [CrossRef]
47. Inoue, J.; Takahashi, M.; Yazaki, Y.; Tsuda, F.; Okamoto, H. Development and validation of an improved RT-PCR assay with nested universal primers for detection of hepatitis E virus strains with significant sequence divergence. *J. Virol. Methods* **2006**, *137*, 325–333. [CrossRef] [PubMed]
48. Kaba, M.; Davoust, B.; Marié, J.-L.; Barthet, M.; Henry, M.; Tamalet, C.; Raoult, D.; Colson, P. Frequent transmission of hepatitis E virus among piglets in farms in Southern France. *J. Med. Virol.* **2009**, *81*, 1750–1759. [CrossRef]
49. Martin-Latil, S.; Hennechart-Collette, C.; Delannoy, S.; Guillier, L.; Fach, P.; Perelle, S. Quantification of Hepatitis E Virus in Naturally-Contaminated Pig Liver Products. *Front. Microbiol.* **2016**, *7*, 1183. [CrossRef]
50. Ahn, J.M.; Rayamajhi, N.; Gyun Kang, S.; Sang Yoo, H. Comparison of real-time reverse transcriptase-polymerase chain reaction and nested or commercial reverse transcriptase-polymerase chain reaction for the detection of hepatitis E virus particle in human serum. *Diagn. Microbiol. Infect. Dis.* **2006**, *56*, 269–274. [CrossRef]
51. Gyarmati, P.; Mohammed, N.; Norder, H.; Blomberg, J.; Belák, S.; Widén, F. Universal detection of hepatitis E virus by two real-time PCR assays: TaqMan®and Primer-Probe Energy Transfer. *J. Virol. Methods* **2007**, *146*, 226–235. [CrossRef]
52. Enouf, V.; Dos Reis, G.; Guthmann, J.P.; Guerin, P.J.; Caron, M.; Marechal, V.; Nicand, E. Validation of single real-time TaqMan®PCR assay for the detection and quantitation of four major genotypes of hepatitis E virus in clinical specimens. *J. Med. Virol.* **2006**, *78*, 1076–1082. [CrossRef]
53. Hennechart-Collette, C.; Fraisse, A.; Guillier, L.; Perelle, S.; Martin-Latil, S. Evaluation of methods for elution of HEV particles in naturally contaminated sausage, figatellu and pig liver. *Food Microbiol.* **2019**, *84*, 103235. [CrossRef]
54. Wenzel, J.J.; Preiss, J.; Schemmerer, M.; Huber, B.; Plentz, A.; Jilg, W. Detection of hepatitis E virus (HEV) from porcine livers in Southeastern Germany and high sequence homology to human HEV isolates. *J. Clin. Virol.* **2011**, *52*, 50–54. [CrossRef]
55. Baechlein, C.; Seehusen, F.; Nathues, H.; Grosse Beilage, E.; Baumgärtner, W.; Grummer, B. Molecular detection of hepatitis E virus in German domestic pigs. *Berl. Munch. Tierarztl. Wochenschr.* **2013**, *126*, 25–31.
56. Boxman, I.L.A.; Verhoef, L.; Dop, P.Y.; Vennema, H.; Dirks, R.A.M.; Opsteegh, M. High prevalence of acute hepatitis E virus infection in pigs in Dutch slaughterhouses. *Int. J. Food Microbiol.* **2022**, *379*, 109830. [CrossRef]
57. Schielke, A.; Sachs, K.; Lierz, M.; Appel, B.; Jansen, A.; Johne, R. Detection of hepatitis E virus in wild boars of rural and urban regions in Germany and whole genome characterization of an endemic strain. *Virol. J.* **2009**, *6*, 58. [CrossRef]
58. Müller, A.; Collineau, L.; Stephan, R.; Müller, A.; Stärk, K.D.C. Assessment of the risk of foodborne transmission and burden of hepatitis E in Switzerland. *Int. J. Food Microbiol.* **2017**, *242*, 107–115. [CrossRef]
59. Cook, N.; D'Agostino, M.; Johne, R. Potential Approaches to Assess the Infectivity of Hepatitis E Virus in Pork Products: A Review. *Food Environ. Virol.* **2017**, *9*, 243–255. [CrossRef]

MDPI
St. Alban-Anlage 66
4052 Basel
Switzerland
www.mdpi.com

Foods Editorial Office
E-mail: foods@mdpi.com
www.mdpi.com/journal/foods